Linear Algebra

Modules for Interactive Learning

Using Maple 6®

Linear Algebra
Modules for Interactive Learning
Using Maple 6®

by The Linear Algebra Modules Project
(LAMP)

Eugene A. Herman
Grinnell College

Michael D. Pepe
Seattle Central Community College

Robert T. Moore
University of Washington

James R. King
University of Washington

Boston San Francisco New York
London Toronto Sydney Tokyo Singapore Madrid
Mexico City Munich Paris Cape Town Hong Kong Montreal

Sponsoring Editor: Laurie Rosatone
Assistant Editor: Susan Laferriere
Managing Editor: Karen Guardino
Associate Production Supervisor: Julie LaChance
Marketing Manager: Michael Boezi
Cover Design Supervisor: Barbara Atkinson
Manufacturing Manager: Evelyn Beaton

Library of Congress Cataloging-in-Publication Data

Linear algebra : modules for interactive learning using MAPLE / by The Linear Algebra Modules Project (LAMP) ; Eugene A. Herman ... [et al.].—1st ed.
    p. cm.
  Includes index.
  ISBN 0-201-44135-7 (alk. paper)
  1. Algebras, Linear—Computer-assisted instruction. 2. Maple (Computer file) I. Title: Linear algebra modules for interactive learning using MAPLE. II. Herman, Eugene A. III. Linear Algebra Modules Project.

QA185.C65 L56 2000
512'.5'02855369—dc21

                                                                                         00-061852

MAPLE is a registered trademark of Waterloo Software, Inc.

Reproduced by Addison-Wesley from camera-ready copy supplied by the authors.

Copyright © 2001 Addison-Wesley.

All rights reserved. No part of this publication may be reproduced, stored in a retrieval system, or transmitted, in any form or by any means, electronic, mechanical, photocopying, recording, or otherwise, without the prior written permission of the publisher. Printed in the United States of America.

ISBN   0-201-44135-7

1 2 3 4 5 6 7 8 9 10 CRS 04 03 02 01 00

Preface

Linear Algebra: Modules for Interactive Learning Using MAPLE 6 is a collection of twenty-nine Maple worksheets (i.e., "modules") produced by the Linear Algebra Modules Project (LAMP). Taken together, these modules cover the entire introductory linear algebra course as taught at most colleges and universities. The modular structure of the collection is designed to permit a variety of styles of use under many different teaching conditions. In designing such a flexible set of materials, we had the advantage of working at three very different institutions with different types of students, computing facilities, and institutional constraints.

We were also fortunate to have had reviewers and class testers at several other institutions who worked through preliminary versions of the modules. This first edition has benefited from their thoughtful feedback and advice.

Goals

The primary objective of these materials is to promote active learning by students in linear algebra courses. In the standard linear algebra course, we have observed, few students find the material genuinely engaging. The computations are not interesting enough to stimulate students intellectually; the theory often strikes them as too abstract or disconnected from their needs; and the applications are often too sketchy to make an impression. Our goal, therefore, has been to engage students in constructing an understanding of the material. We offer them material that is more concrete, familiar, attractive, and useful to them, so they feel encouraged to participate in the learning process.

Other important goals are related to the use of geometry, approach to applications, and attention to development of higher-order thinking.

> **Geometry and intuition** The modules use geometry extensively to help students develop their intuition about the concepts of linear algebra. For example, once students have a clear understanding of the linear geometry of R^2 and R^3, they become much more at ease with discussions of n-space. Also, by interacting with the visualizations presented in the modules and by creating pictures of their own, they become confident that their geometric intuition is solidly grounded.

> **Applications** Each application module is substantial enough that students can develop a solid understanding of how linear algebra is used in that application and what significance that contribution has. Since the students have the power of Maple at hand, they can carry out realistic computations and can bring the results to a satisfying conclusion. For example, in the study of least-squares solutions, students plot the data and the approximating line or curve and compare the results of different choices of approximating curves. In the study of discrete dynamical systems, animations of their trajectories permit students to experiment and conduct "what if" analyses on the models presented.

Higher-order thinking One reason students in a standard course have difficulty appreciating the power of the abstractions of linear algebra is that they spend much of their time doing low-level computations on the entries of a matrix. We have found that, by providing appropriate tools in Maple, we can ask new kinds of questions that encourage students to formulate answers in terms of the higher-level objects that Maple commands act on, such as matrices, vectors, and linear transformations. This stimulates students to think at a higher level of abstraction and to appreciate better the power of the subject.

Structure of the modules

Each module begins with a brief statement of purpose, a list of prerequisite concepts, and a summary of the Maple commands used in that module. These are followed by the two main parts of the module, the Tutorial and the Problems.

Tutorial The Tutorial is designed to present the main topic of the module in a straightforward manner with many concrete examples and exercises to help students grasp the key concepts. Complete solutions for all exercises that appear in the body of the Tutorial are provided in closed (but viewable) sections so students can get immediate feedback and monitor their progress in understanding the material. The Tutorial is designed to be completed by students with minimal assistance from the instructor. After completing the Tutorial, students move on to the Problems.

Problems The Problems are all intended to be fairly substantial, as they provide the work that students will be graded on. Problems include explorations, applications, constructions (e.g., of specified types of matrices, pictures, or animations), counter-examples, short essays, proofs, true/false questions, and, of course, many challenging computations. Since the Problems are both substantial and varied, students do not usually need to be assigned every Problem in the module to master the material.

Active learning with (and without) technology

Using computers and appropriate software, we have found, can provide highly effective tools for promoting active student learning. However, we also believe that every student of linear algebra should master certain fundamental pencil and paper skills and that some topics in linear algebra may not benefit from the use of technology. To facilitate learning away from the computer, especially for students who are not using a traditional textbook, we have included a "Pencil and Paper Tutorial" on Gaussian elimination and another on determinants and cofactors in appendices to this book. Also, Module 3 of Chapter 3 ("Rules of Matrix Algebra") and Module 1 of Chapter 5 ("Subspaces") are similarly designed to be done without Maple. While all these materials are to be done by hand, we have written them in the active style of the on-line tutorials.

Options for using the modules

In designing this collection of modules, we have tried to anticipate the wide variety of expectations that instructors of linear algebra will have for computer-based materials. We are well aware that one linear algebra course may have a quite different syllabus and purpose than another, that a course at one institution may have different scheduling and laboratory constraints than at another, and that one instructor may have different pedagogical goals than another. We hope you will find that our collection of modules is sufficiently flexible to meet your needs. Below we describe three modes in which these materials have been implemented successfully.

Traditional course using modules for computer labs Under this option an instructor assigns a subset of the modules as computer labs to supplement a course that uses a standard linear algebra textbook. The modules are particularly well suited for this kind of use since there is a module for every major topic presented in a traditional course. Within this option, many approaches are possible. Some instructors assign a group of application modules as projects. Others use one or more complete chapters of modules in lieu of traditional lectures on that material. Yet others choose several introductory modules to help students get a good start in every chapter by using the modules to develop intuition. In those colleges where the course is taught in an environment in which the instructor or students have computer access during regular class hours, instructors have found ways to integrate the modules directly into their lectures and class discussions. Many instructors find the special visualization tools that are included in the Lamp library a valuable resource in their lecture demonstrations. Still another mode of use within the context of a traditional course is to have students work through the Tutorials outside of class to reinforce the material from lecture.

The "LAMP" course Instructors who seek a change of pace from the traditional lecture course format can use the complete set of LAMP modules as the electronic textbook for the course. Under this option, almost all class meetings take place in a computer lab where students work through the modules on line. The instructor circulates among the students, answering questions and engaging in brief discussions with individuals or small groups. Although one generally does little or no lecturing in this mode, we recommend holding regular discussions designed to help students see the big picture. We have also provided pencil and paper tutorials for those topics best done without a computer. These materials ensure that students who participate in this technology-dependent option are able to do the essential computations by hand.

Independent study options The LAMP modules have been used successfully for independent study of linear algebra at the college level. Since the modules are self-contained and self-paced and provide substantial feedback for the student, they can be worked through with little outside assistance. The modules are also well-suited for use in a distance-learning environment, since the Tutorials can take the place of on-site lectures.

How to use this book

This book contains a print copy of each of the twenty-nine modules. These print versions include most of the output that you will get when executing a module during a Maple session. Of course, the print versions lack the color one sees on the screen, and the graphics are disappointingly inert. Nevertheless, students find the print copy useful for reviewing the modules after having worked through them. Another use is to preview a module by reading a section or two before working through it on line.

Additional resources provided in the appendices are as follows:

    Paper and Pencil Tutorial: Gaussian Elimination
    Paper and Pencil Tutorial: Determinants and Cofactors
    The Lamp Library (on-line help resources, the Lamp computing environment)
    Glossary of Linear Algebra Definitions (definition of every important linear algebra
        term used in the modules)
    List of Linear Algebra Theorems (a statement of every Theorem in the modules)

The Lamp library

The Lamp library provides more than sixty special-purpose Maple commands that are used in the modules. The library was written with two objectives in mind. First, while Maple 6 includes the powerful new LinearAlgebra package, that package is not designed for the beginning student of linear algebra. Often the syntax and options are not related to the concepts and terminology encountered in an introductory linear algebra course. The Lamp library commands, on the other hand, are designed to be user friendly and to coincide with the needs and expectations of students and instructors in such an academic setting. While some Lamp commands make calls to LinearAlgebra algorithms, the design of the Lamp commands has been tailored to this audience.

The second objective of the Lamp Library is to provide a rich set of visualization tools to help the beginning student "see" and experience the concepts of linear algebra. To this end the authors have developed an extensive set of plotting and animation commands designed to illustrate specific aspects of linear algebra. These commands range from drawing tools for plotting vectors to animations that help one visualize the geometric behavior of matrix transformations, trace the trajectory of a discrete dynamical system, and much more. Each command has been written to be as flexible as possible, with many options to aid users in creating their own pictures.

Software and hardware requirements

You will need Maple 6 to run the modules. Maple 6 runs on all the common hardware platforms—Windows PCs, Macintosh computers, and Unix workstations. The LAMP modules and library work in all these environments.

The LAMP modules and library can be installed from the CD that can be found in the back cover of this book. See the Installation Instructions following the appendix for details on how to install these materials under your particular operating system.

Acknowledgements

We are especially pleased to recognize our excellent student programmers, David Morse, Jeremy Sheeley, Dorene Mboya, and Nathan Corvino, who coded the Lamp library. We also gratefully acknowledge the support of the National Science Foundation, which funded the first two years of our project. We thank our students, who put up with our early efforts, provided crucial feedback, and encouraged us to persevere. We thank our colleagues, especially Sanford Helt, Samuel Rebelsky, John Stone, and Karen Thomson, for their valuable suggestions and their substantial assistance. We thank the participants in our several mini-courses for their suggestions and encouragement, and we thank our reviewers and class testers (listed below) for their informed advice. Finally, we thank our editorial and production team at Addison Wesley Longman.

We also acknowledge the many influences that have helped shape our thinking and our strategies. Especially influential have been our involvement with MAX(the MAtriX Algebra Calculator, the ATLAST project (Augment the Teaching of Linear Algebra through the use of Software Tools) headed by Steven Leon, and the Interactive Mathematics Text Project headed by Gerald Porter and James White. Additionally, we have learned much from the ground-breaking textbooks by David Lay (Linear Algebra and its Applications, 2nd ed., Addison Wesley Longman, 1997), and Gilbert Strang (Linear Algebra and its Applications, 3rd ed., Harcourt Brace Jovanovich, 1976).

Eugene A. Herman, Grinnell College

Michael D. Pepe, Seattle Central Community College

Robert T. Moore and James R. King, University of Washington

October 1, 2000

Reviewers, evaluators, and class testers:

Jacob Appleman, Queensborough Community College
Philip Beckmann, Okanagan University College-Kelowna
Laurie Burton, Central Washington University
Sylvie Desjardins, Okanagan University College-Penticton Mark Farris, Midwestern State University William E. Fenton, Bellarmine College
Christie Gilliland, Green River Community College
Sanford Helt, Seattle Central Community College
David Hill, Temple University
Eugene Johnson, University of Iowa
James P. Marshall, Illinois College
Doug Meade, University of South Carolina
Jonathan Natov, New York City Technical College, CUNY
Clint Lee, Okanagan University College-Vernon James Marshall, Illinois College
Stanley Perrine, Charleston Southern University
Roy Rakestraw, Oral Roberts University
Elsa Newman Schaefer, Marymount University
Carol Scheftic, California Polytechnic State University

To the Student

This book and its on-line version take a quite different approach to linear algebra than you would find in a standard textbook. Because the methods are so different, you will need a different set of study skills. So we offer here a more extensive collection of suggestions and helpful hints than one usually finds in a textbook. We recommend you read this section once now and again more thoroughly after you have used the materials for a while.

A. Working through the tutorial

Learning by doing The Tutorials are designed to engage you in actively doing mathematics rather than passively watching an instructor. So throw yourself into the work by experimenting, anticipating, and writing, as you go through a Tutorial.

As you work through the Tutorial, you will encounter some short Exercises that you should solve before continuing. Their purpose is to ensure that you are comprehending what you are reading and experiencing on the screen. Just open the Student Workspace that follows the Exercise, and do your work there. Resist the temptation to peek at the Answer we have provided, until you have given the question a fair try. Since the Exercises are designed to be straightforward applications of the material, when you cannot answer one you should go back and review the previous example, exploration, or demonstration.

Learning by seeing The Tutorials use geometry extensively to help you develop intuition about the concepts of linear algebra. Test your growing intuition by altering the live pictures and animations in the Tutorial and by creating your own pictures and animations.

Learning a new language A beginning linear algebra student may feel overwhelmed by the large number of definitions to master. To help you keep track of and review these definitions, there is a Glossary of Linear Algebra Definitions in an Appendix to this book. We suggest you highlight each new term as you encounter it and study your "vocabulary list" regularly. Another useful reference is the List of Linear Algebra Theorems.

B. Writing up your solutions to the problems

Organize your screen for efficient work When you have finished the Tutorial and are ready to work on the Problems, you may find the following arrangements helpful.

1. Open the corresponding file of Problems (such as `probc1m1.mws`), which is a duplicate of the Problem set in the module. This file is provided so you can work through the Problems while the Tutorial is simultaneously viewable and also so you can turn in just the Problem file with your solutions.
2. Maximize the screen by clicking on the middle button at the top right corner of the Maple window.
3. Select Horizontal (or Vertical) from the Window menu, which puts the module and the Problems above one another (or side by side).

Annotate your solutions Include some explanatory text before each block of Maple commands or hand computations. Keep the reader informed of the plan of your solution and how you are carrying it out. This does not usually require lengthy prose. A good rule of thumb in deciding how much to write is to write enough so that a classmate who hasn't yet solved the problem could understand what you are doing and why. Your annotations will also help you study from your solutions when you are preparing for an exam (and are having trouble following your own solution).

Explain your conclusions Many problems ask you to "explain" your conclusions, which is an important part of many solutions. When you are asked to draw a conclusion from a computation and the conclusion isn't immediately obvious, explain how you reached your conclusion. Again, this does not usually require lengthy prose, and again you should imagine that your audience is a classmate. Write enough so that a classmate could understand how you reached your conclusion and be persuaded of its validity by the logic and clarity of your reasoning.

By-hand work Many Exercises and Problems ask that some or all of your work be done "by hand." Also, sometimes an Exercise or Problem can by done by hand even when it doesn't say "by hand"; in that case, feel free to do the work by hand or by using Maple. Keep a sheet of paper handy for your by-hand work. When you turn in your Problems, you can simply attach the sheet of paper on which you have done the by-hand work. Be sure to indicate in the worksheet those places where the reader should turn to the attached sheet of paper. Sometimes, by-hand Problems can be left to later, when you are away from the computer lab. Alternatively, you can write your by-hand work in a Maple text region in the worksheet. But if you want formulas to look good in a Maple text region, see the formatting hints in section D below.

Annotate your graphics Although Maple's graphics are good, much is lost when we print in black and white. Draw or write by hand on the printed copy, if this will help the reader understand the picture better. Use a colored pen or pencil for greater clarity.

How to print your solutions First, open just those "Student Workspace" sections that you want to hand in. Be sure to save your work on a hard drive. Then print from the hard drive, not from a diskette (which has proven to be risky).

C. Using Maple effectively: Answers to frequently asked questions

Execute first line You MUST execute the first input region in the module every time you open the module:

```
> restart: with(LinearAlgebra): with(plots): with(Lamp):
  UseHardwareFloats := false: Digits := 6:
```

Many of the commands in the module are in the `LinearAlgebra`, `plots`, or `Lamp library`, and these commands are not available to you until you execute the corresponding `with` commands.

Use help Maple's on-line help is extensive, and the Lamp library is integrated into this on-line help (see Appendix C). To get more information about any command, click on that command, then click on the Help menu in the Maple window, and then select help on that command.

Save often Computers do sometimes crash and lose your work. Save to a hard drive, and save often simply by clicking on the diskette icon. Also, change the file name occasionally, so you have a recent back-up copy.

Semicolons Every Maple command must end in a semicolon or a colon. (The colon tells Maple not to display the output.) If you forget the semicolon or colon, just type one in and execute the command again.

Entering Maple input To get a Maple input prompt (which is required for entering a Maple command), click on the [> icon.

Entering Maple text To get a Maple text region (which is required for entering text in Maple), first get an input prompt as above and then click on the letter T icon.

Copy and paste A quick way to enter a long Maple command is to copy a similar command from elsewhere, paste it into a Maple input region, and edit it. You can also copy output from a Maple command (even a plot) and paste it into a text region.

Colon-equals A common oversight is omitting the colon (:) before the equal sign in a Maple statement. Most Maple statements have the form Name := Expression, which tells Maple to calculate the value of the Expression and assign the result of the calculation to the Name. However, in an equation that you want Maple to solve, you use the equal sign without the colon.

Lists and sets Some Maple commands apply only to lists of expressions (which you recognize by the fact that they are enclosed in square brackets [,]); others apply only to sets of expressions (which you recognize by the fact that they are enclosed in curly braces { , }). So check Maple commands carefully to see which they require.

Maple's forgetfulness When you open a worksheet that contains output, Maple does not actually know the values of any of the variables in the output regions, even though it may look like values have been assigned to some variables. Think of it this way: Maple "forgets" the values assigned in a worksheet as soon as you exit Maple or execute a `restart` command. So if you want Maple to know the values of these variables, you have to execute the commands in the worksheet again. Better yet, before you save a Maple worksheet, remove all the output; do this by going into the Edit menu and selecting Remove Output / From Worksheet.

Maple's lack of forgetfulness Unless you exit Maple or execute a `restart` command or do something else to force Maple to wipe out the value of some variables, Maple will retain these values. So it's a good idea not to use the same variable names repeatedly, since Maple will remember the values of those variables. If you find that you are getting strange responses to Maple commands, the cause may be exactly this unwise repeat of variable names. The cure is to execute the `restart` and `with` commands at the top of the module; then re-execute all the commands you need to get your answer, but change names that you have repeated.

Keep commands in order When you go through a module, execute all the commands and execute them in the order in which they appear. Otherwise, you may get incorrect results. Keep in mind the order in which Maple processes commands: Maple processes each command in the order in which it is executed; Maple pays no attention to the order in which the commands appear in the worksheet.

D. Producing nicely formatted formulas in text regions

You can create mathematical formulas in text regions that look as good as the mathematical formulas in Maple's output. In a text region, type the formula just as you would if it were Maple input. Then highlight the formula, hold down the right mouse button, and select Convert to / Standard Math.

You can also edit a formula: Click on the formula, hold down the right mouse button, and toggle Standard Math. Then edit the formula (which now looks like Maple input), hold down the right mouse button, and again toggle Standard Math.

In memory of Augusta Herman
EAH

*To Margaret, Fiona and Tom, to my parents
and to my teacher Isadore Glaubiger*
MDP

To my wife, Jo
RTM

To Vicki, with love
JRK

CONTENTS

Chapter 1: Systems of Linear Equations — 1
- **Module 1.** Geometric Perspectives on Linear Equations — 2
- **Module 2.** Solving Linear Systems Using MAPLE — 12
- **Module 3.** Some Applications Leading to Linear Systems — 24

Chapter 2: Vectors — 36
- **Module 1.** Geometric Representation of Vectors in the Plane — 37
- **Module 2.** Linear Combinations of Vectors — 48
- **Module 3.** Decomposing the Solution of a Linear System — 63
- **Module 4.** Linear Independence of Vectors — 77

Chapter 3: Matrix Algebra — 93
- **Module 1.** Product of a Matrix and a Vector — 95
- **Module 2.** Matrix Multiplication — 111
- **Module 3.** Rules of Matrix Algebra — 124
- **Module 4.** Markov Chains — An Application — 134
- **Module 5.** Inverse of a Matrix — An Introduction — 150
- **Module 6.** Inverse of a Matrix and Elementary Matrices — 166
- **Module 7.** Determinants — 176

Chapter 4: Matrix Transformations — 196
- **Module 1.** Geometry of Matrix Transformations of the Plane — 198
- **Module 2.** Geometry of Matrix Transformations of 3-Space — 225
- **Module 3.** Computer Graphics — 238

Chapter 5: Vector Spaces — 259
- **Module 1.** Subspaces — 260
- **Module 2.** Basis and Dimension — 268
- **Module 3.** Subspaces Associated with a Matrix — 279
- **Module 4.** Loops and Spanning Trees — An Application — 296

Chapter 6: Eigenvalues and Eigenvectors — 306
- **Module 1.** Introduction to Eigenvalues and Eigenvectors — 308
- **Module 2.** The Characteristic Polynomial — 326
- **Module 3.** Eigenvector Bases and Discrete Dynamical Systems — 340

Module 4. Diagonalization and Similarity 360
Module 5. Complex Eigenvalues and Eigenvectors 380

Chapter 7: Orthogonality 408
Module 1. Orthogonal Vectors and Orthogonal Projections 410
Module 2. Orthogonal Bases 423
Module 3. Least-Squares Solutions 438

Appendix A. Paper and Pencil Tutorial: Gaussian Elimination 456
Appendix B. Paper and Pencil Tutorial: Determinants and Cofactors 474
Appendix C. The Lamp Library 481
Appendix D. Glossary of Linear Algebra Definitions 483
Appendix E. List of Linear Algebra Theorems 490
Index 497
Software Installation Instructions (Windows and Macintosh) 501

Chapter 1: Systems of Linear Equations

Module 1. Geometric Perspectives on Linear Equations

Module 2. Solving Linear Systems Using MAPLE

Module 3. Some Applications Leading to Linear Systems

 Commands used in this chapter

`D(f);` defines the derivative function for the function *f*.
`display([pict1, pict2]);` displays together a group of previously defined pictures.
`Matrix([[a,b],[c,d]]);` defines the matrix with rows $[a, b]$ and $[c, d]$.
`Pivot(M,i,j);` applies elementary row operations to the matrix *M*, pivoting on entry $M_{i,j}$.
`plot(expr,x=a..b);` plots a graph of a function of one variable.
`plot3d(expr, x=a..b, y=c..d);` plots a graph of a function of two variables.
`pointplot([pt1,pt2,...,ptn]);` plots the list of points *pt1, pt2, ..., ptn*, where each point is given as a list of two coordinates.
`solve({eqn1,eqn2},{x,y});` finds exact solutions for the unknowns *x* and *y* in the system of equations given by *eqn1* and *eqn2*.
`spacecurve([expr1,expr2,expr3],t=a..b);` plots a curve in 3-space where *expr1, expr2, expr3* are the component functions and the parameter *t* ranges from *a* to *b*.
`subs({a=3,b=13},expr);` substitutes the values for *a* and *b* in the expression *expr*.

LAMP commands:
`Backsolve(R);` solves the linear system for which *R* is a row echelon form of the augmented matrix.
`Drawlines([eqn1,eqn2]);` draws the graphs of *eqn1* and *eqn2*, which are equations of lines in the plane.
`Drawplanes([eqn1,eqn2]);` draws the graphs of *eqn1* and *eqn2*, which are equations of planes in 3-space.
`Genmatrix([eqn1,eqn2],[x,y]);` produces the augmented matrix for the linear system [*eqn1, eqn2*] in the unknowns *x* and *y*.
`Reduce(M);` produces a row echelon form of the matrix *M*.
`Reduce(M,form=rref);` produces the reduced row echelon form of the matrix *M*.
`Rowop(M,rj <> rk);` interchanges row *j* and row *k* of the matrix *M*.
`Rowop(M,rj <= rj+c*rk);` replaces row *j* by the sum of row *j* and *c* times row *k*.
`Rowop(M,rj <= c*rj);` replaces row *j* by *c* times row *j*.

Linear Algebra Modules Project
Chapter 1, Module 1

Geometric Perspectives On Linear Equations

 Purpose of this module

The purpose of this module is to give you experience interpreting linear equations and their solutions geometrically. You will work with systems of linear equations in two and three unknowns, since their solutions can be pictured as points in the plane and in 3-space. The main question to investigate is: What are the possible solution sets for a system of linear equations in 2 or 3 unknowns?

 Prerequisites

Familiarity with linear equations in two and three unknowns and their representations as lines and planes, respectively.

 Commands used in this module

```
> restart: with(LinearAlgebra): with(plots): with(Lamp):
  UseHardwareFloats := false: Digits := 6:
```

Tutorial

Section 1: Linear Equations in Two Unknowns

We begin by looking at a single linear equation in two unknowns:
$$a x + b y = c$$
This is the equation of a line in the plane, provided the coefficients a and b are not both zero. In Example 1A, we use Maple to solve two such equations simultaneously and to look at their common solutions.

==================

Example 1A: Solve simultaneously the following system of linear equations:
$$x + 4 y = 6$$
$$3 x - y = 5$$
Then plot the two lines, and find the location of the solution on the plot.

Solution: We define the Maple variables eqn1 and eqn2 to be the two equations. Then we use the `solve` command to solve the two equations simultaneously.

```
> eqn1 := x+4*y=6;
```

$$eqn1 := x + 4y = 6$$

```
> eqn2 := 3*x-y=5;
```
$$eqn2 := 3x - y = 5$$

```
> solve({eqn1,eqn2},{x,y});
```
$$\{y = 1, x = 2\}$$

So $(x, y) = (2, 1)$ is the unique solution of the given system of two equations. Next we plot the two lines by first solving each equation for y in terms of x.

```
> y1 := solve(eqn1,y);
  y2 := solve(eqn2,y);
```
$$y1 := -\frac{1}{4}x + \frac{3}{2}$$
$$y2 := 3x - 5$$

```
> plot([y1,y2],x=-5..5,y=-5..5,color=[red,blue]);
```

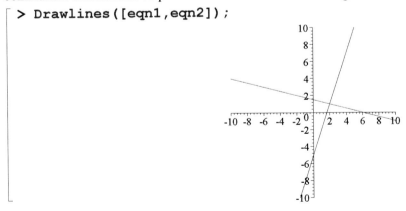

Note that the two lines intersect at the point (2, 1), as we expected. You can also use the special command **Drawlines** to plot the lines without first having to solve each equation for y:

```
> Drawlines([eqn1,eqn2]);
```

==================

Exercise 1.1: (a) Replace the second equation above (the one called eqn2) by a new equation so that the system {eqn1, eqn2} has no solutions. But choose the coefficients of x and y so they are not both zero.

Check your answer by applying the `solve` command to your new system.

(b) How are your two lines related to one another geometrically? How are their coefficients related to one another algebraically?

Maple note: When Maple's `solve` command finds no solutions to a system of equations, it simply returns no output.

 Student Workspace

 Answer 1.1

We can find a new eqn2 by using the same coefficients of x and y as in eqn1 (and hence creating a line with the same slope) but changing the right side:
```
> eqn2 := x + 4*y = 0;
```
$$eqn2 := x + 4y = 0$$
Let's verify that the new system has no solution and that the two lines are parallel:
```
> solve({eqn1,eqn2},{x,y});
```
Maple gave no response to the `solve` command because it found no solutions.
```
> Drawlines([eqn1,eqn2]);
```

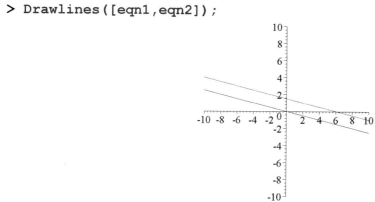

Now let's consider three equations in two unknowns. First, let's go back to the original two equations; then we will consider what can happen when we include a third equation.
```
> eqn1 := x+4*y=6:
  eqn2 := 3*x-y=5:
>
```

Exercise 1.2: Create a third equation (call it eqn3) so the system {eqn1, eqn2, eqn3} has a unique solution, but no two of the three lines are equal.

 Student Workspace

 Answer 1.2

Choose the third line so it goes through the point of intersection (2, 1) of the first two lines. For

example:
```
> eqn3 := x-2*y=0;
```
$$eqn3 := x - 2y = 0$$
```
> solve({eqn1,eqn2,eqn3},{x,y});
```
$$\{x = 2, y = 1\}$$
```
> Drawlines([eqn1,eqn2,eqn3]);
```

Section 2: Linear Equations in Three Unknowns

A single linear equation in three unknowns
$$ax + by + cz = d$$
is the equation of a plane in 3-space, provided the coefficients a, b, and c are not all zero. In Example 2A, we use Maple to solve two such equations simultaneously and to look at their common solutions:

========================

Example 2A: Solve simultaneously the following system of linear equations:
$$x + 2y + 3z = 3$$
$$2x - y - 4z = 1$$

Then plot the two planes, and find the location of the solution on the plot.

Solution:
```
> eqn1 := x+2*y+3*z=3;
  eqn2 := 2*x-y-4*z=1;
```
$$eqn1 := x + 2y + 3z = 3$$
$$eqn2 := 2x - y - 4z = 1$$
```
> solve({eqn1,eqn2},{x,y});
```
$$\{x = 1 + z, y = 1 - 2z\}$$

So the system of equations has infinitely many solutions: $(x, y, z) = (z + 1, -2z + 1, z)$. In this representation of the solutions we refer to z as the "free" variable, because z can have any value and each value of z determines a solution of the system. For example if $z = 4$, then $x = 5$ and $y = -7$, and therefore $(5, -7, 4)$ is a solution to each of the original two equations. Geometrically, the solution set is a parametric representation of the line of intersection of the two planes.

Now let's plot the two planes and look for this line of intersection. First, we solve each equation for z.

```
> z1 := solve(eqn1,z);
  z2 := solve(eqn2,z);
```

$$z1 := -\frac{1}{3}x - \frac{2}{3}y + 1$$

$$z2 := \frac{1}{2}x - \frac{1}{4}y - \frac{1}{4}$$

```
> p1 := plot3d(z1,x=-5..5,y=-5..5,color=blue,style=patchnogrid):
  p2 := plot3d(z2,x=-5..5,y=-5..5,color=green,style=patchnogrid):
  display([p1,p2],scaling=constrained);
```

Click on the picture, and then rotate it in 3-space. Orient the picture so you are looking along the line of intersection. Experiment with some of the plotting options on the toolbar, especially the axes options.

We use Maple's **spacecurve** command to draw the line of intersection. (We changed the name of the parameter from z to the more traditional t.) Adding this to the previous picture of the two planes provides us with a visual check of our solution.

```
> p3 := spacecurve([t+1,-2*t+1,t],t=-4..4,color=black,thickness=2):
  display([p1,p2,p3],scaling=constrained);
```

We can obtain this picture with less effort by using the **Drawplanes** command, which is similar to the **Drawlines** command:

```
> Drawplanes([eqn1,eqn2]);
```

The **Drawplanes** command can also display the planes in another way that is sometimes easier to see:

```
> Drawplanes([eqn1,eqn2],round=on);
```

Here's how to interpret this picture. Imagine a sphere with center at the origin and radius 10, and imagine intersecting each plane with this sphere. You then see each of the circles of intersection. You also see the line of intersection (in black) of each pair of planes. Although representing a plane by a circle is somewhat unconventional, this picture is often easier to interpret than the conventional picture of a parallelogram representing a plane, especially when you have several intersecting planes.

```
>
```

Exercise 2.1: (a) Replace the second equation above (the one called eqn2) by a new equation so that the system {eqn1, eqn2} has no solution. But choose the coefficients of x, y and z so they are not all zero.
(b) How are your two planes related to one another geometrically? How are their coefficients related to one another algebraically?
(Recall that when Maple's **solve** command finds no solutions, it simply returns no output.)

[+] Student Workspace

Answer 2.1

We can find a new eqn2 by using the same coefficients for x, y, and z as in eqn1 (and hence creating a parallel plane) but changing the right side:

```
> eqn2 := x+2*y+3*z=-8;
```
$$eqn2 := x + 2y + 3z = -8$$
```
> solve({eqn1,eqn2},{x,y});
```

Maple gave no response to the **solve** command, because it found no solution. Rotate the figure produced by the **Drawplanes** command below until you can see the two parallel planes.

```
> Drawplanes([eqn1,eqn2]);
```

==========================

Example 2B: Now let's consider three equations in three unknowns. First, let's go back to the original two equations; then we will consider what can happen when we include a third equation.

```
> eqn1 := x+2*y+3*z=3:
  eqn2 := 2*x-y-4*z=1:
> eqn3 := x+y-z=0;
```
$$eqn3 := x + y - z = 0$$
```
> solve({eqn1,eqn2,eqn3},{x,y,z});
```
$$\{z = 1, y = -1, x = 2\}$$

So this system has the unique solution $(x, y, z) = (2, -1, 1)$, and therefore the three corresponding planes must have this single point in common. Look for this point in the two plots below:

```
> Drawplanes([eqn1,eqn2,eqn3]);
```

```
> Drawplanes([eqn1,eqn2,eqn3],round=on):
```
Of course, a system of three equations in three unknowns doesn't always have a unique solution, as the next exercise demonstrates.

==================

Exercise 2.2: (a) Replace the third equation above (the one called eqn3) by a new equation so that the system {eqn1, eqn2, eqn3} has more than one solution, but no two of the three planes are equal.
(b) How is your third plane related geometrically to the line of intersection of the first two planes?

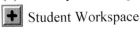 Student Workspace

 Answer 2.2

We find a new eqn3 by finding a plane $a\,x + b\,y + c\,z = d$ that contains the line of intersection $(x, y, z) = (z + 1, -2\,z + 1, z)$. One way to do this is to pick two points that lie on the line of intersection and pick a third point that does not lie on the line and also does not lie on either of the planes defined by eqn1 and eqn2. These three points determine a unique plane. For example, pick $z = 0$ and $z = 1$ to determine the points $(1, 1, 0)$ and $(2, -1, 1)$ on the line, and choose the third point to be $(0, 0, 0)$. The unique plane containing these three points is $-x + y + 3\,z = 0$.

```
> eqn3 := -x+y+3*z=0;
```
$$eqn3 := -x + y + 3\,z = 0$$
```
> solve({eqn1,eqn2,eqn3},{x,y,z});
```
$$\{y = 1 - 2\,z, x = 1 + z, z = z\}$$

Here is the picture:
```
> Drawplanes([eqn1,eqn2,eqn3]);
```

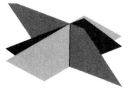

Note the single line of intersection in the picture. This means that each pair of planes intersects in the same line.

Problems

Problem 1: Three equations in two unknowns

Find a system of three linear equations in two unknowns that has no solutions, and choose your equations so that no two of the three lines are parallel. Use {eqn1, eqn2} from Section 1 (given again below) as two of your three equations. Draw the three lines (by using **Drawlines**, for example).
(Recall that when Maple's **solve** command finds no solutions, it simply returns no output.)

```
> eqn1 := x+4*y=6:
  eqn2 := 3*x-y=5:
```

Student Workspace

Problem 2: Three equations in three unknowns

(a) Example 2B gave a system of three linear equations in three unknowns that has a unique solution. In Exercise 2.2, you gave an example of such a system that has an entire line of solutions. We could also give an example of such a system that has no solutions by choosing two of the planes to be parallel, as in Exercise 2.1. In this Problem, you are to find a system of three linear equations in three unknowns that has no solutions but no two of the three planes are parallel. Use {eqn1, eqn2} from Section 2 (given again below) as two of your three equations.
(b) How is your third plane related geometrically to the line of intersection of the first two planes?
(Recall that when Maple's **solve** command finds no solutions, it simply returns no output.)

```
> eqn1 := x+2*y+3*z=3:
  eqn2 := 2*x-y-4*z=1:
```

Student Workspace

Problem 3: Five equations in three unknowns

Find a system of five linear equations in three unknowns that has a line of solutions, and choose your equations so that all five planes are different. Hint: Pick the line of intersection first, and pick one that is easy to work with, such as the z axis.

Student Workspace

Problem 4: Homogeneous equations

(a) Select any system of two homogeneous linear equations in three unknowns, and choose the equations so the two planes are different. (A linear equation is *homogeneous* if the constant on the right side of the equation is zero; that is, the equation has the form $a\,x + b\,y + c\,z = 0$. This is the equation of a plane that goes through the origin.) Find the solution set of this system and describe it geometrically.
(b) No matter how you chose your two planes in (a), you will always get the same kind of solution set

you described in (a). Explain why.

 Student Workspace

Problem 5: Essay on solution sets

(a) Describe all the possible solution sets for a system of linear equations in three unknowns. Be careful to include even the most extreme possibilities: The coefficients of x, y, and z might be all zero; some of the equations might be identical to one another; we might have just one equation, or two equations, or three or more equations. Give some illustrative examples.

(b) How does your answer change if all the equations in the system of linear equations are homogenous?

To help you understand the kind of answer that is desired here, we give a sample answer for systems of linear equations in two unknowns:

(a) If we have just one equation, $ax + by = c$, there are three possibilities: The solution set is a line, or the solution set is empty (for example, if the equation is $0x + 0y = 1$), or the solution set is the entire plane (if the equation is $0x + 0y = 0$). If we have two linear equations, all three of the above solution sets are still possible, and we have the additional possibility that the solution set is a single point. If we have more than two linear equations, all four of these solution sets are still possible.

(b) If all the equations are homogeneous, then the system always has the solution (0, 0). So the solution set is never empty; otherwise, our answer in (a) is unchanged.

 Student Workspace

Problem 6: Find a parabola through three points

Find the equation of the parabola $y = ax^2 + bx + c$ that passes through the points (1, 1), (2, 3), and (3, 9). (Hint: By substituting these three points into the equation $y = ax^2 + bx + c$, you will get three linear equations.) Check your answer by plotting the three points and the parabola; use the commands below, with the quadratic function suitably modified.

```
> p1 := pointplot([[1,1],[2,3],[3,9]],symbol=circle,color=blue):
  p2 := plot(x^2-2*x+3,x=0..4,y=0..10):
  display([p1,p2]):
```

 Student Workspace

Linear Algebra Modules Project
Chapter 1, Module 2

Solving Linear Systems using MAPLE

 Purpose of this module

The purpose of this module is to help you become familiar with some of Maple's many methods for solving a linear system and for reducing a matrix to echelon form. The module will also provide you with practice in solving linear systems without the distraction of doing tedious arithmetic, so you can see the overall strategy more clearly.

 Prerequisites

Consistent versus inconsistent linear systems; the augmented matrix of a linear system; elementary row operations; row equivalent matrices and equivalent linear systems; pivots; reducing a matrix to row echelon form and reduced row echelon form; back substitution. This module assumes that you have learned these concepts and have practiced solving linear systems by hand. (For example, first work through the Appendix "Pencil and Paper Tutorial: Gaussian Elimination" in the accompanying book.)

 Commands used in this module

```
> restart: with(LinearAlgebra): with(Lamp): with(plots):
  UseHardwareFloats := false: Digits := 6:
```

Tutorial

 Section 1: Overview of Methods

Before you work through this module, you might want to review the concepts listed under "Prerequisites"; their definitions can be found in the Appendix at the end of this Tutorial. In addition to these concepts, recall the following two theorems which underlie the techniques for solving linear systems.

Theorem 1: Suppose we are given two systems of linear equations, both with m equations and n unknowns. If their augmented matrices are row equivalent, then the two systems have exactly the same solution set.

Theorem 2: Every matrix is row equivalent to a matrix in row echelon form.

To solve a system of linear equations, we form the augmented matrix of the linear system and reduce it to row echelon form. Theorem 2 assures us that this can always be done, and Theorem 1 says that the linear system corresponding to the reduced matrix has the same solutions as the original linear system. The linear system corresponding to the reduced matrix is much easier to solve than the original, and it can always be solved by back substitution.

Sections 2 - 5 of this module are each devoted to a different Maple method for solving a system of linear equations. In each method, we first construct the augmented matrix of the linear system and then reduce it to row echelon form or reduced row echelon form.

The method in Section 2 is closest to the way you would reduce a matrix by hand: We perform step-by-step row reduction using the three elementary row operations. This will give you a chance to practice row reduction without worrying about the accuracy of the individual calculations, which we gladly leave for Maple to carry out.

In Section 3 we move up a level in computation: We use the **Pivot** command to direct Maple to carry out row reduction an entire column at at time.

Sections 4 and 5 will give you a chance to work with the **Reduce** command which, in one step, finds row echelon and reduced row echelon forms of an augmented matrix.

Section 2: Elementary Row Operations

The first step in solving a linear system is to reduce the augmented matrix of the linear system to row echelon form by using elementary row operations.

================

Example 2A: Use elementary row operations to reduce the augmented matrix of the linear system below to row echelon form.

$$-x_1 - 2x_2 + x_3 = -1$$
$$2x_1 + 4x_2 - 7x_3 = -8$$
$$4x_1 + 7x_2 - 3x_3 = 3$$

Solution: We begin by entering the augmented matrix for the system.
> M := Matrix([[-1,-2,1,-1],[2,4,-7,-8],[4,7,-3,3]]);

$$M := \begin{bmatrix} -1 & -2 & 1 & -1 \\ 2 & 4 & -7 & -8 \\ 4 & 7 & -3 & 3 \end{bmatrix}$$

Maple note: In the **Matrix** command, we enter one row at a time; each row is a list of numbers enclosed in square brackets, and the list of rows is enclosed in another pair of square brackets.

Below are the Maple commands corresponding to the elementary row operations. Take a moment to study the syntax of the **Rowop** command for each of the three types of row operations. In each case, M is the name of the matrix we are operating on.

Rowop(M, rj <= rj+c*rk)   replaces row j by row $j + (c$ times row $k)$
Rowop(M, rk <= c*rk)   replaces row k by c times row k $(c \neq 0)$
Rowop(M, rk <> rj)   interchanges row k and row j

In order to keep track of our progress, we will assign a new name, such as *M1*, *M2*, etc., to each matrix created as the result of applying a row operation. Look closely at each command before executing it. Can you predict what it will do?

```
> M := Matrix([[-1,-2,1,-1],[2,4,-7,-8],[4,7,-3,3]]);
```

$$M := \begin{bmatrix} -1 & -2 & 1 & -1 \\ 2 & 4 & -7 & -8 \\ 4 & 7 & -3 & 3 \end{bmatrix}$$

```
> M1 := Rowop(M,r2 <= r2+2*r1);
```

$$M1 := \begin{bmatrix} -1 & -2 & 1 & -1 \\ 0 & 0 & -5 & -10 \\ 4 & 7 & -3 & 3 \end{bmatrix}$$

```
> M2 := Rowop(M1,r3 <= r3+4*r1);
```

$$M2 := \begin{bmatrix} -1 & -2 & 1 & -1 \\ 0 & 0 & -5 & -10 \\ 0 & -1 & 1 & -1 \end{bmatrix}$$

```
> M3 := Rowop(M2,r2 <> r3);
```

$$M3 := \begin{bmatrix} -1 & -2 & 1 & -1 \\ 0 & -1 & 1 & -1 \\ 0 & 0 & -5 & -10 \end{bmatrix}$$

Note that matrix *M3* is in row echelon form.

================

The linear system whose augmented matrix is *M3* has the same solutions as our original system.

Exercise 2.1: (By hand) Solve the linear system in Example 2A by converting the matrix *M3* into three equations in the variables x_1, x_2, x_3. Then use back substitution to find the solution.

 Student Workspace

 Answer 2.1

The equations whose augmented matrix is *M3* are:

$$-x_1 - 2x_2 + x_3 = -1$$
$$-x_2 + x_3 = -1$$
$$-5x_3 = -10$$

Using back substitution (i.e., solving the last equations first), we find the solution $x_1 = -3, x_2 = 3, x_3 = 2$.

Maple can perform back substitution automatically on any augmented matrix in row echelon form. The command to use is **Backsolve**. Execute the next line to see how it works on matrix *M3*, and compare the result to your solution.

> `Backsolve(M3);`

$$\begin{bmatrix} -3 \\ 3 \\ 2 \end{bmatrix}$$

Exercise 2.2 : Using Maple, solve the following system of three equations by applying row operations to its augmented matrix. When you have reduced the augmented matrix to row echelon form, use **Backsolve** to find the solution.

$$x_1 + 5 x_2 + 2 x_3 = 1$$
$$-x_1 - 4 x_2 + x_3 = 6$$
$$x_1 + 3 x_2 - 3 x_3 = -9$$

> `M := Matrix([[1,5,2,1],[-1,-4,1,6],[1,3,-3,-9]]);`

$$M := \begin{bmatrix} 1 & 5 & 2 & 1 \\ -1 & -4 & 1 & 6 \\ 1 & 3 & -3 & -9 \end{bmatrix}$$

Student Workspace

Answer 2.2

> `M1 := Rowop(M,r2 <= r2+r1);`

$$M1 := \begin{bmatrix} 1 & 5 & 2 & 1 \\ 0 & 1 & 3 & 7 \\ 1 & 3 & -3 & -9 \end{bmatrix}$$

> `M2 := Rowop(M1,r3 <= r3-r1);`

$$M2 := \begin{bmatrix} 1 & 5 & 2 & 1 \\ 0 & 1 & 3 & 7 \\ 0 & -2 & -5 & -10 \end{bmatrix}$$

> `M3 := Rowop(M2,r3 <= r3+2*r2);`

$$M3 := \begin{bmatrix} 1 & 5 & 2 & 1 \\ 0 & 1 & 3 & 7 \\ 0 & 0 & 1 & 4 \end{bmatrix}$$

> `Backsolve(M3);`

$$\begin{bmatrix} 18 \\ -5 \\ 4 \end{bmatrix}$$

Section 3: The Pivot Command

Maple's **Pivot** command enables you to reduce a matrix to echelon form by reducing an entire column in one step. Once you choose the pivot position, Maple applies the appropriate row operations to create zeroes above and below it. Specifically, **Pivot(M,i,j)** treats the (i,j) entry of the matrix *M* as the

Chapter 1 Systems of Linear Equations

pivot for the *j*th column and uses row operations to zero out all the remaining entries of the *j*th column, provided the (i, j) entry is nonzero.

Exercise 3.1: The matrix M below is the augmented matrix of the linear system in Exercise 2.2. Use a sequence of **Pivot** commands to reduce M to row echelon form. Then use **Backsolve** to solve the system. The first pivot command is provided below to get you started.

```
> M := Matrix([[-1,-4,1,6],[1,6,2,1],[1,2,-3,-9]]);
```

$$M := \begin{bmatrix} -1 & -4 & 1 & 6 \\ 1 & 6 & 2 & 1 \\ 1 & 2 & -3 & -9 \end{bmatrix}$$

```
> M1 := Pivot(M,1,1);
```

$$M1 := \begin{bmatrix} -1 & -4 & 1 & 6 \\ 0 & 2 & 3 & 7 \\ 0 & -2 & -2 & -3 \end{bmatrix}$$

+ Student Workspace

− Answer 3.1

```
> M := Matrix([[-1,-4,1,6],[1,6,2,1],[1,2,-3,-9]]):
> M1 := Pivot(M,1,1):
> M2 := Pivot(M1,2,2);
```

$$M2 := \begin{bmatrix} -1 & 0 & 7 & 20 \\ 0 & 2 & 3 & 7 \\ 0 & 0 & 1 & 4 \end{bmatrix}$$

```
> M3 := Pivot(M2,3,3);
```

$$M3 := \begin{bmatrix} -1 & 0 & 0 & -8 \\ 0 & 2 & 0 & -5 \\ 0 & 0 & 1 & 4 \end{bmatrix}$$

```
> Backsolve(M3);
```

$$\begin{bmatrix} 8 \\ \frac{-5}{2} \\ 4 \end{bmatrix}$$

Section 4: Reduce to Row Echelon Form: Reduce(M)

The command **Reduce (M)** reduces the matrix M to row echelon form in one step. In the next example, we use **Reduce** to solve a system of linear equations. However, as the solution set in this example turns out to be infinite, we will have to understand how Maple displays this infinite set.

=================

Example 4A: Use the **Reduce** command to reduce to echelon form the augmented matrix of the system of linear equations below.

Module 1.2 Solving Linear Systems Using MAPLE

$$2x_1 - 6x_2 + 3x_3 - 2x_4 = -1$$
$$-x_1 + 3x_2 - 2x_3 = 4$$
$$3x_1 - 9x_2 + 4x_3 - 4x_4 = 2$$

Solution: As usual, we enter the augmented matrix for the system:

> `M := Matrix([[2,-6,3,-2,-1],[-1,3,-2,0,4],[3,-9,4,-4,2]]);`

$$M := \begin{bmatrix} 2 & -6 & 3 & -2 & -1 \\ -1 & 3 & -2 & 0 & 4 \\ 3 & -9 & 4 & -4 & 2 \end{bmatrix}$$

Now apply the **Reduce** command to M:

> `R := Reduce(M);`

$$R := \begin{bmatrix} 2 & -6 & 3 & -2 & -1 \\ 0 & 0 & \frac{-1}{2} & -1 & \frac{7}{2} \\ 0 & 0 & 0 & 0 & 0 \end{bmatrix}$$

===================

Exercise 4.1: (By hand) Answer the following questions about the linear system defined at the beginning of Example 4A by looking at the row echelon matrix R:
(a) "The system of equations is consistent and has an infinite number of solutions". Justify this statement by referring to pivot columns.
(b) Which are the leading variables (sometimes called "constrained" or "basic" variables) and which are the free variables?
(c) Express the leading variables in terms of the free variables.

 Student Workspace

 Answer 4.1

(a) The system of equations is consistent since the rightmost column of R is not a pivot column. The system of equations has an infinite number of solutions since the number of pivot columns (2) is less than the number of variables (4).
(b) The leading variables (corresponding to the pivot columns) are x_1 and x_3. The free variables are the remaining variables, x_2 and x_4.
(c) $x_1 = 10 + 3x_2 + 4x_4$, $x_2 = x_2$, $x_3 = -7 - 2x_4$, $x_4 = x_4$.

The **Backsolve** command produces the same answer:

> `Backsolve(R);`

$$\begin{bmatrix} 10 + 3X_2 + 4X_4 \\ X_2 \\ -7 - 2X_4 \\ X_4 \end{bmatrix}$$

Maple note: Since Maple cannot tell what names we have given our variables, we have to tell the **Backsolve** command the name it should use; by default it uses capital X. You can choose another name, such as lower-case x:

```
> Backsolve(R,free=x);
```

$$\begin{bmatrix} 10 + 3x_2 + 4x_4 \\ x_2 \\ -7 - 2x_4 \\ x_4 \end{bmatrix}$$

Section 5: Reduce to Reduced Row Echelon Form

The command **Reduce(M,form=rref)** reduces M all the way to <u>reduced</u> row echelon form. (The letters "r-r-e-f" stand for "reduced row echelon form.") We apply this command to the augmented matrix that arose in solving the system in Example 4A:

```
> M := Matrix([[2,-6,3,-2,-1],[-1,3,-2,0,4],[3,-9,4,-4,2]]);
```

$$M := \begin{bmatrix} 2 & -6 & 3 & -2 & -1 \\ -1 & 3 & -2 & 0 & 4 \\ 3 & -9 & 4 & -4 & 2 \end{bmatrix}$$

```
> R := Reduce(M,form=rref);
```

$$R := \begin{bmatrix} 1 & -3 & 0 & -4 & 10 \\ 0 & 0 & 1 & 2 & -7 \\ 0 & 0 & 0 & 0 & 0 \end{bmatrix}$$

And here, for sake of comparison, is what we got earlier using **Reduce(M)**:

```
> G := Reduce(M);
```

$$G := \begin{bmatrix} 2 & -6 & 3 & -2 & -1 \\ 0 & 0 & \frac{-1}{2} & -1 & \frac{7}{2} \\ 0 & 0 & 0 & 0 & 0 \end{bmatrix}$$

Exercise 5.1: (a) (By hand) Compare the echelon matrices R and G above. What do they have in common?
(b) (By Maple or by hand) Use elementary row operations to reduce G to R.

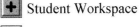 Student Workspace

 Answer 5.1

(a) The two matrices have the same pivot columns and hence the same leading variables (x_1, x_3) and free variables (x_2, x_4). And, of course, both matrices are row equivalent to the original matrix M and hence to each other.
(b)
```
> G1 := Rowop(G,r2 <= -2*r2);
```

$$G1 := \begin{bmatrix} 2 & -6 & 3 & -2 & -1 \\ 0 & 0 & 1 & 2 & -7 \\ 0 & 0 & 0 & 0 & 0 \end{bmatrix}$$

```
> G2 := Rowop(G1,r1 <= r1-3*r2);
```

$$G2 := \begin{bmatrix} 2 & -6 & 0 & -8 & 20 \\ 0 & 0 & 1 & 2 & -7 \\ 0 & 0 & 0 & 0 & 0 \end{bmatrix}$$

```
> G3 := Rowop(G2,r1 <= (1/2)*r1);
```

$$G3 := \begin{bmatrix} 1 & -3 & 0 & -4 & 10 \\ 0 & 0 & 1 & 2 & -7 \\ 0 & 0 & 0 & 0 & 0 \end{bmatrix}$$

Section 6: Pivot Columns and Free Variables

========================

Example 6A: Suppose that the augmented matrix for a linear system reduces to the row echelon matrix shown below.

$$\begin{bmatrix} 4 & -3 & 2 & -4 & 10 \\ 0 & 0 & 1 & 2 & -7 \\ 0 & 0 & 0 & 0 & 0 \end{bmatrix}$$

What information does this echelon matrix provide us about the solutions of the corresponding linear system?

Solution: We know that the system is consistent since there is no pivot in the rightmost column. We also conclude that the solution set has two leading variables, x_1 and x_3, corresponding to the pivot columns, and two free variables, x_2 and x_4. The fact that the solution set has two free variables is an instance of the following theorem (see Problem 8):

> **Theorem 3:** Suppose we are given a consistent linear system with m equations and n unknowns. If a row echelon form of the augmented matrix of this system has exactly r nonzero rows (i.e., rows that are not made up entirely of zeros), then the solution set of the system has exactly $n - r$ free variables.

In our example we have $n = 4$ and $r = 2$, so the number of free variables is $4 - 2 = 2$.

========================

The number r (i.e., the number of nonzero rows of an echelon form of a matrix A) is called the *rank* of A. (We will study the concept of rank in Module 3 of Chapter 5.) So we can restate Theorem 3 in terms of rank:

- The number of free variables in the solution set of a consistent linear system equals the number of unknowns minus the rank of the augmented matrix.

Since the leading entry of each nonzero row in an echelon form matrix identifies a pivot column, we can also restate Theorem 3 in terms of pivots:

- The number of free variables in the solution set of a consistent linear system equals the number of unknowns minus the number of pivot columns in the augmented matrix.

 Appendix: Quick Review of Concepts Related to the Solution of Linear Systems

To solve a system of linear equations, we first form the augmented matrix of the linear system. We then reduce this matrix to row echelon form (or reduced row echelon form) by a sequence of elementary row operations. The steps by which we perform this reduction follow a systematic procedure called "Gaussian elimination." The new matrix is the augmented matrix of an equivalent linear system that is much easier to solve. We then solve the new linear system, which completes our work since this system has the same solutions as the original system. We solve the new system by "back substitution," which means that we first solve the last equation for its leftmost unknown, then the next to last for its leftmost unknown, etc.

Here are the definitions of the concepts used in the above brief review and elsewhere in the Tutorial:

A system of linear equations is *consistent* if it has at least one solution; a system of linear equations is *inconsistent* if it has no solutions.

A matrix is in *row echelon form* if:
(1) all rows that consist entirely of zeros are grouped together at the bottom of the matrix; and
(2) the first (counting left to right) nonzero entry in each nonzero row appears in a column to the right of the first nonzero entry in the preceding row (if there is a preceding row).

A matrix is in *reduced row echelon form* if:
(1) the matrix is in row echelon form; and
(2) the first nonzero entry in each nonzero row is the number 1; and
(3) the first nonzero entry in each nonzero row is the only nonzero entry in its column.

The first nonzero entry in each nonzero row of a matrix in row echelon form is called the *leading entry* of that row; the locations of the leading entries are called the *pivot* positions, and the columns containing the leading entries are called the *pivot columns*.

Two systems of linear equations in n unknowns are *equivalent* if they have the same set of solutions.

Two m by n matrices A and B are *row equivalent* if B can be obtained from A by a sequence of elementary row operations.

The elementary *row operations* performed on a matrix are:
(1) replace row j by row $j + (c$ times row $k)$
(2) replace row k by c times row k $(c \neq 0)$
(3) interchange row k and row j

Problems

Problem 1: Solve by row operations and Pivot

Solve the following linear system twice: First use the **Rowop** command to apply elementary row operations one at a time, and then use the **Pivot** command. In each case, reduce the augmented matrix to echelon form and follow with **Backsolve**. Check that both methods give the same solution set.

$$2 x_1 - 2 x_2 - 4 x_3 = -2$$
$$2 x_1 - x_2 - x_3 = 2$$
$$-3 x_1 + 5 x_2 + 4 x_3 = 3$$
$$-x_1 - 5 x_2 + 4 x_3 = -3$$

Student Workspace

Problem 2: Solve by Reduce with/without form=rref

Solve the following linear system twice: First use **Reduce** without form=rref and then with form=rref. In each case, follow with **Backsolve**. Check that both methods give the same solution set.

$$x_1 + x_2 + 3 x_4 + 3 x_5 = 2$$
$$-2 x_1 + 2 x_2 - x_3 + 2 x_4 + 3 x_5 = 2$$
$$4 x_1 + x_3 + 4 x_4 + 3 x_5 = 2$$
$$-7 x_1 + 5 x_2 - 3 x_3 + 3 x_4 + 6 x_5 = 4$$

Student Workspace

Problem 3: Consistent or inconsistent?

Use any of our four methods to find out whether the following linear system is consistent or inconsistent. Explain why your conclusion follows from the Maple computation.

$$2 x_1 - 6 x_2 + 3 x_3 - 2 x_4 = -1$$
$$-x_1 + 3 x_2 - 2 x_3 = 4$$
$$3 x_1 - 9 x_2 + 4 x_3 - 4 x_4 = 1$$

Student Workspace

Problem 4: Reasoning from the echelon form of the augmented matrix

The matrices A, B, C in the next input region are augmented matrices for three different linear systems. For each matrix, do all of the following:
(a) (By hand) Write down the system of linear equations represented by this matrix.

(b) Use **Reduce** to reduce the matrix, but do not apply **Backsolve** yet. Just use the resulting echelon form to answer the following questions: Is the system consistent? What are the leading and free variables? If the system has a solution, is it unique? Explain your reasoning.
(c) Use back substitution to solve the system.

```
> A := Matrix([[3,-1,5,-1],[-1,4,0,4],[1,1,3,1],[-3,-1,-7,-1]]):
> B := Matrix([[1,0,-1,3,1],[1,-1,-1,4,1],
    [-3,3,3,-2,2],[3,1,-3,0,-1]]):
> C := Matrix([[1,-1,0,1,1],[1,2,4,1,-2],[2,1,0,2,-2],[1,2,1,1,-4]]):
```

 Student Workspace

Problem 5: A system with symbolic constants

Below is a system of three linear equations in three unknowns with two symbolic constants a and b. By changing the values of a and b, we can get not only different solutions but different numbers of solutions. To answer the following questions, you will have to apply **Reduce** without form=rref to the augmented matrix of this system (see matrix B below). Do not use **Backsolve**; just reason from the echelon form, and give explanations.

$$x_1 + 2x_2 + a x_3 = 4$$
$$2x_1 - x_2 + 3x_3 = b$$
$$3x_1 - 4x_2 + 2x_3 = -3$$

For what values of a and b does the system have
(a) no solutions?
(b) a unique solution?
(c) a line of solutions? (i.e., one free variable)
(d) a plane of solutions? (i.e., two free variables)

Suggestion: Check your answers by reducing B with a and b replaced by some specific values.

```
> B := Matrix([[1,2,a,4],[2,-1,3,b],[3,-4,2,-3]]):
```

 Student Workspace

Problem 6 (Challenge problem): Why can we solve for some variables but not others?

Maple has a built-in command, **solve**, for solving systems. In the input region below we have entered the system of equations from Example 4A, followed by a **solve** command. Compare the answer to our earlier results.

```
> eqn1 := 2*x[1] - 6*x[2] + 3*x[3] - 2*x[4]=-1:
> eqn2 := -x[1] + 3*x[2] - 2*x[3] = 4:
> eqn3 := 3*x[1] - 9*x[2] + 4*x[3] - 4*x[4] = 2:
> solve({eqn1,eqn2,eqn3}):
```

In the **solve** command, we can direct Maple to solve for particular variables:
```
> solve({eqn1,eqn2,eqn3},{x[1],x[3]}):
```

Notice that since we asked Maple to solve for x_1 and x_3, the free variables are x_2 and x_4. Direct Maple to solve for other pairs of variables. You will discover that there is one pair that it cannot solve for. Explain why. Hint: Try solving for this pair of variables by hand; that may help you identify the difficulty.

[+] Student Workspace

Problem 7: What echelon forms are possible?

(By hand) Does there exist a consistent linear system of three equations and five unknowns with (a) two free variables? (b) three free variables? (c) one free variable? In each of these three cases, either give an example of such a linear system (or its augmented matrix in echelon form), or explain why none can exist by discussing what echelon forms are possible for a 3 by 6 matrix.

[+] Student Workspace

Problem 8: Explanation of Theorem 3

Explain why Theorem 3 is true. Hint: Ask yourself how many pivot columns and how many non-pivot columns any row echelon form of the augmented matrix must have.

Theorem 3: Suppose we are given a consistent linear system with m equations and n unknowns. If a row echelon form of the augmented matrix of this system has exactly r nonzero rows (i.e., rows that are not made up entirely of zeros), then the solution set of the system has exactly $n - r$ free variables.

[+] Student Workspace

Problem 9: Guaranteed consistency

(a) Consider the linear system whose coefficient matrix is the matrix A below and augmented matrix is the matrix M below. (The last column of M is chosen at random.) Use **Reduce** with form=rref to find the reduced row echelon form of both A and M. Is the linear system whose augmented matrix is M consistent? What are the leading variables and the free variables for the linear system whose augmented matrix is M? Execute several times the command that defines M to see if your answer changes when the last column of M changes.
```
> A := Matrix([[5,-1,-6,5,-1],[0,3,3,0,4],[-9,-3,6,-9,-3]]):
> M := Matrix([A,Randmat(3,1)]):
```
(b) Suppose we are given a linear system of m equations and n unknowns, and suppose that the reduced row echelon form of the coefficient matrix has exactly m nonzero rows. Explain why the linear system must be consistent. (Hint: Looking back at part (a) should help.)

[+] Student Workspace

Linear Algebra Modules Project
Chapter 1, Module 3

Some Applications Leading to Linear Systems

Purpose of this module

The purpose of this module is to provide a few examples of the ways in which linear systems arise and are used in practice.

Prerequisites

Familiarity with linear systems and methods of solving them.

Commands used in this module

```
> restart: with(LinearAlgebra): with(plots): with(Lamp):
  UseHardwareFloats := false: Digits := 6:
```

Tutorial

Section 1: Fitting a Curve to Data

A common problem in many industries is creating elegant, efficient designs for products. Such designs are usually done on computers these days, using techniques that are known as "computer-aided design" (CAD). Typically, the design is made up of curves and surfaces that the designer manipulates on the computer screen. These curves and surfaces are often made up of graphs of fairly simple functions such as polynomials.

Fitting a Polynomial

Here is a simple design problem:

Example 1A: Find a polynomial of degree 2 whose graph goes through the points

$$(1, 2), (2, 6), (3, 4)$$

Solution: We are looking for a polynomial of the form

$$p(x) = a x^2 + b x + c$$

such that $p(1) = 2$, $p(2) = 6$, $p(3) = 4$. These three conditions will give us three linear equations in the three unknowns a, b, c. To find these equations, we define a function p and then substitute the three

points into the function:
```
> p := x -> a*x^2 + b*x + c;
```
$$p := x \to a x^2 + b x + c$$

Maple note: The notation `x -> expr` (where `expr` is an expression in x) defines a function. Since we defined p as a function, we can use standard function notation, such as p(x), to specify values of p.

```
> eqn1 := p(1)=2;
  eqn2 := p(2)=6;
  eqn3 := p(3)=4;
```
$$eqn1 := a + b + c = 2$$
$$eqn2 := 4a + 2b + c = 6$$
$$eqn3 := 9a + 3b + c = 4$$

We solve the system by reducing the augmented matrix of the linear system. Then we substitute the values for a, b, and c back into the polynomial p:

```
> A := Matrix([[1,1,1,2],[4,2,1,6],[9,3,1,4]]):
> Backsolve(Reduce(A));
```
$$\begin{bmatrix} -3 \\ 13 \\ -8 \end{bmatrix}$$

```
> q := subs({a=-3,b=13,c=-8},p(x));
```
$$q := -3 x^2 + 13 x - 8$$

Maple note: The `subs` command above substitutes the values of a, b, and c into the expression p(x).

Finally, let's graph the polynomial and the original three points together in one plot:

```
> p1 := plot(q,x=0..4):
  p2 := pointplot({[1,2],[2,6],[3,4]},symbol=circle,
    symbolsize=15,color=blue):
  display([p1,p2]);
```

Exercise 1.1: Find a polynomial of degree 3 whose graph goes through the points

$$(-1, 20), (1, 12), (2, 5), (3, 16)$$

Graph the polynomial and the four points in a single plot.

Student Workspace

Answer 1.1

We define a function p and determine the values of the unknown coefficients *a*, *b*, *c*, *d*:

```
> p := x -> a*x^3 + b*x^2 + c*x + d;
```

$$p := x \to a x^3 + b x^2 + c x + d$$

```
> eq1 := p(-1)=20;
  eq2 := p(1)=12;
  eq3 := p(2)=5;
  eq4 := p(3)=16;
```

$$eq1 := -a + b - c + d = 20$$
$$eq2 := a + b + c + d = 12$$
$$eq3 := 8a + 4b + 2c + d = 5$$
$$eq4 := 27a + 9b + 3c + d = 16$$

```
> B := Matrix([[-1,1,-1,1,20],[1,1,1,1,12],
  [8,4,2,1,5],[27,9,3,1,16]]):
> s := Backsolve(Reduce(B)):
```

Here is the polynomial whose graph goes through the four points:

```
> q := subs({a=s[1],b=s[2],c=s[3],d=s[4]},p(x));
```

$$q := \frac{5}{2}x^3 - 6x^2 - \frac{13}{2}x + 22$$

```
> p1 := plot(q,x=-2..4):
  p2 := pointplot({[-1,20],[1,12],[2,5],[3,16]},
    symbol=circle,symbolsize=15,color=blue):
  display([p1,p2]);
```

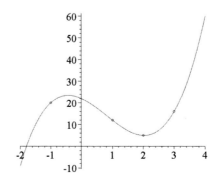

Sometimes the slope of the polynomial is also given at selected points; that too leads to linear equations, as you will see in some of the problems.

==================

Example 1B: Find all polynomials of degree 2 whose graph goes through (1, 2) and (2, 6).
Solution: In Example 1A, we found the unique polynomial of degree 2 whose graph goes through three given points, and in Exercise 1.1, you found the unique polynomial of degree 3 whose graph goes through four given points. In this problem, there are too few points to specify the polynomial completely, which means we can expect lots of solutions to this problem.

```
> p := x -> a*x^2 + b*x + c;
```

$$p := x \to a x^2 + b x + c$$

```
> eqn1 := p(1)=2;
  eqn2 := p(2)=6;
```

$$eqn1 := a + b + c = 2$$
$$eqn2 := 4a + 2b + c = 6$$

```
> A := Matrix([[1,1,1,2],[4,2,1,6]]);
```

$$A := \begin{bmatrix} 1 & 1 & 1 & 2 \\ 4 & 2 & 1 & 6 \end{bmatrix}$$

```
> soln := Backsolve(Reduce(A));
```

$$soln := \begin{bmatrix} 1 + \frac{1}{2} X_3 \\ 1 - \frac{3}{2} X_3 \\ X_3 \end{bmatrix}$$

Let's use the **subs** command to replace X_3 by the simpler name t; the components of the solution are the values for the coefficients a, b, c. Note that we get infinitely many parabolas, $p(x) = a x^2 + b x + c$:

```
> subs(X[3]=t,soln);
```

$$\begin{bmatrix} 1 + \frac{1}{2} t \\ 1 - \frac{3}{2} t \\ t \end{bmatrix}$$

```
> a := 1+t/2:
  b := 1-3*t/2:
  c := t:
> p(x);
```

$$\left(1 + \frac{1}{2} t\right) x^2 + \left(1 - \frac{3}{2} t\right) x + t$$

We graph a few of these parabolas:
```
> p1 := subs(t=2,p(x)):
> p2 := subs(t=-2,p(x)):
> p3 := subs(t=-8,p(x)):
> plot1:=plot([p1,p2,p3],x=0..3,
    color=[red,blue,black],thickness=[3,2,1]):
  plot2 := pointplot([[1,2],[2,6]],
  symbol=circle,symbolsize=20,color=black):
  display([plot1,plot2]);
```

The **animate** command below shows a sequence of parabolas corresponding to integer values of t ranging from -25 to 25, which will give you a feel for the entire continuum of polynomials that pass through the points (1, 2) and (2, 6). To see the animation, execute the commands below, then click on the picture to bring up the animation buttons in the toolbar above, and then click on the play button.

```
> plot1 := animate(p(x),x=0..3,t=-25..25,color=red,frames=51):
  plot2 := pointplot({[1,2],[2,6]},symbol=circle,color=black):
  display([plot1,plot2]):
```

======================

Fitting a Cubic Spline

If we have a large number of points that we want to fit a curve to, choosing a single polynomial is not wise. For such a large number of points, we would need a polynomial of high degree. But a polynomial of high degree usually has a great many maximums and minimums and therefore may have a very bumpy graph. A better choice of curve may be a *cubic spline*, in which we connect consecutive points with cubic (i.e., degree 3) polynomials. For example, here is the graph of the polynomial of degree 7 that goes through a given set of eight points (thin black curve), together with the cubic spline (thick red curve made up of seven adjacent cubics) that goes through the same eight points:

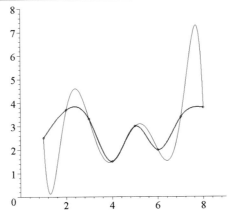

Observe that the graph of the polynomial swings well above and below the given points. On the other hand, the cubic spline not only avoids such wild swings but its adjacent cubics fit together so smoothly that you cannot see where one cubic ends and the next begins. The smoothness of the spline is a consequence of the following "compatibility" condition, which is required of all splines:

- If the cubics p(x) and q(x) both go through a point (a, b), the first derivatives of p and q must be equal at x = a, and their second derivatives must also be equal at x = a.

This condition insures that adjacent cubics have the same tangent lines and same concavity at the points where they meet.

Here is a simple example:

==================

Example 1C: Fit a cubic spline to the three points $P = (1, 3)$, $Q = (2, 7)$ and $R = (3, 4)$.

Solution: First we clear the values of the variables a, b, and c, so we can use these letters again as variables:

```
> a := 'a': b := 'b': c := 'c':
```

Then we introduce two cubic polynomials, one defined on the interval $[1, 2]$ and the other on $[2, 3]$:

```
> p := x -> a*x^3 + b*x^2 + c*x + d;
  q := x -> e*x^3 + f*x^2 + g*x + h;
```

$$p := x \to a x^3 + b x^2 + c x + d$$
$$q := x \to e x^3 + f x^2 + g x + h$$

Notice that there are altogether 8 unknown coefficients in the two polynomials. Our goal is therefore to find a system of 8 linear equations in terms of these unknowns. Solving this system will then give us the coefficients of the two cubics making up the cubic spline.

The polynomial p has to go through the points P and Q. This gives us the first two equations:

```
> eqn1 := p(1)=3;
  eqn2 := p(2)=7;
```

$$eqn1 := a + b + c + d = 3$$
$$eqn2 := 8a + 4b + 2c + d = 7$$

The second polynomial q has to go through the points Q and R.

```
> eqn3 := q(2)=7;
  eqn4 := q(3)=4;
```

$$eqn3 := 8e + 4f + 2g + h = 7$$
$$eqn4 := 27e + 9f + 3g + h = 4$$

The remaining four equations are based on the first and second derivatives of the polynomials. We calculate the first derivatives of p and q using Maple's D operator, naming them Dp and Dq, respectively.

```
> Dp := D(p);
  Dq := D(q);
```

$$Dp := x \to 3ax^2 + 2bx + c$$
$$Dq := x \to 3ex^2 + 2fx + g$$

In a similar way we define the second derivatives, naming them DDp and DDq.

```
> DDp := D(Dp);
  DDq := D(Dq);
```

$$DDp := x \to 6ax + 2b$$
$$DDq := x \to 6ex + 2f$$

Since both p and q go through the point $Q = (2, 7)$, they must satisfy the above compatibility condition; that is, the first derivatives of p and q must be equal at $x = 2$, and their second derivatives must be equal there as well:

```
> eqn5 := Dp(2)=Dq(2);
  eqn6 := DDp(2)=DDq(2);
```

$$eqn5 := 12a + 4b + c = 12e + 4f + g$$
$$eqn6 := 12a + 2b = 12e + 2f$$

So far we have only six equations for our eight unknown coefficients; so we need two more. One convention that is often used to provide the two missing equations is to require that the second derivatives of the spline are zero at the leftmost and rightmost points. So:

```
> eqn7 := DDp(1)=0;
  eqn8 := DDq(3)=0;
```

$$eqn7 := 6a + 2b = 0$$
$$eqn8 := 18e + 2f = 0$$

Now that we have eight equations, we solve the system in the ususal way. However, we introduce a shortcut that saves us from having to type in the 8 by 9 matrix:

```
> M := Genmatrix([eqn1,eqn2,eqn3,eqn4,eqn5,eqn6,eqn7,eqn8],
    [a,b,c,d,e,f,g,h]);
```

$$M := \begin{bmatrix} 1 & 1 & 1 & 1 & 0 & 0 & 0 & 0 & 3 \\ 8 & 4 & 2 & 1 & 0 & 0 & 0 & 0 & 7 \\ 0 & 0 & 0 & 0 & 8 & 4 & 2 & 1 & 7 \\ 0 & 0 & 0 & 0 & 27 & 9 & 3 & 1 & 4 \\ 12 & 4 & 1 & 0 & -12 & -4 & -1 & 0 & 0 \\ 12 & 2 & 0 & 0 & -12 & -2 & 0 & 0 & 0 \\ 6 & 2 & 0 & 0 & 0 & 0 & 0 & 0 & 0 \\ 0 & 0 & 0 & 0 & 18 & 2 & 0 & 0 & 0 \end{bmatrix}$$

Maple note: The **Genmatrix** command generates an augmented matrix from a system of equations. The first argument in the command is the list of equations and the second is the list of variables used in the equations.

```
> s := Backsolve(Reduce(M)):
```

Matching this solution with the list of unknowns [a, b, c, d, e, f, g, h] gives us the values for each of the coefficients. We then graph the cubic spline:

```
> cubic1 := subs({a=s[1],b=s[2],c=s[3],d=s[4]},p(x));
```

$$cubic1 := -\frac{7}{4}x^3 + \frac{21}{4}x^2 + \frac{1}{2}x - 1$$

```
> cubic2 := subs({e=s[5],f=s[6],g=s[7],h=s[8]},q(x));
```

$$cubic2 := \frac{7}{4}x^3 - \frac{63}{4}x^2 + \frac{85}{2}x - 29$$

```
> p1 := plot(cubic1,x=1..2,color=red):
  p2 := plot(cubic2,x=2..3,color=blue):
  p3 := pointplot({[1,3],[2,7],[3,4]},
    symbol=circle,symbolsize=15,color=black):
> display([p1,p2,p3],view=[0..4,0..8]);
```

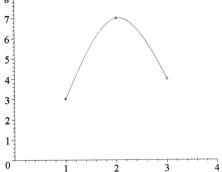

Chapter 1 Systems of Linear Equations

Section 2: Estimating Inside Temperatures From Outside

Suppose we have a thin plate whose temperatures along its outside edges are known, and we want to know the temperatures at its interior points. (We will assume that the temperatures are in a "steady state"; that is, they are not changing with time.)

==================

Example 2A: The rectangular plate below has temperatures of 5 degrees along three edges and 15 degrees along the top edge. Find the temperatures at the four indicated interior points, $t1, t2, t3, t4$.

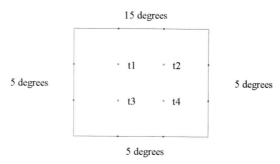

Solution: There are several ways to estimate the temperatures at the interior points, and we will pick the simplest:

- At each interior grid point, assume that the temperature is the average of the temperatures at the four surrounding grid points.

For example, we assume the temperature $t1$ is the average of the temperatures 5 (at left), 15 (above), $t2$ (at right), and $t3$ (below). This method leads to a system of four linear equations in the four unknown interior temperatures:

```
> e1 := t1=(5+15+t2+t3)/4;
  e2 := t2=(t1+15+5+t4)/4;
  e3 := t3=(5+t1+t4+5)/4;
  e4 := t4=(t3+t2+5+5)/4;
```

$$e1 := t1 = 5 + \frac{1}{4} t2 + \frac{1}{4} t3$$

$$e2 := t2 = \frac{1}{4} t1 + 5 + \frac{1}{4} t4$$

$$e3 := t3 = \frac{5}{2} + \frac{1}{4} t1 + \frac{1}{4} t4$$

$$e4 := t4 = \frac{1}{4} t3 + \frac{1}{4} t2 + \frac{5}{2}$$

```
> C := Genmatrix([e1,e2,e3,e4],[t1,t2,t3,t4]):
> Backsolve(Reduce(C)):
```

Therefore our estimates of the interior temperatures are $t1 = 8.75$, $t2 = 8.75$, $t3 = 6.25$, $t4 = 6.25$.

Of course, when this method is used in a practical problem, one usually wants a very large number of interior points for greater accuracy. So the number of equations and unknowns is then very large.

Problems

Problem 1: Polynomial through five points

Find the polynomial of smallest degree whose graph goes through the five points listed below; graph the polynomial and the five points in a single plot.

$$(0, 4), (.5, 2), (1, 1), (1.5, 2.5), (2, 3)$$

 Student Workspace

Problem 2: Given three points and the slope at each

Find the polynomial of degree five whose graph goes through the points (0, 4), (1, 1), (2, 3) and whose slope at $x = 0$ is -1, at $x = 1$ is 2, and at $x = 2$ is -3. Graph the polynomial and the three points in a single plot. (Hint: Use the method of Example 1A, but add equations for the given conditions on the slopes.)

 Student Workspace

Problem 3: Design a ski jump

Design a ski jump that has the following specifications. The ski jump starts at a height of 100 feet and finishes at a height of 10 feet. From start to finish, the ski jump covers a horizontal distance of 120 feet. A skier using the jump will start off horizontally (i.e., with slope = 0) and will fly off the end at a 30 degree angle up from the horizontal. Find a single polynomial whose graph is a side view of the ski jump. [From ATLAST: Computer Exercises for Linear Algebra by Steven Leon, Eugene Herman, and Richard Faulkenberry, Prentice-Hall, 1996, page 8.]

 Student Workspace

Problem 4: Conic section through five points

The equation $a x^2 + b x y + c y^2 + d x + e y + f = 0$, where a, b, and c are not all zero, defines a "conic section" (i.e., an ellipse, a hyperbola, a parabola, or a pair of intersecting lines).

(a) Find the unique conic section through the points (-1, 1), (3, 2), (4, 1/2), (1, 0), and (−2, 2). Which of the four types of conic sections is it? Use the following commands to plot the points and your conic section, with the formula for the conic suitably modified:

34   Chapter 1 Systems of Linear Equations

```
> p1 := pointplot([[-1,1],[3,2],[4,1/2],[1,0],[-2,2]],
    symbol=circle,color=blue):
  p2 := implicitplot(.5*x^2-x*y+3*y^2-2*x-2*y+1=0,
    x=-4..6,y=-2..4,numpoints=500):
  display([p1,p2]):
```
Hint: Use the following function definition to generate the equations:
```
> p := (x,y) -> a*x^2+b*x*y+c*y^2+d*x+e*y+f:
```
(b) Change the point (-1, 1) to (1, 1), and repeat part (a).

 Student Workspace

 Problem 5: Cubic spline through four points

In Example 1C we fit a cubic spline to the three points $P = (1, 3)$, $Q = (2, 7)$ and $R = (3, 4)$. Now we extend that problem by adding one more point. You are to fit a cubic spline to the three points P, Q, R and the additional point $S = (5, 8)$. So you will have to find three cubics and hence will need 12 equations. The first seven equations that we used In Example 1C are applicable here, but eqn8 is not since 3 is not the rightmost x coordinate. Your task is to come up with equations 8 through 12 and then solve the system.

To get you started, all the relevant commands from Example 1C have been entered in the Student Workspace below as has the third cubic polynomial, $r(x) = j x^3 + k x^2 + m x + n$, which will connect points R and S. Your solution should include a picture that shows the cubic spline along with the four given points.

 Student Workspace

 Problem 6: Cubic spline on a large data set

(a) Maple has a function that calculates a cubic spline automatically. Apply it to the points

(0, 0.0), (1, 0.4), (2, 0.4), (3, 0.2), (4, 0.5), (5, -0.5), (6, 0.9), (7, -0.9)

Graph the spline and the points in a single plot. (The commands to do this are given in the Student Workspace. You merely need to execute them.)

(b) Change the sixth point from (5, -0.5) to (5, 0.3), and execute the commands again. Did that alter only the two cubics that go through that point, or did it alter the coefficients of the other cubics as well? Explain.

(c) (By hand) For the problem in (a), Maple had to solve a system of 28 equations in 28 unknowns (seven cubics, each with four unknown coefficients). Write out the row of the augmented matrix that corresponds to the equation which represents the requirement that the graph of the first cubic goes through the second point, (1, 0.4). Suggestion: Denote the 28 unknowns by $a_1, a_2, ..., a_{28}$, and the seven

cubics by $p_1, p_2, ..., p_7$, where
$$p_1 = a_1 x^3 + a_2 x^2 + a_3 x + a_4, \quad p_2 = a_5 x^3 + a_6 x^2 + a_7 x + a_8, \quad \text{etc.}$$

+ Student Workspace

Problem 7: Temperature grid

Estimate the temperatures at the six interior points of the plate pictured below, using the method of Example 2A.

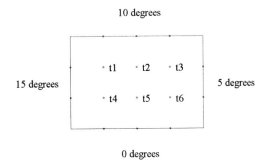

+ Student Workspace

Chapter 2: Vectors

Module 1. Geometric Representation of Vectors in the Plane

Module 2. Linear Combinations of Vectors

Module 3. Decomposing the Solution of a Linear System

Module 4. Linear Independence of Vectors

 Commands used in this chapter

`display([pict1,pict2]);` displays together a group of previously defined pictures.
`Matrix([[a,b],[c,d]]);` defines the matrix with rows $[a, b]$ and $[c, d]$.
`Matrix([u,v,w]);` produces the matrix whose columns are the vectors u, v, w.
`plot3d(expr, x=a..b, y=c..d);` plots a graph of a function of two variables.
`pointplot3d([pt1,pt2,...,ptn]);` plots the list of points $pt1$, $pt2$, ..., ptn, where each point is given as a list of three coordinates.
`spacecurve([expr1,expr2,expr3],t=a..b);` plots a curve in 3-space where $expr1$, $expr2$, $expr3$ are the component functions and the parameter t ranges from a to b.
`subs({a=3,b=13},expr);` substitutes the values for a and b in the expression $expr$.
`Vector([a,b,c]);` defines the vector $\langle a, b, c \rangle$.

LAMP commands:
`Backsolve(R);` solves the linear system for which R is a row echelon form of the augmented matrix.
`Drawmatrix(F);` draws the 2d or 3d figure whose vertices are the columns of the matrix F.
`Drawvec(u, [v,w]);` draws the vector u with tail at the origin and the vector with tail at v and head at w. (Vectors are in 2-space.)
`Drawvec3d(u, [v,w]);` draws the vector u with tail at the origin and the vector with tail at v and head at w. (Vectors are in 3-space.)
`Expand(expr);` expands and evaluates the vector or matrix expression $expr$.
`Gridgame(a,b);` displays $a u + b v$, where u and v are given vectors, and target points are indicated.
`Reduce(M);` produces a row echelon form of the matrix M.
`Unitspan(v,w);` shades in the parallelogram of points $a v + b w$, where a and b range from 0 to 1.
`Vectorgrid(u,v,[a1..a2,b1..b2]);` displays a grid based on the vectors u and v, and shades in all the points $a u + b v$, where a is between $a1$ and $a2$, b is between $b1$ and $b2$.
`Vectorline(v,p);` draws the line through the origin with direction vector v and the parallel line with position vector p.

Linear Algebra Modules Project
Chapter 2, Module 1

Geometric Representation of Vectors in the Plane

Purpose of this module

The purpose of this module is to help you visualize addition and scalar multiplication of vectors in the plane. We then use these ideas to develop the parametric representation of any line in the plane.

Prerequisites

Some familiarity with vectors.

Commands used in this module

```
> restart: with(LinearAlgebra): with(plots): with(Lamp):
  UseHardwareFloats := false: Digits := 6:
```

Tutorial

Section 1: The Geometry of Vector Addition

Geometric Representations of Vectors

Sometimes we represent a vector $\begin{bmatrix} a \\ b \end{bmatrix}$ in R^2 by the point (a, b) in the xy coordinate plane. Other times we represent $\begin{bmatrix} a \\ b \end{bmatrix}$ by the "arrow" (i.e., directed line segment) whose tail is at the origin and head is at the point (a, b). In the figure below, you see the vector $\begin{bmatrix} 3 \\ 1 \end{bmatrix}$ represented both as the point $(3, 1)$ (shown as a small blue circle) and as an arrow from the origin to that point.

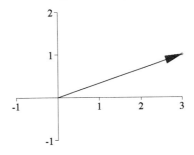

The command **Drawvec** draws vectors as arrows:

Chapter 2 Vectors

```
> v := Vector([3,1]);
```

$$v := \begin{bmatrix} 3 \\ 1 \end{bmatrix}$$

```
> Drawvec(v);
```

You can also draw any number of vectors at once:

```
> u := Vector([1,2]);
```

$$u := \begin{bmatrix} 1 \\ 2 \end{bmatrix}$$

```
> Drawvec(u,v);
```

The **Drawvec** command has several options for customizing a picture. For example, we can specify the head colors:

```
> Drawvec([u,headcolor=blue],[v,headcolor=red]);
```

As we proceed through this chapter and beyond, there will be times when it helps to think of vectors as points and other times when arrows provide a more useful visualization. As much as possible, we will remind you of which interpretation we are using; but try to get in the habit of considering both possibilities when you approach a new situation.

Vector Addition

Recall that we add vectors *componentwise*; that is, we add the corresponding entries (i.e., components) of the vectors together to get their sum. Below, we add the vectors *u* and *v*:

```
> u+v;
```
$$\begin{bmatrix} 4 \\ 3 \end{bmatrix}$$

Plotting two vectors and their sum as arrows provides us with a useful picture of vector addition; in the figure below we show the sum $\begin{bmatrix} 1 \\ 2 \end{bmatrix} + \begin{bmatrix} 3 \\ 1 \end{bmatrix} = \begin{bmatrix} 4 \\ 3 \end{bmatrix}$. Notice that if we form a parallelogram with the vectors u and v as adjacent sides, then the displayed diagonal of the parallelogram gives their sum:

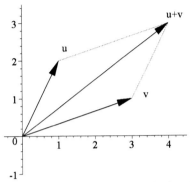

Exercise 1.1: What vector addition is illustrated by the figure below?

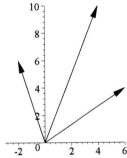

+ Student workspace

− Answer 1.1

$$\begin{bmatrix} -2 \\ 6 \end{bmatrix} + \begin{bmatrix} 6 \\ 4 \end{bmatrix} = \begin{bmatrix} 4 \\ 10 \end{bmatrix}$$

Vector Addition as Translation

If you ask small children to add 8 + 4, they will probably start at 8 and count: 9,10,11,12. In school their teacher will have them do this on a number line: start at 8 and move 4 places to the right. We can think of vector addition in a similar way. If we start with the vector $u = \begin{bmatrix} 1 \\ 2 \end{bmatrix}$ and add to it the vector $v = \begin{bmatrix} 3 \\ 1 \end{bmatrix}$, we have thereby moved 3 units to the right and 1 unit up to arrive at our result $\begin{bmatrix} 4 \\ 3 \end{bmatrix}$. We say that we have *translated* the vector $u = \begin{bmatrix} 1 \\ 2 \end{bmatrix}$ by the vector $v = \begin{bmatrix} 3 \\ 1 \end{bmatrix}$, and we refer to v as a *translation* vector.

One way to visualize this process is to think of the translation vector $v = \begin{bmatrix} 3 \\ 1 \end{bmatrix}$ as if it moves the original point (1, 2) to the resulting point (4, 3). In the figure below, we move the point (1, 2) to the point (4, 3) along a copy of the vector $v = \begin{bmatrix} 3 \\ 1 \end{bmatrix}$ drawn so that its tail begins at the point (1, 2):

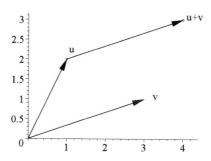

You can also draw this picture using **Drawvec**:

```
> u := Vector([1,2]):
  v := Vector([3,1]):
> Drawvec(u,v,[u,u+v]):
```

The argument [u, $u + v$] in the above **Drawvec** command describes a vector from the head of u to the head of $u + v$.

We could just as easily have taken the vector v as our starting point and added the vector u, tail to head, to again arrive at $u + v$. In this case u would be playing the role of the translation vector. Execute the next command to see this other way of getting to $u + v$, now using distinct head colors:

```
> Drawvec([u,headcolor=blue],[v,headcolor=red],
    [v,u+v,headcolor=black]);
```

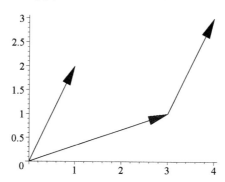

Translation of Geometric Figures

The simple idea of translating (i.e., moving) one point by adding a vector to it can easily be extended to a set of points. In particular, if the set of points defines a geometric figure, adding a vector to each of the points translates the entire figure. For our first example, we will use a collection of four points that forms the letter "T". We draw this picture by connecting the four points (2, 1.5), (2, 4), (3, 4), (1, 4), in that order, with straight line segments. The columns of the matrix T below contain these four points, and the **Drawmatrix** command draws the points and the line segments connecting them:

```
> T := Matrix([[2, 2, 3, 1], [1.5, 4, 4, 4]]);
```

$$T := \begin{bmatrix} 2 & 2 & 3 & 1 \\ 1.5 & 4 & 4 & 4 \end{bmatrix}$$

```
> Drawmatrix(T);
```

Let's add the translation vector $p = \begin{bmatrix} 3 \\ 1 \end{bmatrix}$ to each of the four defining points, which results in four new points. Hence the entire letter T is translated by moving its four defining points the same distance and in the same direction:

```
> p := Vector([3,1]):
  Vectranslate(T,p);
```

For a simpler view of the original and translated points we use "vectors=off":

```
> Vectranslate(T,p,vectors=off):
```

Exercise 1.2: What translation vector *p* is required to move the letter T so that its horizontal bar lies on the *x* axis between 4 and 6? Test your answer by changing the value of *p* below.

```
> p := Vector([2,-1]):
  Vectranslate(T,p):
```

 Student workspace

 Answer 1.2

```
> p := Vector([3,-4]):
  Vectranslate(T,p);
```

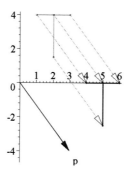

Section 2: The Geometry of Scalar Multiplication

Recall that when we multiply a vector by a real number (i.e., a scalar) we multiply *componentwise*; that is, we multiply each entry (i.e., component) of the vector by the scalar. Below, we define a vector *v* and a scalar *c*, then calculate the scalar multiple *c v*.

```
> v := Vector([3,4]):
> c := 2:
  c*v;
```

$$\begin{bmatrix} 6 \\ 8 \end{bmatrix}$$

The geometric effect of scalar multiplication is easy to see, if we use arrows to display the vectors *v* and *c v*. In the figure below, the head of *v* is in red and the head of the scalar multiple *c v* is in blue.

```
> Drawvec(v,[c*v,headcolor=blue]);
```

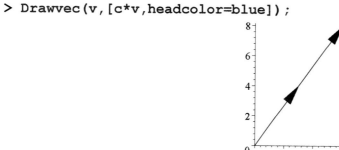

Explore ~~~ Try changing the value of c in **Drawvec** above. Can you predict what you will get when $c = 3$? $c = -2$? $c = .5$?

Exercise 2.1: What do all scalar multiples of a given vector have in common geometrically?

+ Student workspace

− Answer 2.1

They all lie on the line through the origin that contains the given vector.

Section 3: Parametric Representation of a Line

Parametric Representations of Lines Through the Origin: $t\,v$

If we consider the set of all scalar multiples of a given nonzero vector v, we obtain all the points on the line L through the origin that contains v. The vector v is called a *direction vector* for the line L. Here is a picture of the line L (in red) along with its direction vector $v = \begin{bmatrix} 3 \\ 4 \end{bmatrix}$:

```
> v := Vector([3,4]):
> Vectorline(v);
```

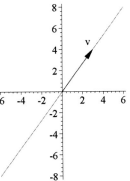

We can describe all the points on L algebraically by writing $t\,v$, where t ranges over all the real numbers. The expression $t\,v$ is called a *parametric representation* (or *vector form*) of the line L, and t is called a parameter. A parametric representation for L is therefore $t\,v = t\begin{bmatrix} 3 \\ 4 \end{bmatrix} = \begin{bmatrix} 3\,t \\ 4\,t \end{bmatrix}$. Another way to express this parametric representation is $x = 3\,t$, $y = 4\,t$. A cartesian equation for L is found by eliminating t between these two equations: $y = \dfrac{4\,x}{3}$.

Exercise 3.1: Find a parametric representation for the line $x + 3\,y = 0$, and give a direction vector for this line. Use the **Vectorline** command to check your answer.

+ Student Workspace

Answer 3.1

Parametric representation: $\begin{bmatrix} -3t \\ t \end{bmatrix}$, where t ranges over the real numbers. Direction vector: $\begin{bmatrix} -3 \\ 1 \end{bmatrix}$.

Many answers are possible. Find several. What do they all have in common?

Parametric Representation of Lines: $tv + p$

At the conclusion of Section 1, we observed that if a translation vector p is added to a set of points, then those points move together in the direction of p. Consider now a line L through the origin as our set of points. So if we add a vector p to each of the points on the line L, we create a new line $L + p$. If the points on L are described parametrically by tv, the points on $L + p$ can be described by the expression $tv + p$, where t ranges over the real numbers. This expression is called a *parametric representation* (or *vector form*) of the line $L + p$.

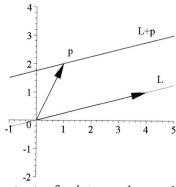

Note: When we use vectors to represent sets of points, we draw only the tips of the vectors. For example, notice that when drawing the line $L + p$ above, we used only the tips of the vectors $tv + p$.

======================

Example 3A: Suppose we are given the line through the origin with direction vector $v = \begin{bmatrix} 4 \\ 2 \end{bmatrix}$ and we are given the translation vector $p = \begin{bmatrix} -3 \\ 4 \end{bmatrix}$. Then the translated line has parametric representation

$$tv + p = t\begin{bmatrix} 4 \\ 2 \end{bmatrix} + \begin{bmatrix} -3 \\ 4 \end{bmatrix} = \begin{bmatrix} 4t - 3 \\ 2t + 4 \end{bmatrix}$$

Alternatively, we can write $x = 4t - 3$, $y = 2t + 4$. The resulting line (thick and blue) is parallel to the direction vector v and passes through the point at the head of the vector p. Because the head of the vector p is a point on the line, p is sometimes referred to as a *position vector* for the line.

```
> v := Vector([4,2]):
  p := Vector([-3,4]):
> Vectorline(v,p);
```

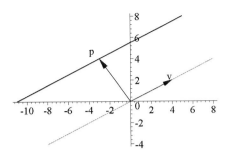

Exercise 3.2: Find a cartesian equation for the blue line above.

 Student Workspace

 Answer 3.2

Solve $x = 4t - 3$ for t in terms of x; then substitute in $y = 2t + 4$. The result is $y = \dfrac{x}{2} + \dfrac{11}{2}$.

Problems

 Problem 1: Vector addition

Six vectors are shown in the figure below. Two of these vectors add up to one of the others. Find this set of three related vectors and write down how they are related, as in $B + C = F$.

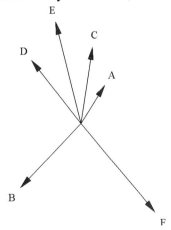

+ Student Workspace

Problem 2: Position and direction vectors from a picture

By simple inspection, find a direction vector *v* for the two lines below, and find a position vector *p* for the thick blue line. Then find parametric equations for the red and blue lines. Check your work by using **Vectorline(v,p)**.

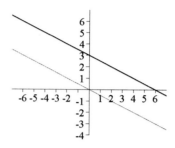

+ Student Workspace

Problem 3: Parametric equations from a cartesian equation

Consider the line $y = 2x - 5$. Do parts (a) through (d) by hand.
(a) Find a direction vector *v* for this line.
(b) Find a position vector *p* for this line.
(c) Find the resulting parametric representation for this line.
(d) There are many answers possible for parts (a) and (b). Choose different answers for both parts and then use them to write another vector form for the line.
(e) Use **Drawvec** to plot all four vectors you found in parts (a), (b), and (d) above. Be sure to note whether they are correctly related to the line $y = 2x - 5$.

+ Student Workspace

Problem 4: Cartesian equation from parametric equations

Consider the line with parametric equations $x = 3t + 5$, $y = -2t + 1$. Do all four parts below by hand.
(a) Find a direction vector *v* for this line.
(b) Find a position vector *p* for this line.
(c) Find a cartesian equation for this line.
(d) Draw the line and the vectors *v* and *p*. (You may use **Vectorline(v,p)** to check your drawing).

+ Student Workspace

Problem 5: Constructing parallelograms

Consider the parallelogram *PQRS*, where $P = (2, 1)$, $Q = (4, 2)$, $R = (3, 6)$, $S = (1, 5)$. Use vector algebra (addition, subtraction, and scalar multiplication) throughout this problem. Do all three parts by hand.

(a) Find the vertices of the translate of the parallelogram *PQRS* in which *P* is translated to the origin.
(b) Find the vertices of the translate of the parallelogram *PQRS* whose center is at the origin. (The "center" of a parallelogram is the common midpoint of its diagonals.)
(c) Find the vertices of the parallelogram that is obtained by rotating *PQRS* 180 degrees about the vertex *P* and making its sides twice as long (keeping *P* as a vertex).

You may check your answers above by using the command **polygonplot**. For example, here is a command to draw the parallelogram *PQRS*:

```
> polygonplot([[2,1],[4,2],[3,6],[1,5]],
    scaling=constrained,view=[0..4,0..6]);
```

+ Student Workspace

Chapter 2 Vectors

Linear Algebra Modules Project
Chapter 2, Module 2

Linear Combinations of Vectors

Purpose of this module

The purpose of this module is to introduce the concept of linear combinations of vectors. In Section 1, we introduce linear combinations of vectors in the plane and 3-space algebraically, and we introduce the related idea of the span of a set of vectors. In Section 2, we consider the same concepts from a geometric perspective. In Section 3, we introduce n-space and linear combinations of vectors in n-space.

Prerequisites

Vector addition, scalar multiplication, and parametric representation of lines.

Commands used in this module

```
> restart: with(LinearAlgebra): with(plots): with(Lamp):
  UseHardwareFloats := false: Digits := 6:
>
```

Tutorial

Section 1: Linear Combinations of Vectors: An Algebraic Approach

======================

Example 1A: Let $u = \begin{bmatrix} 2 \\ 3 \end{bmatrix}$ and $v = \begin{bmatrix} 4 \\ -1 \end{bmatrix}$. We can create a new vector w by adding scalar multiples of u and v. For example, we can create $w = 3u + 2v$:

```
> u := Vector([2,3]);
  v := Vector([4,-1]);
```

$$u := \begin{bmatrix} 2 \\ 3 \end{bmatrix}$$

$$v := \begin{bmatrix} 4 \\ -1 \end{bmatrix}$$

```
> w := 3*u+2*v;
```

$$w := \begin{bmatrix} 14 \\ 7 \end{bmatrix}$$

Here is the calculation displayed on a single line:

$$3\begin{bmatrix} 2 \\ 3 \end{bmatrix} + 2\begin{bmatrix} 4 \\ -1 \end{bmatrix} = \begin{bmatrix} 14 \\ 7 \end{bmatrix}$$

We say that the vector $\begin{bmatrix} 14 \\ 7 \end{bmatrix}$ has been expressed as a *linear combination* of the vectors $\begin{bmatrix} 2 \\ 3 \end{bmatrix}$ and $\begin{bmatrix} 4 \\ -1 \end{bmatrix}$ with *scalar weights* 3 and 2, respectively. In general, if we want to consider all of the possible linear combinations of the vectors u and v, we write $au + bv$ where a and b are any scalar weights (i.e., real numbers).

The set of all linear combinations of the vectors u and v is called the *span* of u and v and is denoted by Span$\{u, v\}$. So the calculation above shows that the vector $\begin{bmatrix} 14 \\ 7 \end{bmatrix}$ is a member of Span$\{u, v\}$.

====================

Maple note: In the command below, we demonstrate a convenient shortcut for entering vectors, `<a,b>`, which is quicker to type than `Vector([a,b])`. We will always use the latter in Maple commands for clarity, although we will sometimes write $\langle a, b \rangle$ in text.

```
> u := <2,3>;
```

$$u := \begin{bmatrix} 2 \\ 3 \end{bmatrix}$$

====================

Example 1B: Suppose we were simply given the three vectors $u = \begin{bmatrix} 2 \\ 3 \end{bmatrix}$, $v = \begin{bmatrix} 4 \\ -1 \end{bmatrix}$ and $w = \begin{bmatrix} 14 \\ 7 \end{bmatrix}$. How could we work backwards to find the particular weights for u and v that produce w?

Solution: We want to express w in the form $au + bv = w$. Thus, we want to solve the vector equation

$$a\begin{bmatrix} 2 \\ 3 \end{bmatrix} + b\begin{bmatrix} 4 \\ -1 \end{bmatrix} = \begin{bmatrix} 14 \\ 7 \end{bmatrix} \text{ which simplifies to } \begin{bmatrix} 2a + 4b \\ 3a - b \end{bmatrix} = \begin{bmatrix} 14 \\ 7 \end{bmatrix}.$$

Equating corresponding components leads us to the familiar problem of solving two linear equations in two unknowns (a and b in this case):

$$2a + 4b = 14$$
$$3a - b = 7$$

We write out the augmented matrix of this linear system and reduce the matrix to find the solution $a = 3$ and $b = 2$:

```
> M := Matrix([[2,4,14],[3,-1,7]]);
```

$$M := \begin{bmatrix} 2 & 4 & 14 \\ 3 & -1 & 7 \end{bmatrix}$$

```
> Backsolve(Reduce(M));
```
$$\begin{bmatrix} 3 \\ 2 \end{bmatrix}$$

Note that the vectors *u*, *v*, *w* are the columns of the matrix *M*. This suggests a more efficient way to construct the matrix *M*: Use the command **Matrix([u,v,w])**, which constructs the matrix whose columns are *u*, *v*, *w*.
```
> M := Matrix([u,v,w]);
```
$$M := \begin{bmatrix} 2 & 4 & 14 \\ 3 & -1 & 7 \end{bmatrix}$$

====================

Exercise 1.1: For the vectors *u*, *v*, and *k* defined below, show that *k* is a member of Span{ *u*, *v* }. Hint: Construct the augmented matrix the quick way demonstrated above.
```
> u := Vector([2,3,4]):
  v := Vector([-4,2,1]):
  k := Vector([-40,4,-8]):
```
+ Student Workspace

− Answer 1.1
```
> M := Matrix([u,v,k]);
```
$$M := \begin{bmatrix} 2 & -4 & -40 \\ 3 & 2 & 4 \\ 4 & 1 & -8 \end{bmatrix}$$
```
> Backsolve(Reduce(M));
```
$$\begin{bmatrix} -4 \\ 8 \end{bmatrix}$$

So $k = -4\,u + 8\,v$.

Exercise 1.2: Show that the vector *m* (defined below) is not a member of Span{ *u*, *v* }, where *u* and *v* are the vectors defined in Exercise 1.1.
```
> m := Vector([20,4,-8]):
```
+ Student Workspace

− Answer 1.2
```
> N := Matrix([u,v,m]):
> Backsolve(Reduce(N));
Error, (in Backsolve) no solutions
```
Since the linear system whose augmented matrix is [*u*, *v*, *m*] is inconsistent, *m* is not a linear combination of *u* and *v*.

Section 2: Linear Combinations of Vectors: A Geometric Approach

Vectors in the Plane

The purpose of this section is to give you some experience visualizing linear combinations of vectors. We begin with a picture of the vectors $u = \begin{bmatrix} 2 \\ 3 \end{bmatrix}$ (in blue) and $v = \begin{bmatrix} 4 \\ -1 \end{bmatrix}$ (in red) from Example 1A along with the vector $w = 3u + 2v = \begin{bmatrix} 14 \\ 7 \end{bmatrix}$ (in black):

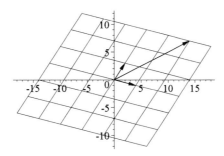

The above grid is based on the vectors u and v. Thus w is the diagonal of the parallelogram whose sides are 3 grid units in the direction of u and 2 grid units in the direction of v; hence $w = 3u + 2v$. (The grid also suggests that the weights 3 and 2 can be thought of as the *coordinates of w* in the coordinate grid determined by u and v; we will exploit this idea in later modules.)

You can generate this picture yourself by using the **Vectorgrid** command. (See below; the argument [3, 2] indicates that we want to multiply the first vector by 3 and the second by 2.)

```
> u := Vector([2,3]):
  v := Vector([4,-1]):
> Vectorgrid(u,v,[3,2]):
```

Exercise 2.1: The figure produced by the **Gridgame** command below has four labelled points, A, B, C and D. Each is a linear combination of $u = \begin{bmatrix} 2 \\ 3 \end{bmatrix}$ (in blue) and $v = \begin{bmatrix} 4 \\ -1 \end{bmatrix}$ (in red); that is, each is of the form $au + bv$ for some choice of a and b. Use the grid to help determine the weights for each; then check your work by changing the values for a and b in the command **Gridgame(a,b)**:

```
> Gridgame(1,2);
```

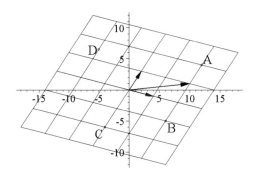

Exercise 2.2: In the figure below, a parallelogram is shown in yellow. Express (by hand) each of the vertices of the parallelogram as a linear combination of the vectors *u* and *v*.

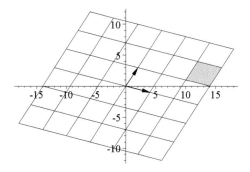

+ Student Workspace

− Answer 2.2

Starting at the bottom-left vertex and going around counter-clockwise:
$$u + 2v, u + 3v, 2u + 3v, 2u + 2v.$$

Now that you are able to picture how a particular choice for the scalar weights *a* and *b* produces a particular linear combination $au + bv$, we next treat *a* and *b* as parameters that can vary. Since we will want to draw many vectors at one time, they will be easier to see if we represent them by points rather than arrows.

======================

Example 2A: Consider the set of points $U = \{au + bv \mid a \text{ is in } [0, 1], \text{ and } b \text{ is in } [0, 1]\}$. In words: U is "the set of all linear combinations of the vectors *u* and *v* with weights ranging from 0 to 1". Clearly U is an infinite set of points, since there are infinitely many real numbers in the interval [0, 1]. We will refer to U as the ***unit span*** of the vectors *u* and *v*.

Explore~~~ (a) Imagine that you pick at random 25 pairs of values for *a* and *b* (all ranging from 0 to 1) and then plot $au + bv$ for each. Try to predict what your picture might look like. You can use **Vectorgrid** below to test your prediction. Maple calculates 25 linear combinations $au + bv$ by choosing random values for *a* between 0 and 1, *b* between 0 and 1.

> `Vectorgrid(u,v,[0..1,0..1],points=25);`

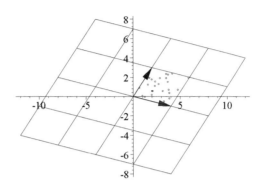

(b) Change the intervals for *a* and *b* in **Vectorgrid**. Can you predict what the new picture will be if *a* is between 0 and 2, and *b* is between 1 and 2?

Alternatively, you can use **Vectorgrid** without the option "points=n", which shades in all the linear combinations $au + bv$ for *a* and *b* in the designated intervals. For example, the command below draws the unit span of *u* and *v* as a yellow parallelogram:

> `Vectorgrid(u,v,[0..1,0..1]);`

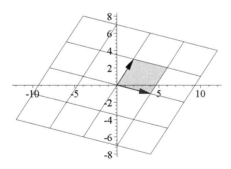

============================

Exercise 2.3: What range of values for *a* and *b* will produce the set of points shown in yellow in the figure below? Check your answer by using the **Vectorgrid** command.

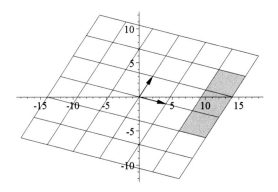

➕ Student Workspace

➖ Answer 2.3

a is between -1 and 2, *b* is between 2 and 3:
```
> Vectorgrid(u,v,[-1..2,2..3]):
```

Vectors in 3-Space

We now consider vectors in 3-space. Below we use the **Drawvec3d** command to picture two vectors *u* (blue) and *v* (red):
```
> u := Vector([5,-1,2]):
  v := Vector([3,7,1]):
> Drawvec3d([u,headcolor=blue],[v,headcolor=red]);
```

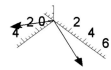

As in the plane, the unit span of two nonparallel vectors in 3-space is the parallelogram defined by the vectors. But now this parallelogram sits in three-dimensional space. Execute the next line to see the unit span for the above vectors *u* and *v*:
```
> Unitspan(u,v);
```

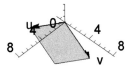

Recall that Span$\{u, v\}$ is defined as the set of <u>all</u> linear combinations of the vectors u and v: $a\,u + b\,v$. So instead of limiting the coefficients a and b to the unit interval $[0,1]$ as we do for the unit span, we allow a and b to range over all the real numbers. Imagine the yellow parallelogram above expanding, creating an infinite plane; then you have a picture of Span$\{u, v\}$.

Explore~~~ Change the values of a and b below to see the linear combination $w = a\,u + b\,v$. By rotating the plot around, confirm visually that w always lies in Span$\{u, v\}$, i.e., lies in the plane containing the unit span.

```
>  a := 2:
   b := 1/2:
>  w := a*u+b*v:
   p1 := Unitspan(u,v):
   p2 := Drawvec3d(w,headcolor=white):
   display([p1,p2]);
```

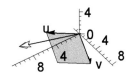

Exercise 2.4: State two vectors whose span is the *yz* plane in 3-space. To make this exercise a little more interesting, use vectors that are not parallel to the *y* or *z* axes.

 Student Workspace

➖ Answer 2.4

$u = \langle 0, 1, 1 \rangle$, and $v = \langle 0, 1, -1 \rangle$, for example. (The first component must be zero, and the vectors must not be multiples of each other.)

Finally, suppose we have three nonzero vectors, *u*, *v*, and *w*, and suppose no two are parallel and that the plane spanned by any two does not contain the third. The unit span of these vectors, $a\,u + b\,v + c\,w$, where the weights a, b, c range over the interval $[\,0,\,1\,]$, is then a solid parallelepiped:

Section 3: Vectors in *n*-Space

The examples we have considered so far have all been limited to linear combinations of vectors with just two or three components -- vectors in R^2 (the plane) or R^3 (3-space). The concepts of linear combinations and span can be applied to vectors with any number of components.

Definition: The set of all vectors with *n* components is denoted by R^n and is called ***n*-space**. Such vectors are added componentwise, and scalars are multiplied by such vectors componentwise.

For example, the vectors *u* and *v* below are in R^4, since they have 4 components. We add them and multiply *u* by the scalar -3:
```
> u := Vector([2,-1,4,0]);
  v := Vector([5,3,-2,-4]);
```
$$u = \begin{bmatrix} 2 \\ -1 \\ 4 \\ 0 \end{bmatrix}, \quad v = \begin{bmatrix} 5 \\ 3 \\ -2 \\ -4 \end{bmatrix}$$

```
> u+v;
> -3*u;
```
$$u + v = \begin{bmatrix} 7 \\ 2 \\ 2 \\ -4 \end{bmatrix}, \quad -3\,u = \begin{bmatrix} -6 \\ 3 \\ -12 \\ 0 \end{bmatrix}$$

The concepts of linear combinations and span can be applied to any finite set of vectors in R^n. Here are the general definitions:

Definition: Given a set of vectors $v_1, v_2, ..., v_k$ in R^n, a ***linear combination*** of these vectors is a vector of the form $c_1\,v_1 + c_2\,v_2 + ... + c_k\,v_k$ (where the scalar weights $c_1, c_2, ..., c_k$ are real numbers). The set of all such linear combinations is called the ***span*** of the vectors $v_1, v_2, ..., v_k$ and is denoted by Span$\{v_1, v_2,...,v_k\}$.

Language note: We also say that the vectors $v_1, v_2, ..., v_k$ **span** the set $\text{Span}\{v_1, v_2,...,v_k\}$; that is, we use "span" as a verb, not just a noun. For example, the vectors $\langle 1, 0 \rangle, \langle 0, 1 \rangle$ span the set R^2.

====================

Example 3A: Determine whether the vector p below is a linear combination of the vectors u, v, w below.

```
> u := Vector([4,5,-2,-5]):
  v := Vector([-1,1,3,1]):
  w := Vector([0,1,-1,-1]):
> p := Vector([-6,0,5,4]):
```

Solution: The vector equation $a\,u + b\,v + c\,w = p$ is the same as the linear system whose augmented matrix has the columns u, v, w, p. So we can solve this system as we did in Example 1B:

```
> M := Matrix([u,v,w,p]);
```

$$M := \begin{bmatrix} 4 & -1 & 0 & -6 \\ 5 & 1 & 1 & 0 \\ -2 & 3 & -1 & 5 \\ -5 & 1 & -1 & 4 \end{bmatrix}$$

```
> Backsolve(Reduce(M));
```

$$\begin{bmatrix} -1 \\ 2 \\ 3 \end{bmatrix}$$

Therefore $-u + 2\,v + 3\,w = p$.

====================

The method we used in Examples 1B and 3A to solve a vector equation will be used so often that it is worth recording. The general form of the method is the following:

- The vector equation $c_1\,v_1 + ... + c_k\,v_k = b$ with unknown scalar weights $c_1, ..., c_k$ can be rewritten as a system of linear equations whose augmented matrix is $[v_1, ..., v_k, b]$; that is, the columns of this matrix are the vectors $v_1, ..., v_k$ and b. The solutions of this linear system are the weights $c_1, ..., c_k$.

An Application of Vectors in R^n

A coffee shop offers two blends of coffees: House and Deluxe. Each is a blend of Brazil, Colombia, Kenya, and Sumatra roasts. The percentages for each blend are shown in the table below.

$$\begin{bmatrix} & House & Deluxe \\ Brazil & 30\% & 40\% \\ Columbia & 20\% & 30\% \\ Kenya & 20\% & 20\% \\ Sumatra & 30\% & 10\% \end{bmatrix}$$

Chapter 2 Vectors

Exercise 3.1: Suppose the store has 36 lbs of Brazil roast, 26 lbs of Colombia roast, 20 lbs of Kenya roast, and 18 lbs of Sumatra roast. How much of the House and Deluxe blends should be made in order to completely use up the stock of coffee on hand? Hint: Let a denote the number of pounds of the House blend and b the number of pounds of the Deluxe blend. Define two vectors, u and v, corresponding to the columns of the table:

```
> u := Vector([.3,.2,.2,.3]):
  v := Vector([.4,.3,.2,.1]):
```

 Student Workspace

 Answer 3.1

The linear combination $a u + b v$, where u and v are the two vectors defined above, gives us the number of pounds of Brazil, Colombia, Kenya, and Sumatra roasts that we use when we make a pounds of House blend and b pounds of Deluxe blend. So we must solve the equation $a u + b v = \langle 36, 26, 20, 18 \rangle$ for a and b. We form the augmented matrix corresponding to this linear system, reduce it and solve:

```
> w := Vector([36,26,20,18]):
  N := Matrix([u,v,w]);
```

$$N := \begin{bmatrix} .3 & .4 & 36 \\ .2 & .3 & 26 \\ .2 & .2 & 20 \\ .3 & .1 & 18 \end{bmatrix}$$

```
> Backsolve(Reduce(N));
```

$$\begin{bmatrix} 40 \\ 60 \end{bmatrix}$$

Therefore we should use 40 pounds of the House blend and 60 pounds of the Deluxe.

Problems

 Problem 1: Linear combinations that produce a parallelogram in R^2

For the vectors u and v below, find the range of values for a and b for which the linear combinations $a u + b v$ produce the yellow parallelogram below. Check your answer by using the `Vectorgrid` command.

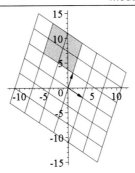

```
> u := Vector([1,3]):
  v := Vector([3,-2]):
```
+ Student Workspace

Problem 2: Linear combinations that span R^2

Show that the vectors u and v below span all of R^2 by showing that, for every vector $\begin{bmatrix} x \\ y \end{bmatrix}$ in R^2, the equation $a\,u + b\,v = \begin{bmatrix} x \\ y \end{bmatrix}$ can be solved for a and b.

```
> u := Vector([2,3]):
  v := Vector([4,-1]):
```
+ Student Workspace

Problem 3: Linear combinations that do not span R^2

The following statement is FALSE: "If u and v are any two nonzero vectors in R^2, then every vector in R^2 can be expressed as a linear combination of u and v."

Give an example of nonzero vectors u and v for which the statement is false. Then modify the first part of the statement to make the statement true.

+ Student Workspace

Problem 4: Seeing linear combinations in R^3

The vectors $i = \langle 1, 0, 0 \rangle$, $j = \langle 0, 1, 0 \rangle$, and $k = \langle 0, 0, 1 \rangle$ are called the standard unit vectors in R^3.
(a) Give a geometric description for each of the following: Span$\{\,i,j\,\}$, Span$\{\,j\,\}$, Span$\{\,j,k\,\}$.
(b) Is the vector $\langle 0, 4, 5 \rangle$ contained in Span $\{\,i+j\,,\,k\,\}$? Give both a geometric and algebraic justification for your answer.

+ Student Workspace

Chapter 2 Vectors

Problem 5: Span of three vectors in R^3

(a) For the vectors u, v, w and p below, show that p is a linear combination of u, v, w. Write your result in the form $p = a\,u + b\,v + c\,w$, using your values of a, b, c.

(b) Is every vector in R^3 a linear combination of these three vectors? Hint: See the method in Problem 2.

(c) What is the span of $\{u, v, w\}$? Explain.

```
> u := Vector([-2,-3,1]):
  v := Vector([4,-2,3]):
  w := Vector([-2,2,3]):
> p := Vector([8,-9,4]):
```

Student Workspace

Problem 6: Concrete application (part 1)

[From ATLAST: Computer Exercises for Linear Algebra by Steven Leon, Eugene Herman, Richard Faulkenberry, Prentice-Hall, 1996, page 81.]

Concrete mix, which is used in jobs as varied as making sidewalks and building bridges, is comprised of five main materials: cement, water, sand, gravel, and fly ash. By varying the percentages of these materials, mixes of concrete can be produced with differing characteristics. For example, the water-to-cement ratio affects the strength of the final mix, the sand-to-gravel ratio affects the "workability" of the mix, and the fly ash-to-cement ratio affects the durability . Since different jobs require concrete with different characteristics, it is important to be able to produce custom mixes.

Assume you are the manager of a building supply company and plan to keep on hand three basic mixes of concrete from which you will formulate custom mixes for your customers. The basic mixes have the following characteristics:

	Super-Strong Type S	All-Purpose Type A	Long-life Type L
cement	20	18	12
water	10	10	10
sand	20	25	15
gravel	10	5	15
fly ash	0	2	8

Each measuring scoop of any mix weighs 60 grams, and the numbers in the table give the breakdown by grams of the components of the mix. Custom mixes are made by combining the three basic mixes. For example, a custom mix might have 10 scoops of Type S, 14 of Type A, and 7.5 of type L. We can represent any mixture by a vector $\langle c, w, s, g, f \rangle$ in R^5 representing the amounts of cement, water, sand, gravel, and fly ash in the final mix. The basic mixes can therefore be represented by the following vectors:

```
> S := Vector([20,10,20,10,0]):
  A := Vector([18,10,25,5,2]):
  L := Vector([12,10,15,15,8]):
```

(a) Give a practical interpretation to the linear combination $3S + 5A + 2L$.

(b) What does Span$\{S, A, L\}$ represent?

(c) A customer requests 6 kilograms (6000 grams) of a custom mix with the following proportions of cement, water, sand, gravel, and fly ash: 16:10:21:9:4. Find the amounts of each of the basic mixes (S, A, and L) needed to create this mix.

(d) Is the solution unique? Explain.

Student Workspace

Problem 7: Convex combinations

Some linear combinations of vectors provide us with useful ways for describing lines, line segments, and convex polygonal regions in the plane. First, work through Illustrations 1 and 2 below; then do the four parts of Problem 7 that follow. Both illustrations will use the vectors u and v defined below:

```
> u := Vector([2,4]):
  v := Vector([5,1]):
```

Note: When we use vectors to represent sets of points, we draw only the tips of the vectors. For example, in Illustration 1 we use only the tips of the vectors $au + bv$ to describe the sets of points A, B, C, D.

Illustration 1: Give a <u>geometric</u> description for each of the sets A, B, C, and D defined below. It may be helpful to refer to the figure below (u in blue, v in red, and $u + v$ in black).

(a) $A = \{au + bv \mid a = 0 \text{ and } 0 < b < 1\}$
(b) $B = \{au + bv \mid b = 1 \text{ and } 0 < a < 1\}$
(c) $C = \{au + bv \mid a = 1 \text{ and } 0 < b < 1\}$
(d) $D = \{au + bv \mid b = 0 \text{ and } 0 < a < 1\}$

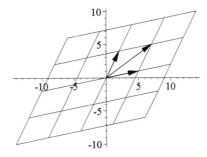

Answer to Illustration 1

(a) A = line segment between origin and the point (5, 1).
(b) B = line segment between the points (5, 1) and (7, 5).

(c) C = line segment between points (2, 4) and (7, 5).
(d) D = line segment between origin and the point (2, 4).

Illustration 2: Let $M = \{ au + bv \mid a + b = 1 \}$
(a) Give a geometric description of the points contained in the set M.
(b) Give an algebraic justification for your answer to part (a). Hint:
$$au + bv = au + (1 - a)v = v + a(u - v)$$

 Answer to Illustration 2

(a) M is the set of points on the line that passes through the heads of the vectors u and v, i.e., the line through the points (5, 1) and (2, 4).
(b) Note that $v + a(u - v)$ is a parametric representation of the line with position vector v and direction vector $u - v$. In particular, when $a = 0$ the line passes through the point $v = (5, 1)$, and when $a = 1$ the line passes through the point $u = (2, 4)$.

Problem 7(a): Let $N = \{ au + bv \mid a + b = 1 \text{ and } 0 \leq a \text{ and } 0 \leq b \}$.
(i) Describe the three points of N that correspond to $a = 0$, $a = .5$, and $a = 1$. How are they related to the vectors u and v?
(ii) Give a geometric description of all the points contained in the set N.
(iii) Give an algebraic justification for your answer to part (ii).
+ Student Workspace

Problem 7(b): Consider the two points $A = (2, 4)$ and $B = (5, 1)$. Describe the set P of all points on and inside the parallelogram $OACB$ (where $C = A + B$ and $O = (0, 0)$) by using suitable linear combinations of the vectors u and v.
+ Student Workspace

Problem 7(c): Consider the three points $A = (2, 4)$, $B = (5, 1)$, $O = (0, 0)$. Describe the set T of all points on and inside the triangle OAB by using suitable linear combinations of the vectors u and v. Hint: Modify your answer to Problem 7(b).
+ Student Workspace

Problem 7(d): (Challenge) Consider the three points $A = (2, 4)$, $B = (5, 1)$, $C = (8, 8)$. Describe the set of all points on and inside the triangle CAB. Your description should express the triangle as a set of linear combinations of the vectors u, v, and w, where $w = \begin{bmatrix} 8 \\ 8 \end{bmatrix}$. Hint: Use a translation to change this problem into one like Problem 7(c).
+ Student Workspace

Linear Algebra Modules Project
Chapter 2, Module 3

Decomposing the Solution of a Linear System

Purpose of this module

The purpose of this module is to enable you to understand how and why one decomposes the solution set of a linear system so it is expressed in terms of a finite number of vectors. To prepare for the discussion of decomposition, we first investigate vector forms of lines and planes in R^3, as these provide visualizations of the decomposition.

Prerequisites

Vector algebra, vector form of lines, linear combinations of vectors, span of vectors.

Commands used in this module

```
> restart: with(LinearAlgebra): with(plots): with(Lamp):
  UseHardwareFloats := false: Digits := 6:
```

Tutorial

Section 1: Vector Representation of Lines in 3-space

Recall that a line in R^2 can be described parametrically in the form $t\,v + p$, where v is a nonzero *direction vector*, p is a *position vector*, and t is a parameter that ranges over all the real numbers. We picture the line (thick and blue) passing through the point p and parallel to the vector v:

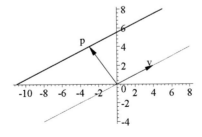

We can understand this vector form by constructing the line in two steps. First form the line through the origin in the direction of v (thin, red). This line is simply all scalar multiples of v and so has the vector form $t\,v$. Next, translate this line by adding the vector p to all the vectors $t\,v$. The point at the origin moves to the head of p, and the remaining points move the same distance and in the same direction to form the line $tv + p$ (thick, blue).

Lines in R^3 can be described in exactly the same vector form: $tv + p$.

Note: When we use vectors to represent sets of points, we draw only the tips of the vectors. For example, notice that when drawing the line above, we used only the tips of the vectors $tv + p$.

=====================

Example 1A: Find a vector form for the line in R^3 that has direction vector $v = \langle 2, -5, 3 \rangle$ and passes through the point $p = (3, 1, -4)$. Also find parametric equations for the line.

Solution: We calculate the expression $tv + p$ using Maple.
Maple note: Maple will not multiply a vector by a symbolic scalar such as t (because it does not know whether the symbol stands for a scalar or a matrix); the **Expand** command forces the scalar multiplication.

```
> v := Vector([2,-5,3]):
  p := Vector([3,1,-4]):
> Expand(t*v+p);
```

$$\begin{bmatrix} 2t+3 \\ -5t+1 \\ 3t-4 \end{bmatrix}$$

Thus we have the parametric equations $x = 2t + 3$, $y = -5t + 1$, $z = 3t - 4$. Here is a picture of the vectors v and p and the line $tv + p$. Try looking at it from different viewpoints, with and without axes displayed. Notice that the two lines and the direction vector remain parallel in all views.

```
> v := Vector([2,-5,3]):
  p := Vector([3,1,-4]):
  Vectorline(v,p);
```

=====================

Maple note: To draw the line by itself, you can use Maple's **spacecurve** command. Notice the syntax for this command: The three entries in the square brackets are the values of x, y, z in the parametric equations, and the next entry, $t = -2 \,..\, 2$, specifies the range for the parameter t.

```
> spacecurve([2*t+3,-5*t+1,3*t-4],t=-2..2,color=blue,axes=normal):
```

Exercise 1.1: (a) What are the coordinates of the two endpoints of the portion of the line drawn using the spacecurve command above?
(b) Does the point (9, -14, 5) lie on the line? Use algebra to answer this question, but also examine the following picture of the line and the point:

```
> p1 := spacecurve([2*t+3,-5*t+1,3*t-4],t=-2..2,
    color=blue,axes=normal):
  p2 :=
  pointplot3d([9,-14,5],symbol=circle,symbolsize=15,color=red):
  display([p1,p2]);
```

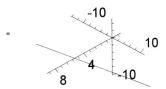

Student Workspace

Answer 1.1

(a) The endpoints have coordinates corresponding to values of $t = -2$ and $t = 2$:

```
> L := [2*t+3,-5*t+1,3*t-4];
```

$$L := [2t+3, -5t+1, 3t-4]$$

```
> subs(t=-2,L);
  subs(t=2,L);
```

$$[-1, 11, -10]$$
$$[7, -9, 2]$$

(b) In order for the point (9, −14, 5) to be on the line, there must be a value for the parameter t which yields this point. Working backwards from the x-coordinate, we have $2t + 3 = 9$ which means that $t = 3$. For this value of t, we get the desired values of y and z: (-5)3+1=-14 and (3)3-4=5. Hence the point (9, −14, 5) does lie on the line. However, since $t = 3$ is not in the interval $[-2, 2]$, the point does not lie on the part of the line that is in the picture. Here's another picture, but with the t range extended to $t = 3$:

```
> p1 := spacecurve([2*t+3,-5*t+1,3*t-4],t=-2..3,
    color=blue,axes=normal):
  p2 := pointplot3d([9,-14,5],symbol=circle,
    symbolsize=15,color=red):
```

```
display([p1,p2]);
```

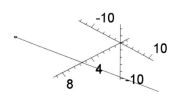

Exercise 1.2: (a) Find a direction vector and a position vector for the line that has the parametric equations

$$x = -4t + 2, \; y = 3t + 5, \; z = t - 4$$

(b) Combine the **spacecurve** command below with a **Drawvec3d** command to display the line and the two vectors together.

```
> p1 :=
  spacecurve([-4*t+2,3*t+5,t-4],t=-1..1,color=blue,axes=normal):
```

Student Workspace

 Answer 1.2

$$\begin{bmatrix} x \\ y \\ z \end{bmatrix} = \begin{bmatrix} -4t+2 \\ 3t+5 \\ t-4 \end{bmatrix} = t\begin{bmatrix} -4 \\ 3 \\ 1 \end{bmatrix} + \begin{bmatrix} 2 \\ 5 \\ -4 \end{bmatrix}. \text{ So } v = \begin{bmatrix} -4 \\ 3 \\ 1 \end{bmatrix} \text{ and } p = \begin{bmatrix} 2 \\ 5 \\ -4 \end{bmatrix}$$

However, there are many other correct answers. Any nonzero multiple of *v* will do for a direction vector, and any point on the line will serve as a position vector.

```
> p2 := Drawvec3d(<-4,3,1>,<2,5,-4>):
  display([p1,p2]);
```

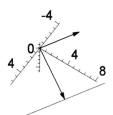

Section 2: Vector Representation of Planes: $s\,u + t\,v$

In Module 2 we saw that the span of two nonzero nonparallel vectors in R^3 is a plane through the origin. For example, here is the "unit span" of two such vectors:

```
> u := Vector([5,-1,2]):
  v := Vector([3,7,1]):
> Unitspan(u,v);
```

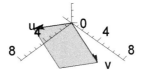

Recall that Span$\{u, v\}$ is defined as the set of all linear combinations of the vectors u and v: $s\,u + t\,v$. So instead of limiting the weights s and t to the unit interval [0,1] as we do for the unit span, we allow s and t to range over all the real numbers. Imagine the yellow parallelogram expanding, creating an infinite plane; then you have a picture of Span$\{u, v\}$. Since the vectors u and v are parallel to this plane and completely determine it, they are referred to as its *direction* vectors.

Recall also that every plane in R^3 can be expressed as the solution set of a linear equation, $a\,x + b\,y + c\,z = d$, where not all the coefficients a, b, c are zero. If the plane goes through the origin, then $d = 0$; so the equation of the plane is then homogeneous: $a\,x + b\,y + c\,z = 0$.

Exercise 2.1: Find coefficients a, b, c for the cartesian equation $a\,x + b\,y + c\,z = 0$ of the plane through the origin that has direction vectors $u = \langle 5, -1, 2 \rangle$ and $v = \langle 3, 7, 1 \rangle$.

 Student Workspace

 Answer 2.1

The vectors u and v are two solutions of the equation $a\,x + b\,y + c\,z = 0$. Substitute them into the equation of the plane and solve for the coefficients a, b, c by forming the augmented matrix of the resulting system of two linear equations:

$$5\,a - b + 2\,c = 0$$
$$3\,a + 7\,b + c = 0$$

```
> M := Matrix([[5,-1,2,0],[3,7,1,0]]);
```

```
                                    M := [ 5  -1  2  0 ]
                                         [ 3   7  1  0 ]
> Backsolve(Reduce(M));
```
$$\begin{bmatrix} -\dfrac{15}{38} X_3 \\ \dfrac{1}{38} X_3 \\ X_3 \end{bmatrix}$$

So if we choose $X_3 = -38$, we have the equation $15x - y - 38z = 0$.

We are more interested in the opposite problem: Given a cartesian equation of a plane through the origin, express the solution set as the span of two direction vectors. In other words, find a vector form $s\,u + t\,v$.

======================

Example 2A: Find the solution set of the plane $x + 2y + z = 0$, and find a vector form $s\,u + t\,v$ of the plane.

Solution: This problem is easy to solve by hand: $x = -2y - z$, and so

$$\begin{bmatrix} x \\ y \\ z \end{bmatrix} = \begin{bmatrix} -2y - z \\ y \\ z \end{bmatrix} = y \begin{bmatrix} -2 \\ 1 \\ 0 \end{bmatrix} + z \begin{bmatrix} -1 \\ 0 \\ 1 \end{bmatrix} = y\,u + z\,v$$

Since y and z can be any real numbers, we have expressed the solution set as Span$\{u, v\}$, where $u = \langle -2, 1, 0 \rangle$ and $v = \langle -1, 0, 1 \rangle$ are our two direction vectors.

Let's reexamine our key step, in which we "decompose" the infinite set of solutions into linear combinations of just two vectors:

$$\begin{bmatrix} -2y - z \\ y \\ z \end{bmatrix} = \begin{bmatrix} -2y - 1z \\ 1y + 0z \\ 0y + 1z \end{bmatrix} = y \begin{bmatrix} -2 \\ 1 \\ 0 \end{bmatrix} + z \begin{bmatrix} -1 \\ 0 \\ 1 \end{bmatrix} = y\,u + z\,v$$

Note that the components of u are just the coefficients of y in the solution set, and the components of v are the coefficients of z.

======================

There are many alternative pairs of direction vectors that we could have used to describe the above plane. In fact, any two nonparallel vectors in the plane will span the plane. We used the procedure above because it demonstrates how the direction vectors arise naturally from the equation of the plane when we "decompose" the solution. This is a process that you will use repeatedly in expressing solutions of linear systems in vector form.

Explore ~~~ We can use **plot3d** or **Drawplanes** to see the plane $x + 2y + z = 0$, and we can then use **Drawvec3d** and **display** to add the vectors u and v to this plot. Since the plane passes through the origin, we expect to see the vectors sitting right in the plane. Move the plane around to confirm this visually.

```
> u := Vector([-2,1,0]):
  v := Vector([-1,0,1]):
> plane := plot3d(-x-2*y,x=-0.5..1,y=-1.0..1.0,
    style=patchnogrid,color=yellow):
> vects := Drawvec3d(u,v,headlength=.5):
> display([plane,vects],axes=none);
```

Exercise 2.2: (a) Find a vector form of the plane through the origin that has the cartesian equation $2x - y + 4z = 0$. Use the method of decomposition, as in Example 2A.
(b) [For students familiar with normal vectors to a plane] Confirm that your vectors u and v are parallel to the plane by showing that they are orthogonal to a normal vector for the plane.

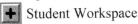 Student Workspace

 Answer 2.2

(a) We can solve for any of the three variables in terms of the other two. Let's solve for y:
$y = 2x + 4z$ and so
$$\begin{bmatrix} x \\ y \\ z \end{bmatrix} = \begin{bmatrix} x \\ 2x+4z \\ z \end{bmatrix} = x \begin{bmatrix} 1 \\ 2 \\ 0 \end{bmatrix} + z \begin{bmatrix} 0 \\ 4 \\ 1 \end{bmatrix} = x\,u + z\,v$$

Therefore the plane is Span$\{\{u, v\}\}$, where $u = \langle 1, 2, 0 \rangle$ and $v = \langle 0, 4, 1 \rangle$.
(b) Recall that, in the equation of a plane $ax + by + cz = d$, the vector whose components are the coefficients a, b, c is a normal vector to the plane; hence $n = \langle 2, -1, 4 \rangle$. Note that the dot product of n with each of u and v is zero.

Chapter 2 Vectors

Section 3: Vector Representation of Planes: $p + s\,u + t\,v$

If we take a vector form of a plane through the origin, $s\,u + t\,v$, and add a position vector p, we get $p + s\,u + t\,v$. Geometrically, the result is a translation of the original plane by the vector p to a new plane that now passes through the point p rather than the origin. The original plane and the translated plane are parallel and therefore have the same direction vectors, namely u and v.

Example 3A: Find a vector form $p + s\,u + t\,v$ for the plane $x + 2\,y + z = 6$ by decomposing the solution.

Solution: We find the general solution of $x + 2\,y + z = 6$ and then decompose this solution. Since $x = 6 - 2\,y - z$, we have

$$\begin{bmatrix} x \\ y \\ z \end{bmatrix} = \begin{bmatrix} 6 - 2y - z \\ y \\ z \end{bmatrix} = \begin{bmatrix} 6 \\ 0 \\ 0 \end{bmatrix} + y \begin{bmatrix} -2 \\ 1 \\ 0 \end{bmatrix} + z \begin{bmatrix} -1 \\ 0 \\ 1 \end{bmatrix}$$

This is our decomposition $p + s\,u + t\,v$ with y in place of s and z in place of t, and where

$$p = \begin{bmatrix} 6 \\ 0 \\ 0 \end{bmatrix},\ u = \begin{bmatrix} -2 \\ 1 \\ 0 \end{bmatrix},\ v = \begin{bmatrix} -1 \\ 0 \\ 1 \end{bmatrix}$$

Exercise 3.1: For the plane $x + 2\,y + z = 6$ in Example 3A, find a vector form by solving the equation for z instead of x.

+ Student Workspace

− Answer 3.1

Since $z = 6 - x - 2\,y$, we have

$$\begin{bmatrix} x \\ y \\ z \end{bmatrix} = \begin{bmatrix} x \\ y \\ 6 - x - 2y \end{bmatrix} = \begin{bmatrix} 0 \\ 0 \\ 6 \end{bmatrix} + x \begin{bmatrix} 1 \\ 0 \\ -1 \end{bmatrix} + y \begin{bmatrix} 0 \\ 1 \\ -2 \end{bmatrix} = p + x\,u + y\,v$$

Note that the vectors p, u, and v here are different from those we found in Example 3A.

Section 4: Decomposing the Solution of a Linear System

In this section, we will decompose the solution set of an arbitrary linear system and not limit ourselves to systems in two and three unknowns. Our approach will draw on our experience with vector forms $p + t\,v$ of a line and $p + s\,u + t\,v$ of a plane. In fact, the decomposition method we will use in this section will be a straightforward extension of the decompositions in the preceding sections.

Example 4A: Solve the following linear system consisting of three equations in four unknowns, and decompose the solution.

$$2x_1 - 6x_2 + 3x_3 - 2x_4 = -1$$
$$-x_1 + 3x_2 - 2x_3 = 4$$
$$3x_1 - 9x_2 + 4x_3 - 4x_4 = 2$$

Solution: Execute the next input region to compute the general solution of this system.

```
> M := Matrix([[2,-6,3,-2,-1],[-1,3,-2,0,4],[3,-9,4,-4,2]]):
> Backsolve(Reduce(M),free=x);
```

$$\begin{bmatrix} 10 + 3x_2 + 4x_4 \\ x_2 \\ -7 - 2x_4 \\ x_4 \end{bmatrix}$$

We decompose this solution as before:

$$\begin{bmatrix} 10 + 3x_2 + 4x_4 \\ x_2 \\ -7 - 2x_4 \\ x_4 \end{bmatrix} = \begin{bmatrix} 10 \\ 0 \\ -7 \\ 0 \end{bmatrix} + x_2 \begin{bmatrix} 3 \\ 1 \\ 0 \\ 0 \end{bmatrix} + x_4 \begin{bmatrix} 4 \\ 0 \\ -2 \\ 1 \end{bmatrix}$$

Thus, the decomposed form of the solution is $p + x_2 u + x_4 v$, where

$$p = \begin{bmatrix} 10 \\ 0 \\ -7 \\ 0 \end{bmatrix}, \quad u = \begin{bmatrix} 3 \\ 1 \\ 0 \\ 0 \end{bmatrix}, \quad v = \begin{bmatrix} 4 \\ 0 \\ -2 \\ 1 \end{bmatrix}$$

Notice that the solution of this linear system has two free variables, x_2 and x_4, which is the same number of free variables we had in Example 3A, where the solution described a plane in R^3. Therefore, even though the solution set in Example 4A is a set of points in R^4, we will find it useful to think of the solution as if it were also a "plane." In particular, in writing the solution in the form $p + su + tv$ (where we have replaced x_2 by s and x_4 by t), we will refer to p as a *position vector* for the solution and will refer to u and v as *direction vectors* for the "parallel plane" through the origin. The vector p is also called a *particular solution* of the linear system.

Here is the most general form of the conclusion that the method of decomposition yields:

Chapter 2 Vectors

- The solution set of a linear system can be decomposed in the form $p + \text{Span}\{v_1, ..., v_k\}$, where p is a particular solution of the linear system and $\text{Span}\{v_1, ..., v_k\}$ is the solution set of the corresponding homogeneous linear system.

======================

Example 4B: In this example, we use an entirely different strategy to decompose the solution of the linear system in Example 4A. For large linear systems, this strategy lets us use Maple to write out the decomposition more quickly.

Solution: We solve the system as usual:

```
> soln := Backsolve(Reduce(M),free=x);
```

$$soln := \begin{bmatrix} 10 + 3x_2 + 4x_4 \\ x_2 \\ -7 - 2x_4 \\ x_4 \end{bmatrix}$$

To find a position vector p, we set all the free variables to 0:

```
> p := subs({x[2]=0,x[4]=0},soln);
```

$$p := \begin{bmatrix} 10 \\ 0 \\ -7 \\ 0 \end{bmatrix}$$

Next, subtract p from the general solution, $p + x_2 u + x_4 v$, which leaves $x_2 u + x_4 v$. This is the solution of the corresponding homogeneous system:

```
> h := soln - p;
```

$$h := \begin{bmatrix} 3x_2 + 4x_4 \\ x_2 \\ -2x_4 \\ x_4 \end{bmatrix}$$

In the solution of the homogeneous system, set one free variable to 1 and the rest to 0. This gives one of the direction vectors. Repeat for each of the free variables.

```
> u := subs({x[2]=1,x[4]=0},h);
  v := subs({x[4]=1,x[2]=0},h);
```

$$u = \begin{bmatrix} 3 \\ 1 \\ 0 \\ 0 \end{bmatrix}, \quad v = \begin{bmatrix} 4 \\ 0 \\ -2 \\ 1 \end{bmatrix}$$

======================

Module 2.3 Decomposing the Solution of a Linear System

Exercise 4.1: Use the method of Example 4B to decompose the solution of $x + 2y + z = 6$.

Student Workspace

Answer 4.1

```
> A := Matrix([[1,2,1,6]]):
> soln := Backsolve(Reduce(A),free=x);
```

$$soln := \begin{bmatrix} 6 - 2x_2 - x_3 \\ x_2 \\ x_3 \end{bmatrix}$$

```
> p := subs({x[2]=0,x[3]=0},soln);
```

$$p := \begin{bmatrix} 6 \\ 0 \\ 0 \end{bmatrix}$$

```
> h := soln - p;
```

$$h := \begin{bmatrix} -2x_2 - x_3 \\ x_2 \\ x_3 \end{bmatrix}$$

```
> u := subs({x[2]=1,x[3]=0},h);
  v := subs({x[3]=1,x[2]=0},h);
```

$$u = \begin{bmatrix} -2 \\ 1 \\ 0 \end{bmatrix}, \quad v = \begin{bmatrix} -1 \\ 0 \\ 1 \end{bmatrix}$$

The decomposition is therefore $p + x_2 u + x_3 v$. (Compare Example 3A.)

Problems

Problem 1: Vector form of a plane

Find a vector form $p + s u + t v$ of the plane $x + 2y + 3z = 8$; decompose the solution by hand.

Student Workspace

Problem 2: Picture of a plane and associated vectors

Use **plot3d**, **Drawvec3d**, and **display** to produce a plot that includes all of the following: the graph of $x + 2y + 3z = 8$, the graph of the parallel plane through the origin, $x + 2y + 3z = 0$, and the vectors p, u, v you found in Problem 1. If you are familiar with the concept of the normal vector to the plane, plot it too. Suggestion: You may find it difficult to orient the plot so it displays both planes and all four vectors in the way you want them to look. In that case, produce two separate plots (but no more than two). If necessary, remove some of the vectors from one of the plots. (Don't spend a lot of time, however, getting the pictures to look just right.)

[+] Student Workspace

Problem 3: Vector form of the intersection of two planes

(a) Below are the equations of two planes that intersect in a line. Solve the system of two equations, and express the solution (i.e., the line of intersection) in vector form, $p + t\,v$; decompose the solution by hand.
(b) [For students familiar with normal vectors to a plane] How is the direction vector of the line related geometrically to the normal vectors of the two planes?

$$2x - y + 3z = 6$$
$$x - 4z = 8$$

[+] Student Workspace

Problem 4: Picture of parallel parallelograms

Use **polygonplot3d** to draw the following two parallelograms in R^3. The first parallelogram has vertices at (1, 2, 0), (0, 0, 0), (2, −3, 1), and the fourth vertex opposite (0, 0, 0). The second parallelogram is parallel to the first one but translated by the vector $p = \langle 3, 0, 2 \rangle$.

Below is a sample of how **polygonplot3d** is used. The main argument in **polygonplot3d** is a list of points; **polygonplot3d** connects consecutive points by line segments and also connects the last point back to the first point. In the sample below, we connect three of the given points to form a triangle.
```
> polygonplot3d([[1,2,0],[0,0,0],[2,-3,1]],color=red,axes=normal):
```
[+] Student Workspace

Problem 5: One solution, many decompositions

(a) Solve the linear system below by applying the **Reduce** and **Backsolve** commands to the augmented matrix, which is also given. Then determine which of the vectors p_1, p_2, p_3, listed below are position vectors and which vectors v_1, v_2, v_3, v_4 listed below are direction vectors for the solution set. Caution: There are many correct answers, not just the ones you would get from decomposing by hand.

$$2x_1 + 3x_2 - x_3 - 3x_4 = 1$$
$$x_1 + 3x_2 + x_3 - 3x_4 = 2$$
$$-3x_1 - 4x_2 + 2x_3 + 4x_4 = -1$$

(b) Write three different decompositions of the solution set by using appropriate vectors from those given below. The number of direction vectors should be the same in each of your decompositions.
```
> A := Matrix([[2,3,-1,-3,1],[1,3,1,-3,2],[-3,-4,2,4,-1]]);
```

$$p_1 := \begin{bmatrix} -1 \\ 1 \\ 0 \\ 0 \end{bmatrix}, \ p_2 := \begin{bmatrix} 1 \\ 1 \\ 1 \\ 1 \end{bmatrix}, \ p_3 := \begin{bmatrix} -1 \\ 1 \\ 0 \\ 1 \end{bmatrix}, \ v_1 := \begin{bmatrix} 0 \\ 1 \\ 0 \\ 1 \end{bmatrix}, \ v_2 := \begin{bmatrix} 2 \\ -1 \\ 1 \\ 0 \end{bmatrix}, \ v_3 := \begin{bmatrix} -1 \\ 1 \\ 0 \\ 1 \end{bmatrix}, \ v_4 := \begin{bmatrix} 2 \\ 0 \\ 1 \\ 1 \end{bmatrix}$$

+ Student Workspace

− Problem 6: Decompose the solution of a linear system I

Solve the following system of equations, and decompose the solution into a linear combination of vectors. (You may use the quick method in Example 4B.) Identify the position and direction vectors that generate the solution.

$$x_1 + x_2 - 2x_3 + x_4 = 3$$
$$2x_1 + 2x_2 - x_3 + 2x_4 = 3$$
$$3x_1 + 3x_2 + 2x_3 + 3x_4 = 1$$

To help you get started, here is the augmented matrix:
```
> A := Matrix([[1,1,-2,1,3],[2,2,-1,2,3],[3,3,2,3,1]]):
```
+ Student Workspace

− Problem 7: Decompose the solution of a linear system II

(a) Solve the following system of equations, and decompose the solution into a linear combination of vectors. (You may use the quick method in Example 4B.) Identify the position and direction vectors that generate the solution.

$$x_1 + x_2 + 3x_4 + 3x_5 = 2$$
$$-2x_1 + 2x_2 - x_3 + 2x_4 + 3x_5 = 2$$
$$4x_1 + x_3 + 4x_4 + 3x_5 = 2$$
$$-7x_1 + 5x_2 - 3x_3 + 3x_4 + 6x_5 = 4$$

To help you get started, here is the augmented matrix:
```
> M := Matrix([[1,1,0,3,3,2],[-2,2,-1,2,3,2],
    [4,0,1,4,3,2],[-7,5,-3,3,6,4]]):
```
(b) Without any further computation, write down the decomposed form of the solution of the corresponding homogeneous system. Describe, in geometric language, how the solution set of the homogeneous system and the solution set of the original system are related. In particular, how are their direction vectors related?

+ Student Workspace

Problem 8: Nonhomogeneous and homogeneous systems

Here is a system of linear <u>homogeneous</u> equations, the coefficient matrix A of the system, and two right-side column vectors b and c:

$$4x_1 + x_2 + 2x_3 - 3x_4 - 2x_5 = 0$$
$$-2x_1 - x_2 - x_3 + 4x_4 + x_5 = 0$$
$$4x_2 - 3x_3 + 2x_4 + 2x_5 = 0$$

```
> A := Matrix([[4,1,2,-3,-2],[-2,-1,-1,4,1],[0,4,-3,2,2]]):
> b := Vector([13,-7,1]):
  c := Vector([6,-7,6]):
```

(a) Solve the homogeneous system and decompose the solution. Do the same for each of the two nonhomogenous systems that have this same coefficient matrix A and whose right sides are b and c, respectively. (The commands in the input region below will help you get started. You may use the quick method of Example 4B.)

(b) How are the three solution sets you found in (a) related to one another? That is, describe in words the similarities and differences among the three solution sets. In your description, use the geometric language we employed in Section 3.

(c) The vector $k = \langle 3, -4, 2, 0, 5 \rangle$ is a solution of the linear system whose coefficient matrix is this same matrix A and whose right-side vector is some vector d. Write down a vector form of the solution set of this linear system. (This should require no calculation; you should not even need to find d.)

```
> zero := Vector(3,0):
  aug1 := Matrix([A,zero]):
> aug2 := Matrix([A,b]):
> aug3 := Matrix([A,c]):
```

+ Student Workspace

Linear Algebra Modules Project
Chapter 2, Module 4

Linear Independence of Vectors

 Purpose of this module

The purpose of this module is to introduce the concepts of linearly dependent and linearly independent vectors.

 Prerequisites

Linear combinations and span of a set of vectors; solution of a system of linear equations.

 Commands used in this module

```
> restart: with(LinearAlgebra): with(plots): with(Lamp):
  UseHardwareFloats := false: Digits := 6:
```

Tutorial

Section 1: Linear Dependence of Vectors: A Geometric Introduction

In Module 2 we observed that any two nonparallel vectors u and v in R^2 span all of 2-space. In other words, every vector in R^2 can be expressed as a linear combination of the vectors u and v. If u and v are nonparallel vectors in R^3, their span is a plane in R^3. So two vectors cannot span all of R^3; we will need at least three vectors in R^3 to span the entire space. Of course, one set of three vectors that spans all of R^3 consists of the standard unit vectors i, j, k.

Will <u>any</u> three nonparallel vectors in R^3 span all of 3-space? Before reading on, take a moment to think about this question and form a preliminary answer.

======================

Example 1A: Consider the set of three vectors $\{u, v, p\}$ defined below. Note that the vectors are nonparallel; that is, they all point in different directions. Does Span$\{u, v, p\} = R^3$?

```
> u := Vector([3,4,1]):
  v := Vector([5,1,2]):
  p := Vector([1,7,0]):
```

Let's start by looking at a picture of the three vectors:

```
> Drawvec3d([u,label="u"],[v,label="v"],[p,label="p"]);
```

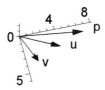

By turning the picture in different directions, you should be able to convince yourself that the three vectors lie in a single plane and hence cannot span R^3. Of course, no such picture can be conclusive.

To determine conclusively that the three vectors lie in a single plane, we need an algebraic method for testing them. Note that if all three vectors do lie in a single plane and no two are parallel, then each of the vectors must be in the span of the other two. Therefore each vector can be expressed as a linear combination of the other two.

So let's attempt to find a solution to the vector equation $p = a u + b v$. We can solve this equation by using the corresponding augmented matrix:

```
> N := Matrix([u,v,p]);
```

$$N := \begin{bmatrix} 3 & 5 & 1 \\ 4 & 1 & 7 \\ 1 & 2 & 0 \end{bmatrix}$$

```
> Backsolve(Reduce(N));
```

$$\begin{bmatrix} 2 \\ -1 \end{bmatrix}$$

A solution exists! In fact, we see that $p = 2u - v$. We prefer to write this equation as $2u - v - p = 0$, which has the advantage of not singling out one of the vectors. An equation of this form is called a *linear dependence relation*. More precisely, a linear dependence relation for the set of vectors $\{u, v, p\}$ is a vector equation of the form $c_1 u + c_2 v + c_3 p = 0$, where at least one of the weights c_1, c_2, c_3 is not zero.

===================

In the next exercise you will see how we can use this linear dependence relation to show that Span$\{u, v, p\}$ is only a plane and not all of R^3.

Exercise 1.1: (a) (By hand) Prove that Span$\{u, v, p\}$ = Span$\{u, v\}$ by demonstrating that every linear combination of the three vectors u, v, p can be rewritten as a linear combination of just the two vectors u and v.

(b) Illustrate the result from part (a) by showing how the vector $4u + 5v + 3p$ can be rewritten as a linear combination of the vectors u and v. Check your work by direct calculation.

Student Workspace

 Answer 1.1

(a) Given any linear combination $q = au + bv + cp$, we can use the equation $p = 2u - v$ to eliminate p and thus express q in terms of just the vectors u and v:
$$au + bv + cp = au + bv + c(2u - v) = (a + 2c)u + (b - c)v$$
(b) $4u + 5v + 3p = 4u + 5v + 3(2u - v) = 10u + 2v$

```
> 4*u+5*v+3*p;
  10*u+2*v;
```

$$4u + 5v + 3p = \begin{bmatrix} 40 \\ 42 \\ 14 \end{bmatrix}, \quad 10u + 2v = \begin{bmatrix} 40 \\ 42 \\ 14 \end{bmatrix}$$

=====================

Example 1B: Consider the set of three vectors $\{u, v, w\}$ defined below. The first two vectors are the same as in Example 1A, but we have replaced p by a new vector w. In particular, we have chosen w so it does not lie in the same plane as u and v.

```
> u := Vector([3,4,1]):
  v := Vector([5,1,2]):
  w := Vector([2,2,4]):
```

The picture below illustrates the fact that the vector w is not in the plane spanned by the vectors u and v.

```
> Drawvec3d([u,label="u"],[v,label="v"],[w,label="w"]);
```

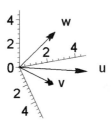

=====================

Exercise 1.2: (a) Prove algebraically that w is not in the plane spanned by u and v.
(b) Explain why the fact that w is not in the plane spanned by u and v implies that: (i) u is not in the plane spanned by the vectors v and w and (ii) v is not in the plane spanned by the vectors u and w.
(c) Explain why no linear dependence relation can exist among the vectors u, v, w. Hint: Show that such a linear dependence relation would lead to a contradiction.

 Student Workspace

Chapter 2 Vectors

Answer 1.2

(a) To show that w is not in the span of u and v, show that the vector equation $a\,u + b\,v = w$ has no solution:

```
> N := Matrix([u,v,w]);
```

$$N := \begin{bmatrix} 3 & 5 & 2 \\ 4 & 1 & 2 \\ 1 & 2 & 4 \end{bmatrix}$$

```
> Backsolve(Reduce(N));
Error, (in Backsolve) no solutions
```

(b) (i) If u were in the plane spanned by v and w, then all three vectors would lie in a single plane, contradicting the result of part (a). (ii) Similar reasoning applies.

(c) If a linear dependence relation existed, we would be able to express one of the vectors as a linear combination of the other two. Hence this vector would be in the span of the other two vectors, contradicting the results from parts (a) and (b).

In Exercise 1.2 you proved that no linear dependence relation exists among the vectors u, v, w. When a set of vectors has this property, we call it a *linearly independent* set. In the next exercise, you will prove that the linearly independent set of vectors $\{u, v, w\}$ spans all of 3-space.

Exercise 1.3: Show that Span$\{u, v, w\} = R^3$. Hint: Take a general vector in R^3, say $q = \langle k, l, m \rangle$, and show that q can be expressed as a linear combination of u, v, w by showing that the vector equation $a\,u + b\,v + c\,w = q$ always has a solution.

Student Workspace

Answer 1.3

```
> q := Vector([k,l,m]):
> N := Matrix([u,v,w,q]);
```

$$N := \begin{bmatrix} 3 & 5 & 2 & k \\ 4 & 1 & 2 & l \\ 1 & 2 & 4 & m \end{bmatrix}$$

```
> Backsolve(Reduce(N));
```

$$\begin{bmatrix} \dfrac{2}{7}l - \dfrac{1}{7}m \\[4pt] -\dfrac{5}{28}l + \dfrac{1}{4}k - \dfrac{1}{28}m \\[4pt] \dfrac{17}{56}m - \dfrac{1}{8}k + \dfrac{1}{56}l \end{bmatrix}$$

Thus, for every vector q in R^3, the equation $a\,u + b\,v + c\,w = q$ can be solved for a, b, c.

Module 2.4 Linear Independence of Vectors

We began this section by asking whether any set of three nonparallel vectors in R^3 will span 3-space. Example 1A demonstrated that it is not enough for the vectors to be merely nonparallel. Example 1B, on the other hand, provided an example of a linearly independent set that does span all of 3-space. We will soon see that linearly independence is, in fact, a necessary and sufficient condition for three vectors to span 3-space. Before we can pursue this topic further, we need to take a closer look at the definitions of linear dependence and independence. The next section presents a systematic algebraic approach to these fundamental concepts.

 Section 2: Linear Dependence of Vectors: An Algebraic Approach

As we have seen, a given set of vectors either satisfies some linear dependence relation, in which case we will say that the set of vectors is *linearly dependent*; or there is no linear dependence relation that it satisfies, in which case we will say that the set of vectors is *linearly independent*. Here are the formal definitions:

Definition: A set of vectors $\{v_1, v_2, \ldots, v_k\}$ in R^n is **linearly dependent** if the vectors satisfy a **linear dependence relation**

$$c_1 v_1 + c_2 v_2 + \ldots + c_k v_k = 0$$

where at least one of the weights $c_1, c_2, \ldots, c_k$ is not zero. (The 0 in the above equation is the zero vector in R^n.) The set of vectors is **linearly independent** if it is not linearly dependent, that is, if there is no linear dependence relation satisfied by the vectors.

Let's take a close look at a set of vectors that is linearly dependent.

==================

Example 2A: Consider the set of three vectors $\{u, v, w\}$ defined below.
```
> u := Vector([2,3]):
  v := Vector([4,-1]):
  w := Vector([14,7]):
```
This set of three vectors satisfies the linear dependence relation $3u + 2v - w = 0$. (Check this by hand.) Therefore $\{u, v, w\}$ is a linearly dependent set of vectors. Notice also that we can solve this equation for any one of the three vectors u, v, w in terms of the other two, which means that each is a linear combination of the other two:

$$w = 3u + 2v, \quad v = \frac{w}{2} - \frac{3u}{2}, \quad u = \frac{w}{3} - \frac{2v}{3}$$

==================

==================

Example 2B: Consider the set of vectors $\{u, v, w\}$ defined below.
```
> u := Vector([2,3]):
```

```
v := Vector([4,6]):
w := Vector([14,7]):
```
This set of vectors satisfies the linear dependence relation $2u - v + 0w = 0$. (Check this by hand.) Therefore $\{u, v, w\}$ is a linearly dependent set of vectors. Unlike the previous example, however, it is not true that each vector is a linear combination of the other two. This time, u is a linear combination of v and w, and v is a linear combination of u and w, but w is not a linear combination of u and v:

$$u = \frac{v}{2} + 0w, \quad v = 2u + 0w$$

Let's look again at the definition of linear dependence. Note that the equation $c_1 v_1 + c_2 v_2 + \ldots + c_k v_k = 0$ can always be solved for $c_1, c_2, \ldots, c_k$ simply by choosing all the weights to be zero. However, the equation $0 v_1 + 0 v_2 + \ldots + 0 v_k = 0$ does not constitute a linear dependence relation. We refer to the solution in which all the weights are zero as the "trivial" solution. So here is another way to express the statement that the set of vectors $\{v_1, v_2, \ldots, v_k\}$ is linearly dependent:

- The set $\{v_1, v_2, \ldots, v_k\}$ is linearly dependent if and only if the equation $c_1 v_1 + c_2 v_2 + \ldots + c_k v_k = 0$ has a nontrivial solution (i.e., a solution in which at least one of the weights is nonzero).

Example 2C: Let's find out whether the set of vectors $\{u, v, w\}$ defined below is linearly dependent or independent.
```
> u := Vector([3,4,1]):
  v := Vector([5,1,2]):
  w := Vector([9,-5,4]):
```
Solution: We need to find out if there exist weights c_1, c_2, c_3 (where at least one is not zero) such that

$$c_1 u + c_2 v + c_3 w = 0$$

We can solve this equation by using the corresponding augmented matrix:
```
> N := Matrix([u,v,w,Vector(3,0)]);
```
$$N := \begin{bmatrix} 3 & 5 & 9 & 0 \\ 4 & 1 & -5 & 0 \\ 1 & 2 & 4 & 0 \end{bmatrix}$$

```
> Backsolve(Reduce(N),free=c);
```
$$\begin{bmatrix} 2 c_3 \\ -3 c_3 \\ c_3 \end{bmatrix}$$

So we have a nontrivial solution of the vector equation $c_1 u + c_2 v + c_3 w = 0$ whenever $c_1 = 2 c_3$,

$c_2 = -3\,c_3$, and $c_3 \neq 0$. Choosing $c_3 = 1$, we have the linear dependence relation $2\,u - 3\,v + w = 0$. Therefore $\{u, v, w\}$ is a linearly dependent set of vectors.

Maple note: The command **Vector(3,0)** specifies a vector with 3 components, each equal to 0.

================

Let's look at these three examples more closely to see what general conclusions we can infer. A set of vectors $\{v_1, v_2, ..., v_k\}$ is defined to be linearly dependent if there is a linear dependence relation $c_1\,v_1 + c_2\,v_2 + ... + c_k\,v_k = 0$, where at least one of the weights $c_1, c_2, ..., c_k$ is not zero. Since at least one of the weights is not zero, we can solve for at least one of the vectors, thus writing that vector as a linear combination of the other vectors. (See Examples 2A and 2B.) However, as we saw in Example 2B, some of the weights can be zero and we still say that the entire set of vectors is linearly dependent. So we have another way to express the statement that the set of vectors $\{v_1, v_2, ..., v_k\}$ is linearly dependent:

- The set of vectors $\{v_1, v_2, ..., v_k\}$ is linearly dependent if and only if at least one of the vectors can be written as a linear combination of the others.

Exercise 2.1: Consider the set of four vectors $\{v_1, v_2, v_3, v_4\}$ in R^5 defined below.
(a) Show that this set of vectors is linearly dependent, and find a linear dependence relation.
(b) Solve for one of the vectors as a linear combination of the others.
(c) Now consider the set consisting of only the three vectors that remain after you remove the vector you solved for in (b). Is this set linearly dependent or independent?

```
> v1 := Vector([4,3,-1,2,1]):
  v2 := Vector([-4,7,-9,0,3]):
  v3 := Vector([2,3,-2,0,1]):
  v4 := Vector([0,2,-2,3,1]):
```

 Student Workspace

Answer 2.1

(a) Form the augmented matrix of the linear system $c_1\,v_1 + c_2\,v_2 + c_3\,v_3 + c_4\,v_4 = 0$, and solve:

```
> N := Matrix([v1,v2,v3,v4,Vector(5,0)]):
> Backsolve(Reduce(N),free=c);
```

$$\begin{bmatrix} -\dfrac{3}{2}c_4 \\ -\dfrac{1}{2}c_4 \\ 2\,c_4 \\ c_4 \end{bmatrix}$$

We have a nontrivial solution of the vector equation $c_1\,v_1 + c_2\,v_2 + c_3\,v_3 + c_4\,v_4 = 0$ whenever c_4 is chosen to be a nonzero number; let's choose $c_4 = 2$ to eliminate the denominators. So we have the linear dependence relation

$$-3 v_1 - v_2 + 4 v_3 + 2 v_4 = 0$$

and therefore the set $\{v_1, v_2, v_3, v_4\}$ is linearly dependent.

(b) We can solve for any of the four vectors, since none of the coefficients is zero. For example,

$$v_4 = \frac{3 v_1}{2} + \frac{v_2}{2} - 2 v_3$$

(c) Repeat the above procedure for the set of vectors $\{v_1, v_2, v_3\}$:

```
> N := Matrix([v1,v2,v3,Vector(5,0)]):
> Backsolve(Reduce(N),free=c);
```

$$\begin{bmatrix} 0 \\ 0 \\ 0 \end{bmatrix}$$

So the set $\{v_1, v_2, v_3\}$ is linearly independent, since the only solution to the equation $c_1 v_1 + c_2 v_2 + c_3 v_3 = 0$ is the trivial solution in which all the weights are zero.

====================

Example 2D -- **The Special Case of a Set of Two Vectors:** If a set consists of only two vectors, say u and v, then $\{u, v\}$ satisfies a linear dependence relation if and only if one of the vectors is a multiple of the other. For example, if $\{u, v\}$ satisfies the linear dependence relation $u - 3 v = 0$ then $u = 3 v$. In general, the linear dependence relation $a u + b v = 0$ is equivalent to $u = \left(-\frac{b}{a}\right) v$, provided $a \neq 0$.

====================

Exercise 2.2: (By hand) For each pair of vectors listed below, determine whether the set consisting of the two vectors is linearly dependent or independent. If it is linearly dependent, find a linear dependence relation.

(a) $v_1 = \begin{bmatrix} 2 \\ 3 \\ 5 \end{bmatrix}$ and $v_2 = \begin{bmatrix} 8 \\ 12 \\ 20 \end{bmatrix}$, (b) $v_1 = \begin{bmatrix} 2 \\ 3 \\ 5 \end{bmatrix}$ and $v_2 = \begin{bmatrix} 8 \\ 12 \\ 25 \end{bmatrix}$, (c) $v_1 = \begin{bmatrix} 2 \\ 3 \\ 5 \end{bmatrix}$ and $v_2 = \begin{bmatrix} 0 \\ 0 \\ 0 \end{bmatrix}$

+ Student Workspace
− Answer 2.2

(a) The set $\{v_1, v_2\}$ is linearly dependent since $v_2 = 4 v_1$ (or $v_1 = \frac{v_2}{4}$ or $4 v_1 - v_2 = 0$).

(b) The set $\{v_1, v_2\}$ is linearly independent since neither vector is a multiple of the other.

(c) The set $\{v_1, v_2\}$ is linearly dependent since $v_2 = 0\, v_1$ (or $0\, v_1 + v_2 = 0$).

===================

Example 2E: Consider the set of four vectors $\{v_1, v_2, v_3, v_4\}$ in R^5 defined below.
(a) Find at least three linearly dependent subsets of $\{v_1, v_2, v_3, v_4\}$.
(b) Find a linearly independent subset that has the same span as Span$\{v_1, v_2, v_3, v_4\}$.

```
> v1 := Vector([0,3,-1,0,1]):
> v2 := Vector([3,1,1,2,-1]):
> v3 := Vector([3,4,0,2,0]):
> v4 := Vector([6,2,2,4,-2]):
```

Solution: (a) Let's find all solutions to $c_1\, v_1 + c_2\, v_2 + c_3\, v_3 + c_4\, v_4 = 0$:

```
> M := Matrix([v1,v2,v3,v4,Vector(5,0)]):
> Backsolve(Reduce(M),free=c);
```

$$\begin{bmatrix} -c_3 \\ -c_3 - 2\, c_4 \\ c_3 \\ c_4 \end{bmatrix}$$

We have a nontrivial solution to $c_1\, v_1 + c_2\, v_2 + c_3\, v_3 + c_4\, v_4 = 0$ whenever c_3 or c_4 has a nonzero value. For example, choosing $c_3 = 1$ and $c_4 = 0$ yields one linear dependence relation, and choosing $c_3 = 0$ and $c_4 = 1$ yields another:

$$-v_1 - v_2 + v_3 = 0 \quad \text{and} \quad -2\, v_2 + v_4 = 0$$

Thus $\{v_1, v_2, v_3\}$ and $\{v_2, v_4\}$ are linearly dependent. We can form other linearly dependent subsets of $\{v_1, v_2, v_3, v_4\}$ by including additional terms in which the coefficients are zero. For example, $\{v_1, v_2, v_4\}$ is linearly dependent since $0\, v_1 - 2\, v_2 + v_4 = 0$.

(b) $\{v_1, v_2\}$ is clearly linearly independent since neither vector is a multiple of the other. (Another way to see that $\{v_1, v_2\}$ must be linearly independent is to note that in part (a) there is no choice of c_3 and c_4 that yields a linear dependence relation for $\{v_1, v_2\}$.) The span of $\{v_1, v_2\}$ is the same as Span$\{v_1, v_2, v_3, v_4\}$, since we can express v_3 and v_4 as linear combinations of v_1 and v_2: $v_3 = v_1 + v_2$ and $v_4 = 2\, v_2$. Other correct answers are $\{v_1, v_3\}$, $\{v_1, v_4\}$, $\{v_2, v_3\}$, and $\{v_3, v_4\}$, but not $\{v_2, v_4\}$.

===================

Section 3: More Examples of Linear Independence and Dependence

If we have a set of vectors that is linearly dependent and we add another vector to the set or remove a vector from the set, can we predict whether the new set is linearly dependent or independent? What if the original set is linearly independent? We examine these questions in this section.

Suppose $\{v_1, \ldots, v_k\}$ is a linearly dependent set of vectors in R^n. Then $\{v_1,\ldots,v_k\}$ must satisfy a linear dependence relation

$$c_1 v_1 + c_2 v_2 + \ldots + c_k v_k = 0$$

where at least one of the weights $c_1, c_2, \ldots, c_k$ is nonzero. So if w is another vector in R^n then, since $0\,w = 0$, we have the linear dependence relation

$$c_1 v_1 + c_2 v_2 + \ldots + c_k v_k + 0\,w = 0$$

with the same coefficients $c_1, c_2, \ldots c_k$ as above and the additional coefficient 0. So we have proved:

- If $\{v_1, \ldots, v_k\}$ is a linearly dependent set of k vectors in R^n and w is another vector in R^n, then the set of $k+1$ vectors $\{v_1, \ldots, v_k, w\}$ must also be a linearly dependent set.

On the other hand, adding another vector to a set of linearly independent vectors may or may not result in a linearly independent set.

Exercise 3.1: The set of vectors $\{u, v\}$ below is linearly independent.
(a) Find vectors p and q such that $\{u, v, p\}$ is linearly independent and $\{u, v, q\}$ is linearly dependent. Use Maple to check your answers.
(b) Explain your answer in geometric terms.

```
> u := Vector([1,-1,0]):
  v := Vector([0,1,-1]):
```

 Student Workspace

 Answer 3.1

The vectors u and v span a plane through the origin, $a\,x + b\,y + c\,z = 0$. To find a, b, and c, substitute u and v into the equation $a\,x + b\,y + c\,z = 0$; the result is the plane $x + y + z = 0$. We must choose p so it <u>is not</u> in this plane and q so that it <u>is</u> in this plane. For example:

```
> p := Vector([1,0,0]):
  q := Vector([1,0,-1]):
```

Similarly, removing a vector from a linearly dependent set may or may not result in a linearly dependent set.

Exercise 3.2: (a) Consider the four vectors in R^3 shown below. This set of four vectors, $\{u, v, x, y\}$, is linearly dependent. Which vector could you remove from this set so the remaining three vectors still form a linearly dependent set? Is there more than one choice? Which vector could you remove from this set so the remaining three vectors form a linearly independent set? Is there more than one choice?

Module 2.4 Linear Independence of Vectors   87

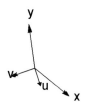

(b) Consider the set of four vectors in R^3 shown below. This set is also linearly dependent. Can you remove a vector from this set so the remaining three form a linearly dependent set? Which vector or vectors could you remove from this set so the remaining three vectors form a linearly independent set?

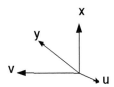

 Student Workspace

 Answer 3.2

(a) The vectors u, v, x all lie in a plane. So if you remove y, the set $\{u, v, x\}$ is still linearly dependent. If, however, you remove any one of the vectors u, v, x, the remaining three form a linearly independent set.

(b) No three of the vectors lie in a plane. So removing any one of them leaves a set of three linearly independent vectors. Furthermore, any three of them span all of R^3, and hence the set of all four vectors is linearly dependent.

Linear Independence of Solutions Produced by Decomposition

===================

Example 3A: In Module 3, we saw how to decompose the solution of a linear system. The general form we obtained from such a decomposition was $p + \text{Span}\{v_1, v_2, ..., v_k\}$, where p is some particular solution of the given linear system and $\text{Span}\{v_1, v_2, ..., v_k\}$ is the set of solutions of the corresponding homogeneous linear system. The result of any decomposition has a geometric interpretation: We can think of the set of solutions of the homogeneous system as analogous to a plane through the origin and the set of solutions of the given system as the translation of this "plane" by the vector p.

In this example, we will use the method of decomposition to find a linearly independent set of vectors whose span is the set of solutions of a given homogeneous linear system:

$$2x_1 - 6x_2 + 3x_3 - 2x_4 = 0$$
$$-x_1 + 3x_2 - 2x_3 = 0$$
$$3x_1 - 9x_2 + 4x_3 - 4x_4 = 0$$

```
> M := Matrix([[2,-6,3,-2,0],[-1,3,-2,0,0],[3,-9,4,-4,0]]):
> Backsolve(Reduce(M),free=x);
```

$$\begin{bmatrix} 3x_2 + 4x_4 \\ x_2 \\ -2x_4 \\ x_4 \end{bmatrix}$$

When we decompose this set of solutions, we see that it is the span of $v_1 = \langle 3, 1, 0, 0 \rangle$ and $v_2 = \langle 4, 0, -2, 1 \rangle$. Note that $\{v_1, v_2\}$ is a linearly independent set of vectors.

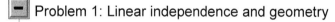

That the method of decomposition in this example produced a <u>linearly independent</u> spanning set was not an accident. You will see in Problem 12 that:

- When the method of decomposition is applied to the solution set of a homogeneous linear system, the vectors produced by the decomposition are always linearly independent.

Problems

Problem 1: Linear independence and geometry

(a) The set of vectors $\{u, v\}$ below is linearly independent and hence spans a plane through the origin. Find the cartesian equation, $ax + by + cz = 0$, for this plane.
(b) Find a vector w so that $\{u, v, w\}$ is linearly independent.
(c) How is w related to the plane in (a)? Does w satisfy the cartesian equation you found? Does your answer depend on your choice of w or will it always be the same, so long as $\{u, v, w\}$ is linearly independent?

```
> u := Vector([3,-1,2]):
  v := Vector([1,4,-3]):
```

Module 2.4 Linear Independence of Vectors   89

[+] Student Workspace

[−] Problem 2: Linear independence and linear combinations

(a) Show that the set of vectors $\{v_1, v_2, v_3, v_4\}$ below is linearly dependent, and find one linear dependence relation for this set of vectors.
(b) Which of the four vectors is a linear combination of the other three? Give all possible answers.

```
> v1 := Vector([3,-1,2,0]):
  v2 := Vector([-5,0,-3,2]):
  v3 := Vector([1,-2,1,2]):
  v4 := Vector([0,4,-2,3]):
```

[+] Student Workspace

[−] Problem 3: Does pairwise independence imply independence?

True or false: "If v_1, v_2, v_3 are vectors in R^3 and the sets $\{v_1, v_2\}$, $\{v_2, v_3\}$ and $\{v_1, v_3\}$ are each linearly independent, then the set $\{v_1, v_2, v_3\}$ must also be linearly independent." If this statement is true, explain why; if it is not, give a counterexample.

[+] Student Workspace

[−] Problem 4: Linear independence depending on a variable

For what value(s) of the variable a is $\{u, v, w\}$ (see below) linearly independent?

```
> u := Vector([-2,-5,5]):
  v := Vector([1,7,-2]):
  w := Vector([3,3,a]):
```

[+] Student Workspace

[−] Problem 5: Linear dependence and independence by inspection

(a) Find, by inspection, a linear dependence relation for the set of vectors $\{u, v, w\}$ below. (No calculation necessary; the relation is fairly easy to spot.)
(b) Find, by inspection, a vector p that is not in Span$\{u, v, w\}$. (Your answer in (a) should give you a hint about how you might construct p.)
(c) What are all the 3-element subsets of $\{u, v, w, p\}$ that are linearly independent?

```
> u := Vector([-1,1,1]):
  v := Vector([-2,3,3]):
  w := Vector([-4,5,5]):
```

[+] Student Workspace

Problem 6: Find the linearly independent subsets

(a) Confirm that the set of vectors $\{v_1, v_2, v_3, v_4, v_5\}$ defined below is linearly dependent. Write down all the linear dependence relations you get by solving the equation

$$c_1 v_1 + c_2 v_2 + c_3 v_3 + c_4 v_4 + c_5 v_5 = 0$$

(b) Which of the subsets $\{v_1, v_2\}$, $\{v_1, v_2, v_3\}$, $\{v_1, v_2, v_5\}$, $\{v_1, v_2, v_3, v_4\}$ are linearly independent? (You should be able to answer this question directly from your answers to (a) without any further computation.)

```
> v1 := Vector([-4, -2, -1, 4, 7]):
  v2 := Vector([7, 1, -2, -6, -8]):
  v3 := Vector([-3, 3, 0, 2, 4]):
  v4 := Vector([-6, -6, -1, -2, 3]):
  v5 := Vector([1, 3, 4, -2, -6]):
> M := Matrix([v1,v2,v3,v4,v5,Vector(5,0)]);
```
Student Workspace

Problem 7: Any four vectors in R^3 are linearly dependent

(a) Demonstrate that the four vectors $\{v_1, v_2, v_3, v_4\}$ in R^3 defined below are linearly dependent.
(b) Explain why every set of four vectors in R^3 must be linearly dependent.

```
> v1 := Vector([1,3,5]):
> v2 := Vector([2,4,1]):
  v3 := Vector([-3,0,3]):
> v4 := Vector([3,1,4]):
```
Student Workspace

Problem 8: Linear dependence and the zero vector

Prove the following statement: If $\{v_1, v_2, ..., v_k\}$ is a set of vectors in R^n that includes the zero vector, the set is linearly dependent.

Student Workspace

Problem 9: Linearly independent subsets of a linearly dependent set

Find a set of four vectors in R^4 that is linearly dependent but every subset of which is linearly independent (except, of course, for the set itself). For example, if the set of four vectors is $\{v_1, v_2, v_3, v_4\}$, then subsets such as $\{v_1, v_2, v_3\}$, $\{v_2, v_3, v_4\}$, $\{v_1, v_2\}$, etc. are all linearly independent sets.

Student Workspace

Problem 10: Linearly dependent subset of a linearly independent set?

Let v_1 and v_2 be two vectors in R^4 defined below. If possible, find two additional vectors v_3 and v_4 such that the set $\{v_1, v_2, v_3\}$ is linearly dependent and the set $\{v_1, v_2, v_3, v_4\}$ is linearly independent. If this is not possible, explain why not.

```
> v1 := Vector([2,3,4,5]):
  v2 := Vector([1,3,5,4]):
```

 Student Workspace

Problem 11: Concrete application (part 2)

This problem follows up on Problem 6 of Module 2. Below is the table that describes the composition of the three basic mixtures of concrete, S, A, L. The corresponding vectors for these mixtures and for two custom mixtures, U and V, have been entered below.

	Super-Strong Type S	All-Purpose Type A	Long-life Type L
cement	20	18	12
water	10	10	10
sand	20	25	15
gravel	10	5	15
fly ash	0	2	8

	Type U	Type V
cement	12	15
water	12	10
sand	12	20
gravel	12	10
fly ash	12	5

```
> S := Vector([20,10,20,10,0]):
> A := Vector([18,10,25,5,2]):
> L := Vector([12,10,15,15,8]):
> U := Vector([12,12,12,12,12]):
> V := Vector([15,10,20,10,5]):
```

(a) Show that $\{S, A, L\}$ is a linearly independent set of vectors. What practical advantage does that have?
(b) Show that we can make the custom mix V but not the custom mix U from the three basic mixes S, A, L. What does this say about the linear independence or dependence of the sets $\{S, A, L, U\}$ and $\{S, A, L, V\}$?
(c) Explain why any combination of S, A, L and V can also be achieved by a combination of just S, A and L. For example, show how to make the custom mix $3\,S + 4\,A + 2\,L + 3\,V$ using only S, A and L.
(d) Define a fifth basic mix W to add to $\{S, A, L, U\}$ such that any custom mixture can be expressed as a

linear combination of the set of mixes $\{S, A, L, U, W\}$.

(e) Why will there still be mixes that cannot be physically produced from this set of five basic mixes? (Hint: Consider the signs of the scalar weights.) Give an example of such a mix.

 Student Workspace

 Problem 12: Linear independence and decomposition

(a) Use decomposition to express the set of solutions of the homogeneous linear system below in the form $\text{Span}\{v_1, v_2, ..., v_k\}$. By hand, check that the set of vectors $\{v_1, v_2, ..., v_k\}$ is linearly independent, using the following quick method: When you write out the equation of the form $c_1 v_1 + c_2 v_2 + ... + c_k v_k = 0$ and combine the terms on the left, note that this equation is exactly the equation you would get if you were to set the result of the **Backsolve** command below equal to 0. Then look at the components that correspond to the free variables in the solution of the linear system.

(b) (Challenge) Explain why the method of decomposition, when applied to the solution set of a homogeneous linear system, always yields a linearly independent set of vectors whose span is the set of solutions. Hint: Use the quick method in (a) that helped you check linear independence by hand.

$$-2x_1 + 3x_2 - x_3 + x_4 + 4x_5 + x_6 + 4x_7 = 0$$
$$x_1 - 3x_2 - 2x_4 - 3x_5 + x_7 = 0$$
$$-5x_1 + x_2 + 3x_3 - 4x_4 - 2x_5 - 3x_6 = 0$$

```
> M := Matrix([[-2,3,-1,1,4,1,4,0],[1,-3,0,-2,-3,0,1,0],
    [-5,1,3,-4,-2,-3,0,0]]);
> Backsolve(Reduce(M),free=x):
```

 Student Workspace

Chapter 3: Matrix Algebra

Module 1. Product of a Matrix and a Vector

Module 2. Matrix Multiplication

Module 3. Rules of Matrix Algebra

Module 4. Markov Chains -- An Application

Module 5. Inverse of a Matrix -- An Introduction

Module 6. Inverse of a Matrix and Elementary Matrices

Module 7. Determinants

 Commands used in this Chapter

`Matrix([[a,b],[c,d]]);` defines the matrix with rows $[a, b]$ and $[c, d]$.
`Matrix([u,v,w]);` produces the matrix whose columns are the vectors u, v, w.
`Row(A,i);` selects the ith row of the matrix A.
`subs({a=3,b=13},expr);` substitutes the values for a and b in the expression *expr*.
`Transpose(M);` produces the transpose of the matrix M.
`Vector([a,b,c]);` defines the vector $\langle a, b, c \rangle$.

LAMP commands:
`Backidmat(n);` produces the n by n backwards identity matrix.
`Backsolve(R);` solves the linear system for which R is a row echelon form of the augmented matrix.
`Componentplot(A,x0,points=n);` plots (in an animation) the components of the vectors $A^k x_0$ for k from 0 to n.
`Copyinto(A,B,m,n);` copies the matrix A into the matrix B with the $[1, 1]$ entry of A going into the $[m, n]$ entry of B.
`Det(M);` calculates the determinant of the matrix M.
`Diagmat([a,b,c]);` produces a diagonal matrix with diagonal entries a, b, c.
`Idmat(n);` produces the n by n identity matrix.
`Inverse(M);` produces the inverse of the matrix M.
`Jordanmat(n);` produces the n by n matrix with 1's along the diagonal just above the main diagonal and zeros elsewhere.

LetterL(n); produces the *n* by *n* matrix of 0's and 1's in the shape of the letter L.
Matsolve(A,b,free=t); solves the matrix-vector equation $Ax = b$ for *x*. The solution is expressed in terms of the free variable *t*.
Randmat(m,n); produces an *m* by *n* matrix with random integer entries.
Reduce(M); produces a row echelon form of the matrix *M*.
Reduce(M,form=rref); produces the <u>reduced</u> row echelon form of the matrix *M*.
Rowop(M,rj <> rk); interchanges row *j* and row *k* of the matrix *M*.
Rowop(M,rj <= rj+c*rk); replaces row *j* by the sum of row *j* and *c* times row *k*.
Rowop(M,rj <= c*rj); replaces row *j* by *c* times row *j*.

Linear Algebra Modules Project
Chapter 3, Module 1

Product of a Matrix and a Vector

Purpose of this module

The purpose of this module is to acquaint you with the definition of matrix-vector multiplication, its properties, and its uses.

Prerequisites

Algebra of vectors, linear independence, span.

Commands used in this module

```
> restart: with(LinearAlgebra): with(Lamp): with(plots):
  UseHardwareFloats := false: Digits := 6:
```

Tutorial

Section 1: Definition of the Matrix-Vector Product $A\,x$

Definition: Suppose A is a matrix with m rows and n columns, and x is a vector with n components. The **matrix-vector product** $A\,x$ is defined to be the linear combination of the columns of A in which the weights are the components of x.

We can express this definition in symbols: Let $A_1, A_2, \ldots, A_n$ denote the columns of the matrix A, and $x_1, x_2, \ldots, x_n$ denote the components of the vector x. Then

$$A\,x = [A_1, A_2, \ldots, A_n] \begin{bmatrix} x_1 \\ \cdot \\ \cdot \\ \cdot \\ x_n \end{bmatrix} = x_1 A_1 + x_2 A_2 + \ldots + x_n A_n$$

Example 1A: Find the matrix-vector product $A\,x$, where

$$A = \begin{bmatrix} 1 & 3 & -2 \\ 2 & -1 & 4 \end{bmatrix} \quad \text{and} \quad x = \begin{bmatrix} 5 \\ 3 \\ 2 \end{bmatrix}$$

Chapter 3 Matrix Algebra

Solution: The columns of A are $A_1 = \begin{bmatrix} 1 \\ 2 \end{bmatrix}$, $A_2 = \begin{bmatrix} 3 \\ -1 \end{bmatrix}$, $A_3 = \begin{bmatrix} -2 \\ 4 \end{bmatrix}$, and the components of x are $x_1 = 5$, $x_2 = 3$, $x_3 = 2$. Using the components of x as weights and the columns of A as vectors, we calculate the product $A\,x$ as the linear combination:

$$5 \begin{bmatrix} 1 \\ 2 \end{bmatrix} + 3 \begin{bmatrix} 3 \\ -1 \end{bmatrix} + 2 \begin{bmatrix} -2 \\ 4 \end{bmatrix} = \begin{bmatrix} 10 \\ 15 \end{bmatrix}$$

=====================

The Maple commands to compute this product are as follows:

```
> A := Matrix([[1,3,-2],[2,-1,4]]);
  x := Vector([5,3,2]);
```

$$A := \begin{bmatrix} 1 & 3 & -2 \\ 2 & -1 & 4 \end{bmatrix}$$

$$x := \begin{bmatrix} 5 \\ 3 \\ 2 \end{bmatrix}$$

```
> A.x;
```

$$\begin{bmatrix} 10 \\ 15 \end{bmatrix}$$

Maple note: Maple's multiplication symbol for matrix multiplication is . instead of the more usual *.

Exercise 1.1: Compute the matrix-vector product $M\,u$ using the matrix M and vector u below. First do this by hand to practice the definition above. Then check your answer by having Maple calculate the product.

```
> M := Matrix([[3,-2],[-4,2],[8,5]]);
> u := Vector([1,2]);
```

$$M := \begin{bmatrix} 3 & -2 \\ -4 & 2 \\ 8 & 5 \end{bmatrix}$$

$$u := \begin{bmatrix} 1 \\ 2 \end{bmatrix}$$

+ Student Workspace

− Answer 1.1

$$1 \begin{bmatrix} 3 \\ -4 \\ 8 \end{bmatrix} + 2 \begin{bmatrix} -2 \\ 2 \\ 5 \end{bmatrix} = \begin{bmatrix} -1 \\ 0 \\ 18 \end{bmatrix}$$

```
> M.u;
```

$$\begin{bmatrix} -1 \\ 0 \\ 18 \end{bmatrix}$$

Exercise 1.2: Below we define vectors u, v, w and compute the linear combination $p = 2u + 11v + 7w$. Find a matrix A and vector x such that $p = Ax$.

```
> u := Vector([2,3,8]);
  v := Vector([5,-6,0]);
  w := Vector([7,4,-2]);
```

$$u := \begin{bmatrix} 2 \\ 3 \\ 8 \end{bmatrix}, \; v := \begin{bmatrix} 5 \\ -6 \\ 0 \end{bmatrix}, \; w := \begin{bmatrix} 7 \\ 4 \\ -2 \end{bmatrix}$$

```
> p := 2*u+11*v+7*w;
```

$$p := \begin{bmatrix} 108 \\ -32 \\ 2 \end{bmatrix}$$

+ Student Workspace

− Answer 1.2

```
> A := Matrix([u,v,w]);
```

$$A := \begin{bmatrix} 2 & 5 & 7 \\ 3 & -6 & 4 \\ 8 & 0 & -2 \end{bmatrix}$$

```
> x := Vector([2,11,7]);
```

$$x := \begin{bmatrix} 2 \\ 11 \\ 7 \end{bmatrix}$$

```
> A.x, p;
```

$$\begin{bmatrix} 108 \\ -32 \\ 2 \end{bmatrix}, \begin{bmatrix} 108 \\ -32 \\ 2 \end{bmatrix}$$

Exercise 1.3: (a) If the vector x is simple enough, we can calculate Ax by inspection. Vectors that have 1's and 0's are especially easy to work with. By inspection, multiply the matrix A below by each of the vectors u, v, w, z below. (Just describe each linear combination in words.)

$$A = \begin{bmatrix} a & b & c \\ d & e & f \end{bmatrix}, \; u = \begin{bmatrix} 1 \\ 0 \\ 0 \end{bmatrix} \; v = \begin{bmatrix} 0 \\ 0 \\ 1 \end{bmatrix} \; w = \begin{bmatrix} 0 \\ 1 \\ 1 \end{bmatrix} \; z = \begin{bmatrix} 0 \\ 0 \\ 2 \end{bmatrix}$$

(b) Suppose N is a 8 by 5 matrix and you want to add the odd-numbered columns together to form a single 8 by 1 column vector. Describe how you could accomplish this using a matrix-vector product.

+ Student Workspace

− Answer 1.3

(a) $A\,u$ = the first column of A.
$A\,v$ = the last column of A.
$A\,w$ = the sum of the second and third columns of A.
$A\,z$ = two times the last column of A.
(b) Multiply N by the vector $x = \langle 1, 0, 1, 0, 1 \rangle$.

Dot Product Method for Calculating $A\,x$

Another method for computing matrix-vector products uses dot products.

Definition: The **dot product** of two vectors in R^n is the sum of the products of corresponding components.

For example, here is the dot product of two vectors in R^4:

$$\langle 2, -1, 4, 0 \rangle \cdot \langle 5, 3, -2, -4 \rangle = 2(5) - 1(3) + 4(-2) + 0(-4) = -1$$

Let's take another look at the matrix-vector product that we calculated in Example 1A:

$$\begin{bmatrix} 1 & 3 & -2 \\ 2 & -1 & 4 \end{bmatrix} \begin{bmatrix} 5 \\ 3 \\ 2 \end{bmatrix} = \begin{bmatrix} 10 \\ 15 \end{bmatrix}$$

Notice that the first component of the product $A\,x$ is simply the dot product of the first row of A and the vector x, and similarly for the second component:

$$5\begin{bmatrix}1\\2\end{bmatrix} + 3\begin{bmatrix}3\\-1\end{bmatrix} + 2\begin{bmatrix}-2\\4\end{bmatrix} = \begin{bmatrix} 5(1)+3(3)+2(-2) \\ 5(2)+3(-1)+2(4) \end{bmatrix} = \begin{bmatrix} 10 \\ 15 \end{bmatrix}$$

So we now have a second way to compute the product $A\,x$:

- The ith component of $A\,x$ is the dot product of the ith row of A with the vector x.

Exercise 1.4: (By hand) Use the dot product method to calculate the following matrix-vector products:

$$\text{(a)} \begin{bmatrix} 1 & 2 \\ 4 & 0 \\ -3 & 5 \end{bmatrix} \begin{bmatrix} 7 \\ 3 \end{bmatrix} \qquad \text{(b)} \begin{bmatrix} 5 & -2 & 3 \\ 7 & 10 & 4 \end{bmatrix} \begin{bmatrix} x_1 \\ x_2 \\ x_3 \end{bmatrix}$$

 Student Workspace

Answer 1.4

(a) $\begin{bmatrix} 1(7)+2(3) \\ 4(7)+0(3) \\ -3(7)+5(3) \end{bmatrix} = \begin{bmatrix} 13 \\ 28 \\ -6 \end{bmatrix}$

(b) $\begin{bmatrix} 5x_1 - 2x_2 + 3x_3 \\ 7x_1 + 10x_2 + 4x_3 \end{bmatrix}$

Section 2: Vector Equations, Matrix Equations and Linear Systems

We begin this section by reconsidering a problem that we first solved in Chapter 2. We will briefly repeat our earlier solution and then introduce a new way of looking at this familiar problem by using matrix-vector products.

========================

Example 2A: For the vectors u, v and b below, express b as a linear combination of u and v.

```
> u := Vector([2,3,4]);
  v := Vector([-4,2,1]);
```

$$u := \begin{bmatrix} 2 \\ 3 \\ 4 \end{bmatrix}, \quad v := \begin{bmatrix} -4 \\ 2 \\ 1 \end{bmatrix}$$

```
> b := Vector([-40,4,-8]);
```

$$b := \begin{bmatrix} -40 \\ 4 \\ -8 \end{bmatrix}$$

Solution method 1: To express b as a linear combination of u and v requires us to find weights x_1 and x_2 that satisfy the vector equation $x_1 u + x_2 v = b$:

$$x_1 \begin{bmatrix} 2 \\ 3 \\ 4 \end{bmatrix} + x_2 \begin{bmatrix} -4 \\ 2 \\ 1 \end{bmatrix} = \begin{bmatrix} -40 \\ 4 \\ -8 \end{bmatrix} \qquad [1]$$

Recall that in Chapter 2, Module 2, we solved a vector equation by writing it as a linear system:

$$\begin{aligned} 2x_1 - 4x_2 &= -40 \\ 3x_1 + 2x_2 &= 4 \\ 4x_1 + x_2 &= -8 \end{aligned} \qquad [2]$$

This linear system can then be solved by reducing its augmented matrix:

```
> M := Matrix([u,v,b]);
  Backsolve(Reduce(M));
```

$$M := \begin{bmatrix} 2 & -4 & -40 \\ 3 & 2 & 4 \\ 4 & 1 & -8 \end{bmatrix}$$

$$\begin{bmatrix} -4 \\ 8 \end{bmatrix}$$

So we have found that $b = -4\,u + 8\,v$.

Solution method 2: The left side of the vector equation [1] can be expressed as a matrix-vector product:

$$x_1 \begin{bmatrix} 2 \\ 3 \\ 4 \end{bmatrix} + x_2 \begin{bmatrix} -4 \\ 2 \\ 1 \end{bmatrix} = \begin{bmatrix} 2 & -4 \\ 3 & 2 \\ 4 & 1 \end{bmatrix} \begin{bmatrix} x_1 \\ x_2 \end{bmatrix}$$

Therefore equation [1] can be written in the form:

$$\begin{bmatrix} 2 & -4 \\ 3 & 2 \\ 4 & 1 \end{bmatrix} \begin{bmatrix} x_1 \\ x_2 \end{bmatrix} = \begin{bmatrix} -40 \\ 4 \\ -8 \end{bmatrix} \qquad [3]$$

which we will refer to as a "matrix-vector equation." Equation [3] can be written more succinctly yet in the form $A\,x = b$, where $A = \begin{bmatrix} 2 & -4 \\ 3 & 2 \\ 4 & 1 \end{bmatrix}$ and $x = \begin{bmatrix} x_1 \\ x_2 \end{bmatrix}$. Thus, to solve the given problem, we simply need to solve $A\,x = b$ for the unknown vector x, which is accomplished by using the command `Matsolve(A,b)`:

> `A := Matrix([u,v]);`

$$A := \begin{bmatrix} 2 & -4 \\ 3 & 2 \\ 4 & 1 \end{bmatrix}$$

> `Matsolve(A,b);`

$$\begin{bmatrix} -4 \\ 8 \end{bmatrix}$$

Maple note: The command `Matsolve(A,b)` is equivalent to the combination of commands `Backsolve(Reduce(Matrix([A,b])))`. However, there is no need to think of `Matsolve` in such a complicated way; it simply finds all solutions x to the equation $A\,x = b$.

=====================

In the next exercise, we ask you to practice moving between the three equivalent problems: a vector equation [1], a linear system [2], and a matrix-vector equation [3].

Exercise 2.1: Consider the vector equation $x_1\,u + x_2\,v = b$, where u, v, b are defined below:

> `u := Vector([5,3]);`

```
v := Vector([3,2]);
b := Vector([7,2]);
```

$$u := \begin{bmatrix} 5 \\ 3 \end{bmatrix}, \quad v := \begin{bmatrix} 3 \\ 2 \end{bmatrix}, \quad b := \begin{bmatrix} 7 \\ 2 \end{bmatrix}$$

(a) Write out the corresponding linear system.
(b) Write out the corresponding matrix-vector equation $A\,x = b$, and solve this equation using the **Matsolve** command.
(c) Check that the solutions you found in (b) solve the vector equation and the linear system.

➕ Student Workspace

➖ Answer 2.1

(a) $5x_1 + 3x_2 = 7$
$3x_1 + 2x_2 = 2$

(b)
```
> A := Matrix([u,v]);
```

$$A := \begin{bmatrix} 5 & 3 \\ 3 & 2 \end{bmatrix}$$

```
> Matsolve(A,b);
```

$$\begin{bmatrix} 8 \\ -11 \end{bmatrix}$$

(c)
```
> 8*u-11*v = b;
```

$$\begin{bmatrix} 7 \\ 2 \end{bmatrix} = \begin{bmatrix} 7 \\ 2 \end{bmatrix}$$

```
> 5*8+3*(-11)=7;
  3*8+2*(-11)=2;
```

$$7 = 7$$
$$2 = 2$$

The following theorem gives the general statement of the equivalence of vector equations, matrix-vector equations, and linear systems:

Theorem 1: Suppose A is a m by n matrix with columns $A_1, A_2, \ldots, A_n$ and b is a vector in R^m. Then
- the vector equation $x_1 A_1 + x_2 A_2 + \ldots + x_n A_n = b$,
- the matrix-vector equation $A\,x = b$, and
- the linear system with coefficient matrix A and right-side vector b

all have the same set of solutions.

In the remainder of this section, we use matrix-vector products to review and deepen our understanding of the fundamental concepts of span and linear independence introduced in Chapter 2.

Span and the Equation $Ax = b$

Recall that a vector b is in the *span* of the vectors $v_1, \ldots, v_k$ if there are scalars $x_1, \ldots, x_k$ such that $x_1 v_1 + \ldots + x_k v_k = b$. By Theorem 1, this vector equation is the same as the matrix-vector equation $Ax = b$, where A is the matrix whose columns are $v_1, \ldots, v_k$ and x is the column vector with entries $x_1, \ldots, x_k$. Thus:

- The vector b is in the span of the columns of a matrix A if and only if the matrix-vector equation $Ax = b$ has a solution.

Exercise 2.2: Use the **Matsolve** command to solve the equations $Ax = b$ and $Ax = c$, where the matrix A and vectors b and c are defined below.
(a) Is b in the span of the columns of A? If so, write b as a linear combination of the columns of A.
(b) Is c in the span of the columns of A? If so, write c as a linear combination of the columns of A.

```
> A := Matrix([[-3,0,4], [2,-1,5],[-4,5,2],[2,4,6]]);
```

$$A := \begin{bmatrix} -3 & 0 & 4 \\ 2 & -1 & 5 \\ -4 & 5 & 2 \\ 2 & 4 & 6 \end{bmatrix}$$

```
> b := Vector([10, 27,-15,16]);
> c := Vector([10, 34,-15,16]);
```

$$b := \begin{bmatrix} 10 \\ 27 \\ -15 \\ 16 \end{bmatrix}, \quad c := \begin{bmatrix} 10 \\ 34 \\ -15 \\ 16 \end{bmatrix}$$

 Student Workspace

Answer 2.2

```
> Matsolve(A,b);
```

$$\begin{bmatrix} 2 \\ -3 \\ 4 \end{bmatrix}$$

Therefore b is a linear combination of the columns of A, and $b = 2 A_1 - 3 A_2 + 4 A_3$, where A_1, A_2, A_3 denote the columns of A.

```
> Matsolve(A,c);
Error, (in Matsolve) no solutions
```

Therefore c is not a linear combination of the columns of A.

Linear Independence and the Equation $A\,x = 0$

Recall that a set of vectors $\{v_1, ..., v_k\}$ is *linearly independent* if the only solution of the vector equation $c_1 v_1 + ... + c_k v_k = 0$ is the trivial solution where all the weights c_i are zero. By Theorem 1, solving the equation $c_1 v_1 + ... + c_k v_k = 0$ is equivalent to solving the matrix-vector equation $A\,x = 0$, where the columns of A are the vectors $v_1, ..., v_k$. Thus:

- The columns of a matrix A are linearly independent if and only if the matrix-vector equation $A\,x = 0$ has only the trivial solution $x = 0$.

Exercise 2.3: Determine whether the columns of the matrix A below are linearly independent by using **Matsolve** to solve $A\,x = 0$.

```
> A := Matrix([[4,1,3],[2,3,-1],[1,-1,2]]);
```

$$A := \begin{bmatrix} 4 & 1 & 3 \\ 2 & 3 & -1 \\ 1 & -1 & 2 \end{bmatrix}$$

```
> z := Vector(3,0);
```

$$z := \begin{bmatrix} 0 \\ 0 \\ 0 \end{bmatrix}$$

 Student Workspace

 Answer 2.3

```
> Matsolve(A,z);
```

$$\begin{bmatrix} -X_3 \\ X_3 \\ X_3 \end{bmatrix}$$

Since the matrix-vector equation $A\,x = 0$ has a nontrivial solution, the columns of A are linearly dependent.

In fact, by choosing $x_3 = 1$, we see that $A\,x = 0$ has the nontrivial solution $x_1 = -1$, $x_2 = 1$, $x_3 = 1$. Therefore we have the linear dependence relation

$$-\begin{bmatrix} 4 \\ 2 \\ 1 \end{bmatrix} + \begin{bmatrix} 1 \\ 3 \\ -1 \end{bmatrix} + \begin{bmatrix} 3 \\ -1 \\ 2 \end{bmatrix} = \begin{bmatrix} 0 \\ 0 \\ 0 \end{bmatrix}$$

Maple note: The default name of the free variables produced by **Matsolve** is X; so you should not assign a value to X. Or, you can switch to another name for the free variables, as shown below. Also, you can use the special word "zero" for the right-side vector, rather than defining a specific zero vector.

```
> Matsolve(A,z,free=t);
```

$$\begin{bmatrix} -t_3 \\ t_3 \\ t_3 \end{bmatrix}$$

```
> Matsolve(A,zero);
```

$$\begin{bmatrix} -X_3 \\ X_3 \\ X_3 \end{bmatrix}$$

Section 3: Rules of Algebra for Matrix-Vector Products

So far matrix-vector products have provided a convenient notation and a new viewpoint. However, what makes them especially powerful is that they satisfy rules of algebra that allow us to manipulate them in familiar ways. Here are the two key rules:

- $$A(x+y) = Ax + Ay \quad \text{(distributive law)}$$

- $$A(cx) = c(Ax) \quad \text{(scalars factor out)}$$

where A is any m by n matrix, x and y are column vectors with n entries, and c is any real number.

==================

Example 3A: Use Maple to verify the distributive law for the matrix A and the vectors x and y below.

```
> A := Matrix([[2,3],[5,7]]);
> x := Vector([2,3]);
> y := Vector([5,-1]);
```

$$A := \begin{bmatrix} 2 & 3 \\ 5 & 7 \end{bmatrix}$$

$$x := \begin{bmatrix} 2 \\ 3 \end{bmatrix}, \quad y := \begin{bmatrix} 5 \\ -1 \end{bmatrix}$$

Solution: We calculate $Ax + Ay$ and $A(x+y)$ and observe that they are equal:

```
> A.x+A.y;
```

$$\begin{bmatrix} 20 \\ 49 \end{bmatrix}$$

```
> A.(x+y);
```

$$\begin{bmatrix} 20 \\ 49 \end{bmatrix}$$

==================

In the next exercise, you are asked to prove the two rules of algebra above for the special case when the matrix A has 3 columns. The general case is proved in the same way.

Exercise 3.1: (By hand) Let A be an m by 3 matrix with columns A_1, A_2, A_3, and let x and y be vectors with 3 entries:

$$A = [A_1, A_2, A_3], \quad x = \begin{bmatrix} x_1 \\ x_2 \\ x_3 \end{bmatrix}, \quad y = \begin{bmatrix} y_1 \\ y_2 \\ y_3 \end{bmatrix}$$

(a) Use vector algebra to prove the distributive law, $A(x+y) = Ax + Ay$. Hint: Use the definition of a matrix-vector product to expand the products Ax, Ay and $A(x+y)$.

(b) Use vector algebra to prove the second law, $A(cx) = c(Ax)$.

[+] Student Workspace

[−] Answer 3.1

(a) By definition,

$$Ax = x_1 A_1 + x_2 A_2 + x_3 A_3 \text{ and } Ay = y_1 A_1 + y_2 A_2 + y_3 A_3 \text{ and}$$

$$A(x+y) = (x_1 + y_1) A_1 + (x_2 + y_2) A_2 + (x_3 + y_3) A_3$$

By using the distributive law for scalars and vectors, we can expand the right side of the last equation:

$$A(x+y) = x_1 A_1 + x_2 A_2 + x_3 A_3 + y_1 A_1 + y_2 A_2 + y_3 A_3$$

which is just $Ax + Ay$.

(b) Since $cx = \begin{bmatrix} cx_1 \\ cx_2 \\ cx_3 \end{bmatrix}$, $A(cx) = cx_1 A_1 + cx_2 A_2 + cx_3 A_3 = c(x_1 A_1 + x_2 A_2 + x_3 A_3) = c(Ax)$.

Solution Set of a Linear System $Ax = b$

We can use the above two rules of algebra to prove a central fact about the solution set of a linear sytem. By Theorem 1, we can express any linear system in the compact form $Ax = b$, where A is the coefficient matrix of the system and b is the right-side vector. The corresponding homogeneous linear system is then $Ax = 0$. As we saw in Chapters 1 and 2, the solution sets of $Ax = b$ and $Ax = 0$ are closely related:

Theorem 2: If p is a particular solution of the linear system $Ax = b$ and Span$\{v_1, ..., v_k\}$ is the solution set of the corresponding homogeneous system $Ax = 0$, then $p + $ Span$\{v_1, ..., v_k\}$ is the solution set of $Ax = b$.

For example, in Chapter 2, we saw that the solution set of $A\,x = 0$ can be decomposed into a span of a set of vectors $\{v_1, ..., v_k\}$ and visualized as a plane through the origin (if $k = 2$). Also we saw that the solution set of $A\,x = b$ can be visualized as the translate by a vector p of the solution set of $A\,x = 0$.

The following proof of Theorem 2 for $k = 2$ can be extended easily to the general case.

========================

Example 3B: Use the rules of matrix-vector algebra to prove Theorem 2 for the special case when $k = 2$.
Solution: Suppose p is a solution of $A\,x = b$ and Span$\{v_1, v_2\}$ is the solution set of the corresponding homogeneous equation $A\,x = 0$. To prove Theorem 2, we must do both of the following:
(a) Show that $p + c_1 v_1 + c_2 v_2$ is a solution of $A\,x = b$ for all choices of the weights c_1, c_2.
(b) If q is any solution of $A\,x = b$, show that q can be written in the form $p + c_1 v_1 + c_2 v_2$.

(a) By the rules of algebra, $A\,(p + c_1 v_1 + c_2 v_2) = A\,p + c_1\,A\,v_1 + c_2\,A\,v_2$. But we know that $A\,p = b$ and $A\,v_1 = 0$ and $A\,v_2 = 0$. Therefore $A(p + c_1 v_1 + c_2 v_2) = b + 0 + 0 = b$, and therefore $p + c_1 v_1 + c_2 v_2$ is a solution of $A\,x = b$.
(b) Since $A\,q = b$, by the rules of algebra $A\,(q - p) = A\,q - A\,p = b - b = 0$. Therefore $q - p$ is a solution of the homogeneous system, and so we can write $q - p$ in the form $q - p = c_1 v_1 + c_2 v_2$. Therefore $q = p + c_1 v_1 + c_2 v_2$.

========================

Exercise 3.2: (a) For the matrix A below, use **Matsolve** to solve $A\,x = 0$. Then decompose the solution set and write it in the form Span$\{v_1, ..., v_k\}$.
(b) Use matrix-vector multiplication to check that $A\,p = b$, where p and b are also below. With no further computation, use Theorem 2 to write down the entire solution set of $A\,x = b$.

```
> A := Matrix([[4,1,2,-3,-2],[-2,-1,-1,4,1],[0,4,-3,2,2]]);
```

$$A := \begin{bmatrix} 4 & 1 & 2 & -3 & -2 \\ -2 & -1 & -1 & 4 & 1 \\ 0 & 4 & -3 & 2 & 2 \end{bmatrix}$$

```
> p := Vector([1,0,-2,2,-3]);
> b := Vector([0,5,4]);
```

$$p := \begin{bmatrix} 1 \\ 0 \\ -2 \\ 2 \\ -3 \end{bmatrix}, \quad b := \begin{bmatrix} 0 \\ 5 \\ 4 \end{bmatrix}$$

➕ Student Workspace

Answer 3.2

```
> x := 'x':
  Matsolve(A,zero,free=x);
```

$$\begin{bmatrix} -\dfrac{25}{6}x_4 + \dfrac{1}{6}x_5 \\ 5x_4 \\ \dfrac{22}{3}x_4 + \dfrac{2}{3}x_5 \\ x_4 \\ x_5 \end{bmatrix}$$

(a) So the solution set of $Ax = 0$ is

$$x_4 \begin{bmatrix} -\dfrac{25}{6} \\ 5 \\ \dfrac{22}{3} \\ 1 \\ 0 \end{bmatrix} + x_5 \begin{bmatrix} \dfrac{1}{6} \\ 0 \\ \dfrac{2}{3} \\ 0 \\ 1 \end{bmatrix} = x_4 v_1 + x_5 v_2$$

and therefore is Span$\{v_1, v_2\}$, where v_1 and v_2 are given above.

```
> A.p, b;
```

$$\begin{bmatrix} 0 \\ 5 \\ 4 \end{bmatrix}, \begin{bmatrix} 0 \\ 5 \\ 4 \end{bmatrix}$$

(b) The above Maple computation shows that $Ap = b$. So, by Theorem 2, the solution set of $Ax = b$ is $p + \text{Span}\{v_1, v_2\}$, where v_1 and v_2 are given above. That is, the solution set is:

$$\begin{bmatrix} 1 \\ 0 \\ -2 \\ 2 \\ -3 \end{bmatrix} + x_4 \begin{bmatrix} -\dfrac{25}{6} \\ 5 \\ \dfrac{22}{3} \\ 1 \\ 0 \end{bmatrix} + x_5 \begin{bmatrix} \dfrac{1}{6} \\ 0 \\ \dfrac{2}{3} \\ 0 \\ 1 \end{bmatrix}$$

Problems

Problem 1: Linear combinations written as matrix-vector products

Suppose M is a 7 by 5 matrix and you want to subtract the sum of columns 2 and 4 from the sum of columns 1, 3, and 5. Find a vector x such that Mx is this linear combination. Check your answer by multiplying it by the random 7 by 5 matrix below.

```
> M := Randmat(7,5);
```

+ Student Workspace

Problem 2: Linear systems and linear combinations

Solve the linear system $Ax = b$ using **Matsolve**, where A and b are defined below. Use the solutions to express b as a linear combination of the columns A_1, A_2, A_3, A_4, A_5 of A. Find at least three different linear combinations of these columns that equal b.

```
> A := Matrix([[1,3,3,-3,-4],[-3,0,-3,4,-3],[-4,2,-2,-4,4]]);
> b := Vector([1,3,4]);
```

+ Student Workspace

Problem 3: Geometric explanation of consistency

(a) For the matrix A below, check that the linear system $Ax = b$ is consistent for every right-side column vector b and that the solution has one free variable. Hint: Apply **Reduce** to the augmented matrix below, which is formed by augmenting the coefficient matrix with a completely general right-side vector b.
(b) Give two separate geometric explanations for why this system is consistent and the solution has a free variable. Base your first explanation on the observation that each of the two equations in the linear system can be thought of as an equation of a plane. Base your second explanation on the observation that the columns of A are vectors in R^2 that are not all multiples of one another.

```
> A := Matrix([[1,3,-2],[2,-4,1]]);
> aug := Matrix([A,Vector([b1,b2])]);
```

+ Student Workspace

Problem 4: Geometric explanation of inconsistency

(a) For the matrix A below, check that the linear system $Ax = b$ is inconsistent for some right-side column vectors b. Hint: Apply **Reduce** to the augmented matrix below, which is formed by augmenting the coefficient matrix A with a completely general right-side vector b. Look closely at the last row of the reduced matrix.

```
> A := Matrix([[1,2],[3,-4],[-2,1]]);
> aug := Matrix([A,Vector([b1,b2,b3])]);
```

(b) Give an example of a right-side vector b for which $A\,x = b$ is inconsistent. Also give an example of a nonzero right-side vector b for which $A\,x = b$ is consistent. Hint: look at the echelon form of the augmented matrix *aug*.

(c) Give a geometric explanation for why this system is inconsistent for some right-side vectors b. Base your explanation on the fact that the columns of A are vectors in R^3.

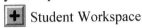 Student Workspace

Problem 5: Linear systems and matrix-vector equations

For the linear system below, do the following:

(a) Express this system in the form $A\,x = b$, and use **Matsolve** to solve this matrix-vector equation.

(b) Write down the vector equation that corresponds to this linear system. What vector have you thereby expressed as a linear combination of other vectors?

(c) Use **Matsolve** to solve the corresponding homogeneous system $A\,x = 0$.

(d) Write down the vector equation that corresponds to this homogeneous system. What set of vectors have you thereby shown is linearly independent or linearly dependent? Which is it?

$$2 x_1 - 2 x_2 - 4 x_3 = -2$$
$$2 x_1 - x_2 - x_3 = 2$$
$$-3 x_1 + 5 x_2 + 4 x_3 = 3$$
$$-x_1 - 5 x_2 + 4 x_3 = -3$$

Student Workspace

Problem 6: Linear independence and dependence

Check that the set of vectors $\{v_1, v_2, v_3, v_4, v_5\}$ is linearly dependent. Which of the subsets $\{v_1, v_2\}$, $\{v_1, v_2, v_3\}$, $\{v_1, v_2, v_5\}$, $\{v_1, v_2, v_3, v_4\}$ is linearly independent? Answer all these questions by solving the matrix-vector equation $A\,x = 0$, where $A = [v_1, v_2, v_3, v_4, v_5]$. Explain, using the definitions of linear independence and dependence, why your answers are correct.

```
> v1 := Vector([-4, -2, -1, 4, 7]):
  v2 := Vector([7, 1, -2, -6, -8]):
  v3 := Vector([-3, 3, 0, 2, 4]):
  v4 := Vector([-6, -6, -1, -2, 3]):
  v5 := Vector([1, 3, 4, -2, -6]):
```

Student Workspace

Problem 7: Structure of solution sets

Use only matrix-vector multiplication to check that p is a solution of the linear system $Ax = b$, and that u, v are solutions of the corresponding homogeneous system $Ax = 0$. (The coefficient matrix A, the right-side vector b, and the solutions p, u, v are given below.)

```
> A := Matrix([[-4,-4,-1,-4,3],[-4,2,2,0,1],
    [2,4,2,2,-2]]);
> b := Vector([2,5,2]):
> p := Vector([-1,-3,5,0,-3]):
> u := Vector([1,-6,8,3,0]):
  v := Vector([2,3,-2,0,6]):
```

In answering the questions below, use only Theorem 2; do not use Maple, except to compute matrix-vector products to check your answers.

(a) Write down three more solutions of $Ax = 0$, and use matrix-vector multiplication to check that they are indeed solutions.

(b) Write down three corresponding solutions of $Ax = b$, and use matrix-vector multiplication to check that they are indeed solutions.

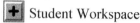 Student Workspace

Problem 8: Rules of algebra and linear combinations

(a) For the matrix A and vectors u_1, u_2, u_3, v_1, v_2, v_3 below, check that $A u_1 = v_1$, $A u_2 = v_2$, and $A u_3 = v_3$.

(b) Using only the rules of algebra, not doing any computation, rewrite the product $A(3 u_1 - u_2 + 2 u_3)$ in terms of v_1, v_2, v_3. Show your use of the rules of algebra explicitly.

```
> A := Matrix([[0,4,-3],[2,-3,2],[3,-1,1],[-1,0,4]]);
> u1 := Vector([4,3,1]):
  u2 := Vector([3,0,2]):
  u3 := Vector([5,3,2]):
  v1 := Vector([9,1,10,0]):
  v2 := Vector([-6,10,11,5]):
  v3 := Vector([6,5,14,3]):
```

Student Workspace

Problem 9: Linear dependence relations and rules of algebra

Suppose $\{v_1, ..., v_k\}$ is a linearly dependent set of vectors in R^n and A is an m by n matrix. Use the rules of algebra to prove that $\{A v_1, ..., A v_k\}$ is a linearly dependent set of vectors in R^m.

Student Workspace

Linear Algebra Modules Project
Chapter 3, Module 2

Matrix Multiplication

Purpose of this module

The purpose of this module is to acquaint you with the definition and properties of matrix multiplication.

Prerequisites

The product of a matrix and a vector.

Commands used in this module

```
> restart: with(LinearAlgebra): with(Lamp): with(plots):
  UseHardwareFloats := false: Digits: 6:
```

Tutorial

Section 1: The Product of Two Matrices

In Module 1 we defined the product of a matrix and a vector. Now we use that definition to define the product of two matrices.

Definition: Suppose A is a matrix with m rows and n columns and B is a matrix with n rows and r columns. The ***matrix product*** $A\,B$ is defined column-by-column as follows. The first column of $A\,B$ is the matrix A times the first column of B; the second column of $A\,B$ is A times the second column of B; and so on. Thus each column of $A\,B$ is the matrix-vector product formed by multiplying the matrix A by the corresponding column of B.

We can express this definition in symbols: Let $B_1, ..., B_r$ denote the columns of the matrix B. Then

$$A\,B = A\,[B_1, B_2, ..., B_r] = [A\,B_1, A\,B_2, ..., A\,B_r]$$

========================

Example 1A: Find the product $A\,B$ for the matrix A and matrix B below.

$$A = \begin{bmatrix} 1 & 3 & -2 \\ 2 & -1 & 4 \end{bmatrix} \quad B = \begin{bmatrix} 4 & 3 \\ 1 & 5 \\ 2 & 6 \end{bmatrix}$$

Solution: We calculate the product directly from the definition, where $B = [B_1, B_2]$:

$$AB_1 = \begin{bmatrix} 1 & 3 & -2 \\ 2 & -1 & 4 \end{bmatrix} \begin{bmatrix} 4 \\ 1 \\ 2 \end{bmatrix} = \begin{bmatrix} 3 \\ 15 \end{bmatrix} \text{ and } AB_2 = \begin{bmatrix} 1 & 3 & -2 \\ 2 & -1 & 4 \end{bmatrix} \begin{bmatrix} 3 \\ 5 \\ 6 \end{bmatrix} = \begin{bmatrix} 6 \\ 25 \end{bmatrix}$$

and hence

$$AB = A[B_1, B_2] = [AB_1, AB_2] = \begin{bmatrix} 3 & 6 \\ 15 & 25 \end{bmatrix}$$

We confirm using Maple:
```
> A := Matrix([[1,3,-2],[2,-1,4]]);
  B := Matrix([[4,3],[1,5],[2,6]]);
> A.B;
```
$$\begin{bmatrix} 3 & 6 \\ 15 & 25 \end{bmatrix}$$

Maple note: Maple's multiplication symbol for matrix multiplication is . instead of the more usual *.

========================

Exercise 1.1: (a) Compute the product MN of the following two matrices in your head. Just imagine the product of M times each column of N, and think of each of these products as a linear combination of the columns of M.

$$M = \begin{bmatrix} 2 & -1 & 3 & -2 \\ 1 & 4 & -1 & -3 \\ -3 & 1 & 2 & 3 \end{bmatrix} \quad N = \begin{bmatrix} 0 & 1 & 1 \\ 0 & 1 & 0 \\ 0 & 1 & 1 \\ 0 & 1 & 0 \end{bmatrix}$$

(b) How would you change the matrix N so that all of the columns of MN are identical to the third column of M?

 Student Workspace

 Answer 1.1

(a) $MN = \begin{bmatrix} 2 & -1 & 3 & -2 \\ 1 & 4 & -1 & -3 \\ -3 & 1 & 2 & 3 \end{bmatrix} \begin{bmatrix} 0 & 1 & 1 \\ 0 & 1 & 0 \\ 0 & 1 & 1 \\ 0 & 1 & 0 \end{bmatrix} = \begin{bmatrix} 0 & 2 & 5 \\ 0 & 1 & 0 \\ 0 & 3 & -1 \end{bmatrix}$

M times a column of 0's is a column of 0's; M times a column of 1's adds the columns of M; and M times a column of alternately 1's and 0's adds the odd-numbered columns of M.

(b) $N = \begin{bmatrix} 0 & 0 & 0 \\ 0 & 0 & 0 \\ 1 & 1 & 1 \\ 0 & 0 & 0 \end{bmatrix}$

Dot Product Method for Calculating AB

If the columns of B are not as simple as in the preceding exercise, computing the product of two matrices by hand can be a bit tedious. Using the dot product method for each column product AB_i may be faster:

- If A is a matrix with m rows and n columns, and B is a matrix with n rows and r columns, then the (i, j) entry of AB is the dot product of the ith row of A and the jth column of B.

Exercise 1.2: (By hand) (a) For the matrices A and B below, calculate the product AB using the dot product method.

$$A = \begin{bmatrix} 4 & 3 \\ 2 & 1 \end{bmatrix} \qquad B = \begin{bmatrix} 7 & -2 \\ 1 & 5 \end{bmatrix}$$

(b) Calculate the product BA. Since the matrix B is now on the left, we calculate the (i, j) entry of BA by taking the dot product of the ith row of B and the jth column of A. Notice that $AB \neq BA$.

(c) For the matrices G and H below, find the $(2,1)$ and $(3,2)$ entries of GH.

$$G = \begin{bmatrix} a & b & c \\ d & e & f \\ g & h & i \end{bmatrix} \qquad H = \begin{bmatrix} j & k \\ l & m \\ n & o \end{bmatrix}$$

 Student Workspace

 Answer 1.2

(a)
$$\begin{bmatrix} 4(7)+3(1) & 4(-2)+3(5) \\ 2(7)+1(1) & 2(-2)+1(5) \end{bmatrix} = \begin{bmatrix} 31 & 7 \\ 15 & 1 \end{bmatrix}$$

(b)
$$\begin{bmatrix} 7(4)-2(2) & 7(3)-2(1) \\ 1(4)+5(2) & 1(3)+5(1) \end{bmatrix} = \begin{bmatrix} 24 & 19 \\ 14 & 8 \end{bmatrix}$$

(c) The $(2,1)$ entry of GH is the dot product of the second row of G and first column of H: $dj + el + fn$. The $(3,2)$ entry is the dot product of the third row of G and the second column of H: $gk + hm + io$.

Section 2: Explorations with Matrix Products

Products of Special Matrices

In this section we look at several special types of matrices and their products. Sometimes we will want to explore the behavior of matrices with random integer entries; for this we use the **Randmat** command, which produces a different matrix every time you execute it.

Example 2A (Identity Matrix): What is the product $A\,I$, where A is a random matrix and I is the special matrix below?

```
> A := Randmat(3,3);
  I := Matrix([[1,0,0],[0,1,0],[0,0,1]]);
```

$$A := \begin{bmatrix} 8 & 4 & -5 \\ -5 & 3 & -5 \\ 8 & 5 & -1 \end{bmatrix},\quad I := \begin{bmatrix} 1 & 0 & 0 \\ 0 & 1 & 0 \\ 0 & 0 & 1 \end{bmatrix}$$

When we multiply A times the first column of I, we get the first column of A; similarly, A times the second column of I is the second column of A; and so on. That is, $A\,I = A$. What is the product $I\,A$?

```
> A.I, I.A;
```

$$\begin{bmatrix} 8 & 4 & -5 \\ -5 & 3 & -5 \\ 8 & 5 & -1 \end{bmatrix} \begin{bmatrix} 8 & 4 & -5 \\ -5 & 3 & -5 \\ 8 & 5 & -1 \end{bmatrix}$$

No matter what the entries of A, the product of A and the special matrix with 1's along the main diagonal and 0's elsewhere is always A. So this special matrix is called the *identity matrix*, and is usually denoted by the letter I. (Although we say "the" identity matrix, there is an n by n identity matrix for every n.) The property we have just observed is similar to the defining property of the number 1:

$$A\,I = A \text{ and } I\,A = A$$

===================

Try the following command, and change the number 3 to other positive integers.

```
> I := Idmat(3);
```

$$I := \begin{bmatrix} 1 & 0 & 0 \\ 0 & 1 & 0 \\ 0 & 0 & 1 \end{bmatrix}$$

===================

Example 2B: Now let's create a matrix P by interchanging the first and second columns of the above identity matrix, and let's compute the product $A\,P$, where A is a random matrix:

```
> A := Randmat(3,3);
  P := Matrix([[0,1,0],[1,0,0],[0,0,1]]);
```

$$A := \begin{bmatrix} 0 & 3 & -4 \\ -5 & -2 & -1 \\ -9 & 5 & 8 \end{bmatrix},\quad P := \begin{bmatrix} 0 & 1 & 0 \\ 1 & 0 & 0 \\ 0 & 0 & 1 \end{bmatrix}$$

```
> A.P;
```

$$\begin{bmatrix} 3 & 0 & -4 \\ -2 & -5 & -1 \\ 5 & -9 & 8 \end{bmatrix}$$

Note that the product $A\,P$ interchanges the first and second columns of A. P is called a permutation matrix. (A *permutation matrix* is a matrix obtained by rearranging (i.e., permuting) the columns of the identity matrix.)

The matrix Q below is the permutation matrix obtained by moving each column of the 3 by 3 identity matrix to the right one column, except for the last column, which is moved to the first column:

```
> Q := Matrix([[0,1,0],[0,0,1],[1,0,0]]);
```

$$Q := \begin{bmatrix} 0 & 1 & 0 \\ 0 & 0 & 1 \\ 1 & 0 & 0 \end{bmatrix}$$

Can you predict what we will get when we compute $A\,Q$, where A is a random matrix?

```
> A := Randmat(3,3);
```

$$A := \begin{bmatrix} 0 & 3 & -4 \\ -5 & -2 & -1 \\ -9 & 5 & 8 \end{bmatrix}$$

```
> A.Q;
```

$$\begin{bmatrix} -4 & 0 & 3 \\ -1 & -5 & -2 \\ 8 & -9 & 5 \end{bmatrix}$$

So the reordering of the columns from A to $A\,Q$ is identical to that from I to Q.

==================

Exercise 2.1: The matrix N below is the result of permuting the columns of the matrix A in a particular way. Find a permutation matrix R such that $A\,R = N$.

```
> A := Matrix([[2,3,7,6],[8,2,1,5]]);
```

$$A := \begin{bmatrix} 2 & 3 & 7 & 6 \\ 8 & 2 & 1 & 5 \end{bmatrix}$$

```
> N := Matrix([[7,6,2,3],[1,5,8,2]]);
```

$$N := \begin{bmatrix} 7 & 6 & 2 & 3 \\ 1 & 5 & 8 & 2 \end{bmatrix}$$

+ Student Workspace

− Answer 2.1

Permute the columns of the 4 by 4 identity matrix so that the first column is third, the second column is fourth, the third column is first, and the fourth column is second:

```
> R := Matrix([[0,0,1,0],[0,0,0,1],[1,0,0,0],[0,1,0,0]]);
```

$$R := \begin{bmatrix} 0 & 0 & 1 & 0 \\ 0 & 0 & 0 & 1 \\ 1 & 0 & 0 & 0 \\ 0 & 1 & 0 & 0 \end{bmatrix}$$

```
> A.R, N;
```

$$\begin{bmatrix} 7 & 6 & 2 & 3 \\ 1 & 5 & 8 & 2 \end{bmatrix} \begin{bmatrix} 7 & 6 & 2 & 3 \\ 1 & 5 & 8 & 2 \end{bmatrix}$$

In the next module, we will learn about the rules of algebra that apply to matrices. Most of the rules of the algebra of real numbers that you are familiar with are also valid for matrix algebra, as we will see. However, unlike multiplication of real numbers, matrix multiplication is not commutative! That is, $A B = B A$ is not always true for matrices. You saw an example of this in Exercise 1.2 at the end of Section 1. In fact, it is so rarely true that a matrix product commutes that if we multiply two random matrices A and B, we will almost always observe that $A B \neq B A$. Execute the following commands several times:

```
> A := Randmat(2,2);
  B := Randmat(2,2);
```

$$A := \begin{bmatrix} 2 & 5 \\ -5 & 9 \end{bmatrix}, \quad B := \begin{bmatrix} -9 & 4 \\ 0 & -8 \end{bmatrix}$$

```
> A.B;
```

$$\begin{bmatrix} -18 & -32 \\ 45 & -92 \end{bmatrix}$$

```
> B.A;
```

$$\begin{bmatrix} -38 & -9 \\ 40 & -72 \end{bmatrix}$$

Occasionally, however, a matrix product does commute, as in $A I = I A$ (see Example 2A).

A *square* matrix is a matrix with the same number of rows as columns. A *diagonal* matrix is a square matrix whose nondiagonal entries are all zero:

```
> K := Diagmat([2,-3,0,4]);
```

$$K := \begin{bmatrix} 2 & 0 & 0 & 0 \\ 0 & -3 & 0 & 0 \\ 0 & 0 & 0 & 0 \\ 0 & 0 & 0 & 4 \end{bmatrix}$$

Maple note: Maple won't let us use the name D for this diagonal matrix; it has reserved D as the name of the differentiation operator for functions.

Exercise 2.2: Let A and K (K diagonal) be the matrices shown below. How is the product $A K$ related to A? Explain why this pattern must always be true. Hint: Analyze $A K$ one column at a time.

```
> A := Matrix([[1,2,3],[6,2,5],[-3,4,8]]);
> K := Diagmat([-2,3,5]);
```

$$A := \begin{bmatrix} 1 & 2 & 3 \\ 6 & 2 & 5 \\ -3 & 4 & 8 \end{bmatrix}, \quad K := \begin{bmatrix} -2 & 0 & 0 \\ 0 & 3 & 0 \\ 0 & 0 & 5 \end{bmatrix}$$

```
> A.K;
```

$$\begin{bmatrix} -2 & 6 & 15 \\ -12 & 6 & 25 \\ 6 & 12 & 40 \end{bmatrix}$$

Student Workspace

Answer 2.2

The first column of the product $A\,K$ is the linear combination of the columns of A whose weights are the entries of the first column of K. Since all but the first entry of this column of K are zero, the first column of $A\,K$ is the first column of A times the first diagonal entry of K. Likewise, the second column of $A\,K$ is the second column of A times the second diagonal entry of K, etc.

Vector-Matrix Products: $y\,A$

In Module 1, we saw that when we multiply a matrix A on the right by a column vector, we get a linear combination of the columns of A. Similarly, when we multiply A on the left by a row vector (i.e., a matrix with just one row), we get a linear combination of the rows of A. Here is an example.

===================

Example 2C: For the matrices y and A below, we compute the product $y\,A$ by computing the linear combination of the rows of A in which the weights are the components of y:

$$[2 \quad -3 \quad 4]\begin{bmatrix} a & b \\ c & d \\ e & f \end{bmatrix} = 2\,[a \quad b] - 3\,[c \quad d] + 4\,[e \quad f] = [2a - 3c + 4e \quad 2b - 3d + 4f]$$

We confirm using Maple:
```
> A := Matrix([[a,b],[c,d],[e,f]]);
> y := Matrix([[2,-3,4]]);
```
$$A := \begin{bmatrix} a & b \\ c & d \\ e & f \end{bmatrix}$$
$$y := [2 \quad -3 \quad 4]$$
```
> y.A;
```
$$[2a - 3c + 4e \quad 2b - 3d + 4f]$$

===================

The general rule (which is proved in Module 3) is:

- The vector-matrix product $y\,A$ is the linear combination of the rows of A in which the weights are the components of y.

Exercise 2.3: Suppose A is a 3 by n matrix and $y = [1, 0, 0]$ and $z = [1, 0, 1]$. Describe the products $y\,A$ and $z\,A$ in terms of the rows of A.

Student Workspace

Answer 2.3

yA is the first row of A, since $yA = 1$ (row 1 of A) + 0 (row 2 of A) + 0 (row 3 of A).
zA is the sum of the first and third rows of A by similar reasoning.

Exercise 2.4: Let A and K (K diagonal) be the matrices shown below. How is the product KA related to A? Explain why this pattern will always be true. Hint: Analyze KA one row at a time, and think of each of these rows as a linear combination of rows of A.

```
> K := Diagmat([-2,3,5]);
> A := Matrix([[1,2,3],[6,2,5],[-3,4,8]]);
```

$$K := \begin{bmatrix} -2 & 0 & 0 \\ 0 & 3 & 0 \\ 0 & 0 & 5 \end{bmatrix}, \quad A := \begin{bmatrix} 1 & 2 & 3 \\ 6 & 2 & 5 \\ -3 & 4 & 8 \end{bmatrix}$$

```
> K.A;
```

$$\begin{bmatrix} -2 & -4 & -6 \\ 18 & 6 & 15 \\ -15 & 20 & 40 \end{bmatrix}$$

Student Workspace

Answer 2.4

The first row of the product KA is the linear combination of the rows of A whose weights are the entries of the first row of K. Since all but the first entry of this row of K are zero, the first row of KA is the first row of A times the first diagonal entry of K. Likewise, the second row of KA is the second row of A times the second diagonal entry of K, etc. (Compare Exercise 2.2.)

Exercise 2.5: Find the 3 by 3 permutation matrix P so that the product PA permutes the rows of A (defined below) by moving each row of A down one row, except the last row, which is moved to the top.

```
> A := Matrix([[1,2,3],[6,2,5],[-3,4,8]]);
```

$$A := \begin{bmatrix} 1 & 2 & 3 \\ 6 & 2 & 5 \\ -3 & 4 & 8 \end{bmatrix}$$

Student Workspace

Answer 2.5

Form the permutation matrix P by permuting the rows of the 3 by 3 identity matrix in the same way we want to permute the rows of A:

```
> P := Matrix([[0,0,1],[1,0,0],[0,1,0]]);
```

$$P := \begin{bmatrix} 0 & 0 & 1 \\ 1 & 0 & 0 \\ 0 & 1 & 0 \end{bmatrix}$$

> P.A;

$$\begin{bmatrix} -3 & 4 & 8 \\ 1 & 2 & 3 \\ 6 & 2 & 5 \end{bmatrix}$$

Exercise 2.6: Since multiplying on the left of a matrix allows us to manipulate the rows of the matrix, we ask you to carry out an elementary row operation by matrix multiplication. Specifically, suppose we want to replace row 2 of M by the sum of row 2 and -2 times row 1 (see below). Find a matrix R so that this row operation is carried out in the product $R\,M$. Hint: We want R to leave the first and third rows of M unchanged and to replace the second row by a certain linear combination of the rows of M.

> M := Matrix([[1,2,5],[2,7,3],[4,3,9]]);

$$M := \begin{bmatrix} 1 & 2 & 5 \\ 2 & 7 & 3 \\ 4 & 3 & 9 \end{bmatrix}$$

> M1 := Matrix([[1,2,5],[0,3,-7],[4,3,9]]);

$$M1 := \begin{bmatrix} 1 & 2 & 5 \\ 0 & 3 & -7 \\ 4 & 3 & 9 \end{bmatrix}$$

 Student Workspace

Answer 2.6

The first and third rows of R should be [1, 0, 0] and [0, 0, 1], respectively, so they leave the first and third rows of M unchanged, The second row of R should be [−2, 1, 0], since the product of this row vector and M is

-2 (row 1 of M) + 1 (row 2 of M) + 0 (row 3 of M)

> R := Matrix([[1,0,0],[-2,1,0],[0,0,1]]);

$$R := \begin{bmatrix} 1 & 0 & 0 \\ -2 & 1 & 0 \\ 0 & 0 & 1 \end{bmatrix}$$

> R.M, M1;

$$\begin{bmatrix} 1 & 2 & 5 \\ 0 & 3 & -7 \\ 4 & 3 & 9 \end{bmatrix}, \begin{bmatrix} 1 & 2 & 5 \\ 0 & 3 & -7 \\ 4 & 3 & 9 \end{bmatrix}$$

Powers of Matrices

Definition: If A is a square matrix, we define A^2 to be the product $A\,A$. Higher powers are defined similarly: A^n is the product of n copies of A.

Exercise 2.7: Let J be the n by n matrix whose entries are all ones. (The command below constructs the 4 by 4 matrix of all ones. Be sure to change the command to get matrices of other sizes.)

> J := Matrix(4,4,1);

$$J := \begin{bmatrix} 1 & 1 & 1 & 1 \\ 1 & 1 & 1 & 1 \\ 1 & 1 & 1 & 1 \\ 1 & 1 & 1 & 1 \end{bmatrix}$$

(a) What is J^2 ? (Your answer will depend on n.)
(b) Explain why your answer is correct for <u>all</u> values of n.

 Student Workspace

 Answer 2.7

(a) If J is n by n, the entries of J^2 are all n's. Thus $J^2 = n\,J$. What is J^3? Hint: $J^3 = J^2\,J$.
(b) The dot product of n 1's with another n 1's is the sum of 1 with itself n times, which is n.

Exercise 2.8: Let K be the matrix whose entries are all 1's along the diagonal just above the main diagonal and 0's elsewhere. (Such a matrix is called a Jordan matrix. Below is a command that constructs the 4 by 4 Jordan matrix. Change the number 4 to get other sizes.)

> K := Jordanmat(4);

$$K := \begin{bmatrix} 0 & 1 & 0 & 0 \\ 0 & 0 & 1 & 0 \\ 0 & 0 & 0 & 1 \\ 0 & 0 & 0 & 0 \end{bmatrix}$$

(a) What is K^2?
(b) What is K^n for every positive integer power n?
(c) Explain why your answer to (b) is correct for <u>all</u> values of n.

 Student Workspace

 Answer 2.8

(a) The matrix K^2 is all 0's except for 1's along the diagonal that is two above the main diagonal.
(b) The matrix K^3 is all 0's except for 1's along the diagonal three above the main diagonal, and so on, until we get to K^n (where K is n by n). This power of K is all 0's, and so are all higher powers.
(c) Suppose you have computed some power of K (which we will denote by B) and are about to compute the next higher power. We can do this by multiplying B on the right by K. B times the first

Problems

Problem 1: Permute the components of a vector

Let u and v be the vectors defined below. Find the permutation matrix P such that $Pu = v$. (Thus P permutes the components of u to get the components of v.)

```
> u := Vector([1,2,3,4]);
  v := Vector([4,1,3,2]);
```

Student Workspace

Problem 2: Zero columns

Suppose A and B are matrices whose product AB is defined.
(a) If the first column of B is all zeros, what is the first column of AB? Explain. Is a similar statement true for other columns of B and AB?
(b) If the first column of AB is all zeros but no column of B is all zeros, what can you say about the columns of A? (Hint: Think of the first column of AB as a linear combination of the columns of A.) Is a similar statement true for other columns of AB?

Student Workspace

Problem 3: Commuting matrices

The 3 by 3 diagonal matrix K below commutes with all 3 by 3 diagonal matrices; that is, $AK = KA$ whenever A is diagonal. (In fact, diagonal matrices always commute with one another.) However, because of its special structure, K also commutes with some nondiagonal matrices. Find all the 3 by 3 matrices that commute with K. Hint: The upper-left 2 by 2 corner of K is a particularly simple matrix.

```
> K := Diagmat([2,2,3]);
```

Student Workspace

Problem 4: Inverse of a matrix

The command **Matsolve(A,b)** only works when b is a column vector. But we can solve $AX = B$ for the matrix X when B has many columns by applying **Matsolve** to one column of B at a time.
(a) Use this idea to solve the equation $AX = I$, where A is the 3 by 3 matrix below and I is the 3 by 3

identity matrix. (The matrix X is called the "inverse" of the matrix A.)

```
> A := Matrix([[1,1,-1],[4,-2,-3],[-2,2,1]]);
> I := Idmat(3);
```

(b) Check your answer by computing $A\,X$. Also compute $X\,A$. (Note that the relationship between X and A is similar to the relationship between a real number and its reciprocal.)

(c) Every nonzero real number has a reciprocal, but not every nonzero matrix has an inverse. Find a nonzero 3 by 3 matrix that has no inverse. (Hint: Choose A so that $A\,x = b$ has no solution, where b is one of the columns of I.)

Student Workspace

Problem 5: Elementary row operations

(a) For the matrix A below, find the matrices R, S, and T that perform these row operations on A:

- $R\,A$ replaces row 2 of A by the sum of row 2 and -3 times row 1;

- $S\,A$ replaces row 3 of A by the sum of row 3 and 2 times row 1;

- $T\,A$ replaces row 3 of A by the sum of row 3 and 4 times row 2;

Check your answers by computing $R\,A$, $S\,A$, and $T\,A$.

(b) Some of the products $R\,S$, $R\,T$, $S\,T$ commute, and some do not. Determine which do and which do not, and explain. Hint: Think of what each of these matrices does to A.

```
> A := Matrix([[2,-1,3,0],[6,-3,0,2],[-4,1,-1,3]]);
```

Student Workspace

Problem 6: Jordan matrices

(a) Let K be a Jordan matrix, as in Exercise 2.8. If A is any square matrix, describe $A\,K$ and $K\,A$. That is, what does K do to A when it is multiplied on the right of A? on the left of A? Experiment with matrices of various sizes, such as 4 by 4, 5 by 5, and 6 by 6. Your answer must be valid for Jordan matrices of every possible size. (Use the **Jordanmat** command.)

(b) Explain why your answer in (a) is correct by explaining what multiplying by K does to the columns and rows of A. Hint: Think of each column of $A\,K$ as a linear combination of the columns of A, and think of each row of $K\,A$ as a linear combination of the rows of A.

Student Workspace

Problem 7: Backward identity matrices

Let B be a "backward identity" matrix, which is a square matrix with 1's along the antidiagonal from the upper-right to lower-left corner and 0's elsewhere. For any square matrix A, experiment with products

$A\,B$ and $B\,A$ for A and B of various sizes, such as 4 by 4, 5 by 5, and 6 by 6, to help you answer the following questions. Your answers to the questions must be valid for backward identity matrices of every possible size. (To construct B, use the **Backidmat** command below, but change the argument 4 to other values; to construct A, use the **Randmat** command.)

```
> B := Backidmat(4);
```

(a) Describe the products $A\,B$ and $B\,A$. That is, what does B do to A when it is multiplied on the right of A? on the left of A?

(b) Explain why your answers to (a) are correct. Hint: B is a permutation matrix.

(c) What is B^2?

(d) What is B^n for every positive integer n?

[+] Student Workspace

Problem 8: Letter L matrices

Let L be a "Letter L" matrix, which is a square matrix of 0's and 1's in which the 1's form the shape of the letter L. For any square matrix A, experiment with the product $A\,L$ for A and L of various sizes, such as 4 by 4, 5 by 5, and 6 by 6, to help you answer the following questions. Your answers to the questions must be valid for letter L matrices of every possible size. (To construct L, use the **LetterL** command below, but change the number 4 to other values; to construct A, use the **Randmat** command.)

```
> L := LetterL(4);
```

(a) Describe the product $A\,L$. That is, what does L do to A when it is multiplied on the right of A? Explain.

(b) What is L^2?

(c) What is L^n for every positive integer n?

[+] Student Workspace

Linear Algebra Modules Project
Chapter 3, Module 3

Rules of Matrix Algebra

 Special instructions

The Tutorial for this module is designed to be done by hand without the use of Maple. Likewise, most of the Problems can and should be done without the use of Maple. The answers to the Exercises are all at the end of the Tutorial.

 Purpose of this module

The purpose of this module is to acquaint you with the rules of matrix algebra, give you practice using them, and help you appreciate their limitations.

 Prerequisites

Some familiarity with matrices, especially the product of two matrices.

```
> restart: with(LinearAlgebra): with(Lamp): with(plots):
  UseHardwareFloats := false: Digits := 6:
```

Tutorial

 Section 1: Familiar Rules of Algebra that Apply to Matrices

The rules of matrix algebra are much like the rules for the ordinary algebra of numbers but with a couple of significant differences. First let's look at the rules of matrix algebra that are exactly like the rules of ordinary algebra.

Theorem 3: Matrix algebra obeys the following rules (where uppercase letters denote matrices and lowercase letters denote scalars; also, the letter O denotes the zero matrix, and the letter I denotes the identity matrix):

$$A + B = B + A \quad \text{(commutative law for addition)}$$
$$(A + B) + C = A + (B + C) \quad \text{(associative law for addition)}$$
$$(A\,B)\,C = A\,(B\,C) \quad \text{(associative law for multiplication)}$$
$$A\,(B + C) = A\,B + A\,C \quad \text{(first distributive law)}$$
$$(A + B)\,C = A\,C + B\,C \quad \text{(second distributive law)}$$
$$A + O = A \quad \text{(identity law for addition)}$$

$$AI = IA = A \quad \text{(identity law for multiplication)}$$
$$A - A = O \quad \text{(inverse law for addition)}$$

And here are rules involving scalar multiplication:
$$(c\,d)\,A = c\,(d\,A)$$
$$c(A\,B) = (c\,A)\,B = A\,(c\,B)$$
$$(c + d)\,A = c\,A + d\,A$$
$$c\,(A + B) = c\,A + c\,B$$

Proofs of these rules can be found in Section 4.

==================

Example 1A: Perhaps the most familiar use of rules of algebra is in solving equations. For the matrices

$$A = \begin{bmatrix} -3 & 4 & 3 \\ -4 & -2 & 2 \\ 0 & 1 & -2 \end{bmatrix} \text{ and } b = \begin{bmatrix} 4 \\ 3 \\ -2 \end{bmatrix}$$

solve the equation $4\,x = 2\,b - A\,x$ for the unknown column vector x.

Solution: We write out all the steps in detail, making explicit use of the rules in Theorem 3.

- (1) Rearrange terms so all terms with x are on the left. Specifically, add $A\,x$ to both sides and then use the associative, inverse, and identity laws for addition:

$$4\,x + A\,x = 2\,b$$

- (2) Write x as $I\,x$ so we can factor out x; then use the second distributive law to factor:

$$4\,I\,x + A\,x = 2\,b$$
$$(4\,I + A)\,x = 2\,b$$

- (3) The last equation has the familiar form of a matrix-vector equation $C\,x = d$, where

$$C = 4\,I + A = \begin{bmatrix} 1 & 4 & 3 \\ -4 & 2 & 2 \\ 0 & 1 & 2 \end{bmatrix} \text{ and } d = 2\,b = \begin{bmatrix} 8 \\ 6 \\ -4 \end{bmatrix}$$

- (4) Solve $C\,x = d$ (using, for example, **Matsolve(C,d)**) and get $x = \begin{bmatrix} -1 \\ 6 \\ -5 \end{bmatrix}$.

==================

Exercise 1.1: Multiply out and simplify the following product:
$$(I+A+A^2)(I-A)$$

 Student Workspace

Section 2: Familiar Rules of Algebra that Do Not Apply to Matrices

Some rules that are true for the ordinary algebra of numbers are false for matrix algebra. In Module 2, for example, we observed that usually $AB \neq BA$ when A and B are matrices. This fact is usually put into words as follows: Although multiplication of numbers is commutative, matrix multiplication is not commutative. Here is an example from Module 2 of a pair of matrices that do not commute:

$$\begin{bmatrix} 4 & 3 \\ 2 & 1 \end{bmatrix}\begin{bmatrix} 7 & -2 \\ 1 & 5 \end{bmatrix} \neq \begin{bmatrix} 7 & -2 \\ 1 & 5 \end{bmatrix}\begin{bmatrix} 4 & 3 \\ 2 & 1 \end{bmatrix}$$

While in general matrix multiplication is not commutative, examples of matrices that do commute are fairly easy to find. We have already seen that the identity matrix commutes with any matrix. The next two examples consider some other special matrices that commute.

========================

Example 2A: For the diagonal matrix $K = \begin{bmatrix} 2 & 0 \\ 0 & -3 \end{bmatrix}$, what are the matrices A such that $KA = AK$? Hint: Think of how K changes A when it is multiplied on the left and right of A.

Solution: We saw in Module 2 that diagonal matrices always commute. So $KA = AK$ for every diagonal matrix A. Do any nondiagonal matrices commute with K? KA multiplies the first row of A by 2 and the second row by -3, and AK multiplies the first column of A by 2 and the second column by -3. So $KA = AK$ if and only if both operations produce the same matrix. This is true <u>only</u> when A is also a diagonal matrix.

========================

Exercise 2.1: For the permutation matrix $P = \begin{bmatrix} 0 & 1 \\ 1 & 0 \end{bmatrix}$, what are the matrices A such that $PA = AP$? Hint: Think of how P changes A when it is multiplied on the left and right of A.

 Student Workspace

Here is another way in which the rules of matrix algebra differ from those of ordinary algebra:

Example 2B: If $A = \begin{bmatrix} 1 & 1 \\ 1 & 1 \end{bmatrix}$, find a 2 by 2 matrix B such that $A\,B = O$ (where O denotes the zero matrix) but B is not the zero matrix.

Solution: One answer is $B = \begin{bmatrix} 1 & 1 \\ -1 & -1 \end{bmatrix}$. How can you find other answers? Hint: Solve the equation

$$\begin{bmatrix} 1 & 1 \\ 1 & 1 \end{bmatrix} \begin{bmatrix} a & b \\ c & d \end{bmatrix} = \begin{bmatrix} 0 & 0 \\ 0 & 0 \end{bmatrix}$$

by converting this equation into four linear equations in the four unknowns a, b, c, d.

So the familiar cancellation law of ordinary algebra is false for matrices: $A\,B = O$ does <u>not</u> imply that $A = O$ or $B = O$.

Summary: The following are the main differences between the rules of matrix algebra and the rules of ordinary algebra:

- Matrix multiplication is not commutative.
- There is no cancellation law for matrix products.

Section 3: Rules for the Transpose of a Matrix

The *transpose* of a matrix A is the matrix obtained from A by making its rows into its columns (and hence its columns into its rows). The Maple command for the transpose of A is **Transpose(A)**; the mathematical symbol for it is A^T. Here are some matrices and their transposes:

$$A = \begin{bmatrix} 2 & 3 & 4 \\ 6 & 8 & 9 \end{bmatrix}, \quad A^T = \begin{bmatrix} 2 & 6 \\ 3 & 8 \\ 4 & 9 \end{bmatrix}$$

$$B = \begin{bmatrix} 12 & 35 & 0 \\ 68 & 94 & -56 \\ 3 & 6 & 8 \end{bmatrix}, \quad B^T = \begin{bmatrix} 12 & 68 & 3 \\ 35 & 94 & 6 \\ 0 & -56 & 8 \end{bmatrix}$$

Note that for a square matrix (same number of rows as columns), its transpose can found by reflecting the entries of the matrix across the main diagonal.

Most of the rules of algebra for the transpose are straightforward and easy to use, but the fourth rule below may surprise you:

Theorem 4: The transpose operation for matrices obeys the following rules of algebra (where uppercase letters denote matrices and the lowercase letter is a scalar):

$$(A+B)^T = A^T + B^T$$
$$(cA)^T = cA^T$$
$$(A^T)^T = A$$
$$(AB)^T = B^T A^T$$

Note that in the fourth rule the order of the matrices A and B is switched when you take the transpose of their product. Thus, in general, $(AB)^T \neq A^T B^T$.

Exercise 3.1: Suppose A is a 2 by 2 matrix and B is a 2 by 3 matrix. Explain how, without doing any calculating, you can tell that $(AB)^T$ could not possibly equal $A^T B^T$ for these particular matrices. (Hint: look at the size of each of the matrices.)

 Student Workspace

===================

Example 3A: Solve the equation $2(Ax)^T = x^T M + b^T A^T$ for the column vector x, given the matrices A, b, M below.

$$A = \begin{bmatrix} 2 & 3 \\ 5 & 7 \end{bmatrix}, \quad b = \begin{bmatrix} -3 \\ 2 \end{bmatrix}, \quad M = \begin{bmatrix} 3 & 5 \\ 4 & 2 \end{bmatrix}$$

Solution:

- (1) Take the transpose of both sides and apply all four parts of Theorem 4:
$$2Ax = M^T x + Ab$$
- (2) Collect all terms with x on one side:
$$2Ax - M^T x = Ab$$
- (3) Factor out x by using the second distributive law in Theorem 3:
$$(2A - M^T)x = Ab$$
- (4) We now have a matrix equation in the familiar form $Cx = d$, where
$$C = 2A - M^T = \begin{bmatrix} 1 & 2 \\ 5 & 12 \end{bmatrix}, \quad d = Ab = \begin{bmatrix} 0 \\ -1 \end{bmatrix}$$
- (5) Solve $Cx = d$ (using, for example, **Matsolve(C,d)**) and get $x = \begin{bmatrix} 1 \\ -\frac{1}{2} \end{bmatrix}$.

===================

Example 3B: Recall the assertion made in Module 2: The product yA, where y is a row vector and A is a matrix, can be written as the linear combination of the rows of A in which the weights are the components of y. Here is an example in which we compute yA first using dot products, then using linear combinations of rows:

$$yA = [2 \ -3 \ 4] \begin{bmatrix} a & b \\ c & d \\ e & f \end{bmatrix} = [2a - 3c + 4e \quad 2b - 3d + 4f]$$

$$yA = 2[a \ b] - 3[c \ d] + 4[e \ f] = [2a - 3c + 4e \quad 2b - 3d + 4f]$$

We can use the rules of transposes to prove the above assertion in general. First use the rule for the transpose of a product:

$$(yA)^T = A^T y^T$$

The product $A^T y^T$ is a matrix-vector product and is therefore the linear combination of the columns of A^T (i.e., the rows of A) in which the weights are the components of y^T (i.e., the components of y). Since yA is the transpose of $A^T y^T$, yA is also the linear combination of the rows of A in which the weights are the components of y; but now the rows of A are written horizontally, since the transpose operation changes column vectors to row vectors.

Section 4: How to Prove the Rules of Algebra

We can prove the rules of algebra in Theorem 3 fairly easily by using the two rules of algebra for matrix-vector products introduced in Module 1:

$$A(x + y) = Ax + Ay \quad \text{(distributive law)}$$
$$A(cx) = c(Ax) \quad \text{(scalars factor out)}$$

where A is any m by n matrix, x and y are column vectors with n entries, and c is any real number.

The following observation shows why rules for matrix-vector products are useful in proving rules for matrix-matrix products: The jth column of the matrix-matrix product AB is the matrix-vector product of A times the jth column of B. We use this observation to prove several rules in Theorem 3:

First distributive law: The jth column of $A(B + C)$ is, by the preceding observation, the matrix-vector product $A(B+C)_j$, which is the same as $A(B_j + C_j)$, where B_j is the jth column of B and C_j is the jth column of C. By the above distributive law for matrix-vector products,

$$A(B_j + C_j) = AB_j + AC_j$$

The right side of the above equation is the jth column of $AB + AC$. Therefore

$$A(B+C) = AB + AC$$

Second distributive law: This law can be proved in a similar way, but it requires a different distributive law for matrix-vector products:

$$(A+B)x = Ax + Bx$$

Exercise 4.1: Prove the second distributive law, $(A+B)C = AC + BC$, by using the above distributive law for matrix-vector products.

 Student Workspace

Associative law for multiplication: This law requires yet another new rule for matrix-vector products, namely the following associative law:

$$(AB)x = A(Bx)$$

Here's a proof of this rule, where x_j denotes the jth component of the vector x and B_j denotes the jth column of B and P_j denotes the jth column of the product AB:

$$(AB)x = x_1 P_1 + \ldots + x_k P_k = x_1 A B_1 + \ldots + x_k A B_k = A(x_1 B_1 + \ldots + x_k B_k) = A(Bx)$$

Note especially the third equal sign in the above proof; at that step we used both of the two rules of algebra for matrix-vector products stated at the beginning of this subsection. From this new rule for matrix-vector products, we deduce the associative law $(AB)C = A(BC)$ by applying the new rule with x replaced by the jth column of C.

We conclude with an exercise that shows why the surprising fourth rule for the transpose is true (see Theorem 4).

Exercise 4.2: (a) What entry of A equals the (i, j) entry of A^T? (Hint: First consider a specific entry of A, such as the (1, 2) entry.)

(b) How is the (i, j) entry of $(AB)^T$ calculated from the entries of A and B? (Hint: The (i, j) entry of AB is the dot product of the ith row of A and the jth column of B.) Use your answer to justify the rule $(AB)^T = B^T A^T$ in Theorem 4.

 Student Workspace

Answers to Exercises

Answer 1.1:
$(I+A+A^2)(I-A) = (I+A+A^2)I - (I+A+A^2)A = I+A+A^2 - (A+A^2+A^3) = I - A^3$

Answer 2.1:
PA swaps the rows of A, and AP swaps the columns of A. So $PA = AP$ if and only if swapping the rows of A and swapping the columns of A produce the same matrix. That is true only when A has the form

$$A = \begin{bmatrix} a & b \\ b & a \end{bmatrix}$$

Answer 3.1:
A^T is 2 by 2 and B^T is 3 by 2. So the product $A^T B^T$ is not even defined.

Answer 4.1:
The jth column of $(A+B)C$ is the matrix-vector product $(A+B)C_j$, where C_j is the jth column of C. By the above distributive law for matrix-vector products,

$$(A+B)C_j = AC_j + BC_j$$

The right side of the above equation is the jth column of $AC + BC$. Therefore

$$(A+B)C = AC + BC$$

Answer 4.2:
(a) The (i, j) entry of A^T is the (j, i) entry of A.
(b) The (i, j) entry of $(AB)^T$ is the (j, i) entry of AB, which is computed as the dot product of the jth row of A and the ith column of B. But that is the same as the dot product of the ith row of B^T and the jth column of A^T. Note that, by this discussion, we have proved the rule $(AB)^T = B^T A^T$.

Problems

 Problem 1: Use matrix algebra to solve an equation

Express the matrix equation $AX + 5A = AB + 4X$ in the form $CX = K$, where C and K are expressions in A and B.

 Student Workspace

Problem 2: Factor a matrix expression

Factor the matrix expression $3A - AB + 12C - 4CB$ into a product of two matrices MN.

+ Student Workspace

Problem 3: Counter-examples in matrix algebra I

These are examples that demonstrate again how different matrix algebra can be from ordinary algebra. All of the statements below are FALSE. For each statement, use legitimate rules of matrix algebra to rewrite (by hand) the given equation in a form that shows more clearly why the statement is false. Then use an example from either this module or the module on "Matrix Multiplication" to demonstrate that the statement is false.

(a) $(A - B)(A + B) = A^2 - B^2$
(b) $(A + B)^2 = A^2 + 2AB + B^2$
(c) If $AC = AB$ and $A \neq O$, then $C = B$.

+ Student Workspace

Problem 4: Counter-examples in matrix algebra II

These are examples that demonstrate again how different matrix algebra can be from ordinary algebra. All of the statements below are FALSE. For each part below, give an example of a matrix that fails to satisfy the stated property, hence demonstrating that the statement is false. (Suggestions: Try examples from either this module or the module on "Matrix Multiplication", or experiment with some simple 2 by 2 matrices whose entries are small integers, such as 0, 1, and -1.)

(a) If $A^2 = O$, then $A = O$ (O denotes the zero matrix)
(b) If $A^2 = I$, then $A = I$ or $A = -I$.
(c) The entries of A^2 are all greater than or equal to 0.

+ Student Workspace

Problem 5: Use rules of transposes to solve a matrix equation

With the help of the rules of matrix algebra, solve the matrix equation $3C + 2x^T = x^T B$ for x, where B and C are given below. That is, by hand write the equation in a form that can be solved using the **Matsolve** command. Then solve it by any method.

```
> B := Matrix([[-1,3,2],[-1,3,3],[0,2,-1]]):
  C := Matrix([[1,-1,4]]):
```

$$B := \begin{bmatrix} -1 & 3 & 2 \\ -1 & 3 & 3 \\ 0 & 2 & -1 \end{bmatrix}, \quad C := \begin{bmatrix} 1 & -1 & 4 \end{bmatrix}$$

Student Workspace

Problem 6: Symmetric matrices

A matrix A is said to be *symmetric* if $A^T = A$. In other words, each row of A has the same entries as the corresponding column of A. Also note that when a symmetric matrix is reflected across its main diagonal, the matrix is unchanged. A symmetric matrix is always square.

(a) Give an example of a 3 by 3 symmetric matrix. Check that it is symmetric by applying the **Transpose** command to it. Can you find an example of a 3 by 3 symmetric matrix with 6 different entries? with 7?

(b) Use the rules of algebra for the transpose operation to prove that, for every matrix A, the matrix $A A^T$ is symmetric. Similarly, show that $A^T A$ is symmetric.

(c) Use the rules of algebra for the transpose operation to prove that if A and B are symmetric n by n matrices, so is $s A + t B$ for all scalars s and t.

(d) Give an example of 2 by 2 symmetric matrices A and B such that $A B$ is not symmetric. Hint: Use the rules of algebra to find out how A and B would have to be related if the product $A B$ were symmetric.

 Student Workspace

134 Chapter 3 Matrix Algebra

Linear Algebra Modules Project
Chapter 3, Module 4

Markov Chains -- An Application

 Purpose of this module

The purpose of this module is to introduce an important class of applications that involve products and powers of matrices. We will return to these applications in later modules where you will be able to use your expanding knowledge of linear algebra to derive even more insights than we can at this early stage.

 Prerequisites

Matrix algebra; solution of linear systems of equations.

 Commands used in this module

```
> restart: with(LinearAlgebra): with(Lamp): with(plots):
  UseHardwareFloats := false: Digits := 6:
```

Tutorial

 Section 1: A Typical (though simplified) Markov Chain Problem

In this section you will investigate a simple mathematical model for population migration. Our focus will be on the movement of a country's population between urban and rural areas. The table below shows current urban/rural breakdowns for a sampling of countries.

$$\begin{bmatrix} & Urban & Rural \\ Argentina & 87\% & 13\% \\ China & 29\% & 71\% \\ Ireland & 57\% & 43\% \\ Lampland & 45\% & 55\% \\ Nigeria & 16\% & 84\% \\ Sri\ Lanka & 22\% & 78\% \\ Sweden & 83\% & 17\% \\ United\ States & 75\% & 25\% \end{bmatrix}$$

While there is clearly a wide variation among countries in the relative size of their urban populations, the population of the world as a whole is steadily moving towards the cities. In 1800, only 3% of the world population lived in cities. By 1900 that number had grown to 10%, and currently about 50% of the world population lives in urban areas.

We now turn our attention to the newly formed country of Lampland. Our goal is to build a mathematical model that we can use to predict how the urban population for Lampland will change over the next 100 years. Suppose we have collected enough data to justify the following assumptions about the Lampland population:

- (1) Each year 6% of the urban population moves to rural areas and the remaining 94% stay in urban areas.
- (2) Each year 9% of the rural population moves to urban areas and the remaining 91% stay in rural areas.
- (3) At the beginning of our study, the Lampland population is 45% urban and 55% rural.

We can diagram assumptions (1) and (2) as follows, where U denotes the urban population and R the rural population:

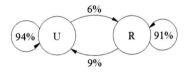

This is obviously a highly simplified mathematical model of a real problem. You might ask yourself what the most serious limitations of this model are. That is, in what ways is it unrealistic, and what might we want to add to our model to overcome its most serious limitations? However, when constructing a mathematical model of a real problem, starting with a very simple model is a useful strategy. And, as we will see, even this simple model provides substantial insights.

Here are the questions we will investigate:

- (a) What will happen to the population in the long run? For example, will all of the population end up in the cities? Will the urban and rural populations eventually settle down, or will they oscillate back and forth endlessly?
- (b) How does the long-term distribution of the two populations depend on the initial 45-55 distribution? For example, if the urban population had started out at 75%, how different would the long-term distribution be?

We will use u_k to denote the percent of the total population that is in urban areas after k years and r_k to denote the percent of the population that is in rural areas after k years. So, in particular, by our assumption (3), we have

$$u_0 = .45 \quad \text{and} \quad r_0 = .55$$

We can use our assumptions (1) and (2) to write equations that describe the population percentages in year $k+1$ in terms of the population percentages in year k:

$$u_{k+1} = .94\, u_k + .09\, r_k$$
$$r_{k+1} = .06\, u_k + .91\, r_k$$

We will refer to these as our "migration equations."

Derivation of the Migration Equations

Let the populations for the urban and rural areas after k years be U_k and R_k, respectively. Then, by our assumptions (1) and (2), the populations for these areas a year later will be the sum of those who stay plus those who move in from the other area:

$$U_{k+1} = .94\, U_k + .09\, R_k$$
$$R_{k+1} = .06\, U_k + .91\, R_k$$

To convert to population percentages, we divide the equations by the total population T (which we assume stays constant):

$$\frac{U_{k+1}}{T} = .94\, \frac{U_k}{T} + .09\, \frac{R_k}{T}$$

$$\frac{R_{k+1}}{T} = .06\, \frac{U_k}{T} + .91\, \frac{R_k}{T}$$

These are our migration equations, which we see from the fact that $u_k = \frac{U_k}{T}$ and $r_k = \frac{R_k}{T}$, and similarly for u_{k+1} and r_{k+1}.

The migration equations are called *recursion* equations, which means that they describe how to get the next values of the variables from the previous values. We will find it easier to work with these two recursion equations if we change them into a single recursion equation using matrices and vectors. Let x_k be the vector with components u_k and r_k:

$$x_k = \begin{bmatrix} u_k \\ r_k \end{bmatrix}$$

We will call x_k the kth *state vector*.

Exercise 1.1: Write the two migration equations above as a single recursion equation of the form

$$x_{k+1} = M x_k$$

for some 2 by 2 matrix M. That is, find the matrix M. Check that your matrix M is correct by computing u_1 and r_1 two ways: by computing the matrix-vector product $M x_0$, and by using our original two recursion equations with $k = 0$.

+ Student Workspace

− Answer 1.1

$$\begin{bmatrix} u_{k+1} \\ r_{k+1} \end{bmatrix} = \begin{bmatrix} .94 & .09 \\ .06 & .91 \end{bmatrix} \begin{bmatrix} u_k \\ r_k \end{bmatrix}$$

Thus
```
> M := Matrix([[.94,.09],[.06,.91]]);
```
$$M := \begin{bmatrix} .94 & .09 \\ .06 & .91 \end{bmatrix}$$

In Section 2, we will see how to use the matrix-vector form of the recursion equations, $x_{k+1} = M x_k$, to investigate the questions we posed above.

Section 2: Analysis of the Problem

Let's restate the mathematical setup from the end of Section 1. We have a 2 by 2 matrix M and a sequence of state vectors, $x_0, x_1, x_2, \ldots$, that are related by the recursion equation

$$x_{k+1} = M x_k$$

where
```
> x0 := Vector([.45,.55]);
  M := Matrix([[.94,.09],[.06,.91]]);
```
$$x0 := \begin{bmatrix} .45 \\ .55 \end{bmatrix}$$

$$M := \begin{bmatrix} .94 & .09 \\ .06 & .91 \end{bmatrix}$$

Therefore, by the recursion equation, we can compute the value of x_1 from x_0, and x_2 from x_1, and so on:
```
> x1 := M.x0;
> x2 := M.x1;
> x3 := M.x2;
```
$$x1 := \begin{bmatrix} .4725 \\ .5275 \end{bmatrix}, \quad x2 := \begin{bmatrix} .491625 \\ .508375 \end{bmatrix}, \quad x3 := \begin{bmatrix} .507882 \\ .492118 \end{bmatrix}$$

Notice that we can compute x_2 another way:

$$x_2 = M x_1 = M M x_0 = M^2 x_0$$

That is, x_2 can be computed directly from x_0 and M by simply multiplying x_0 by M^2.

Exercise 2.1: (a) (By hand) Produce a similar formula for computing x_3 directly from x_0 and M. Do the same for x_k.

(b) (Using Maple) Check that your new formula for computing x_3 yields the same answer as we found above for x_3. Use your new formula to compute x_9.

 Student Workspace

 Answer 2.1

(a) $x_3 = M^3 x_0$. More generally, $x_k = M^k x_0$ for every value of k.
(b)
```
> (M^3).x0;
  (M^9).x0;
```

$$\begin{bmatrix} .507881 \\ .492118 \end{bmatrix}, \begin{bmatrix} .565257 \\ .434743 \end{bmatrix}$$

Investigation of Question (a)

- (a) What will happen to the population in the long run? For example, will all of the population end up in the cities? Will the urban vs. rural populations eventually settle down, or will they oscillate back and forth endlessly?

Do you see any trend yet? Formulate a hypothesis (that is, make a guess!) about the long-term trend before you go on to Exercise 2.2. Enter your hypothesis here and compare yours with another student's:

 Student Workspace

Exercise 2.2: (a) Use the formula in your answer to Exercise 2.1 to compute x_k for some larger values of k, such as $k = 25, 50, 100$.
(b) Describe in words the long-term trend. That is, answer question (a).

 Student Workspace

 Answer 2.2

```
> (M^25).x0;
  (M^50).x0;
  (M^100).x0;
```

$$\begin{bmatrix} .597421 \\ .402579 \end{bmatrix}, \begin{bmatrix} .599957 \\ .400043 \end{bmatrix}, \begin{bmatrix} .600000 \\ .400000 \end{bmatrix}$$

(b) The urban and rural populations seem to be stabilizing at 60% and 40%, respectively.

Here is a less tedious way to see the long-term trend. The following short computer program (in fact, a simple "loop") computes the population percentages over a 150 year period in ten-year increments:

```
> for k from 10 to 150 by 10
    do (M^k).x0 end do;
```

Maple note: You might want to delete the output from this command when you are done looking at it, since the volume of output will clutter up your worksheet. To do so, highlight any portion of the output, and then select "Remove Output / From Selection" in the Edit menu.

Another way to see the long-term trend is to plot each component of the state vectors x_k as a function of the subscript k, which we can think of as time. Below, the red circles give the values of the first component (the urban population) as a function of time; the blue crosses give the values of the second component (the rural population). We see that they tend toward 0.6 and 0.4, respectively. The following command is an animation; so click on the figure and then on the play button.

```
> Componentplot(M,x0,points=30);
```

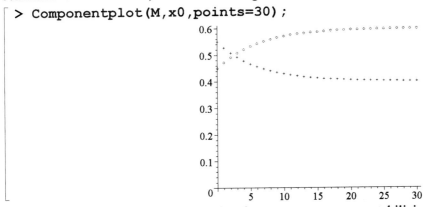

A mathematical way in which to say that the state vectors x_k are stabilizing as k gets larger is to say that the state vectors x_k are converging to a *limit vector* x. Judging from our above computations of x_k for large k's, we might guess that the limit vector x is approximately $\begin{bmatrix} .60 \\ .40 \end{bmatrix}$. This limit vector is called a *steady-state* vector for the following reason. As the computation in the next input region shows, the limit vector x satisfies the equation $Mx = x$. This equation says that if the urban and rural population percentages should happen to equal 60% and 40% (the components of x), these percentages will remain unchanged (i.e., "steady") from that point on.

```
> x := Vector([.6,.4]);
```

$$x := \begin{bmatrix} .6 \\ .4 \end{bmatrix}$$

```
> M.x;
```

$$\begin{bmatrix} .600 \\ .400 \end{bmatrix}$$

Exercise 2.3: Explain why the limit vector x should satisfy the equation $Mx = x$.

Student Workspace

 Answer 2.3

Since the state vectors x_k are approaching a limit, they must not be changing much as k gets large. That is, the population percentages are not changing much from year to year as k gets large. So we would expect that the limiting percentages, which are the components of x, shouldn't be changing at all. In other words, we would expect that $Mx = x$.

A second method for finding the steady-state vector x is to solve the equation $Mx = x$. This equation, however, has infinitely many solutions, which leads to the following question: How can we pick out the one solution that we want, the steady-state vector? The answer comes from an important observation, which you may have already made. The components of all the state vectors add to 1, since they are percentages which must add to 100%. So their limit vector, x, must have this same property -- its components add to 1.

========================

Example 2A: Solve the equation $Mx = x$. Also pick out the steady-state vector from the infinitely many solutions of this equation.

Solution: $Mx = x$ is the same as $Mx = Ix$ which is also $(M - I)x = 0$. Therefore we can solve for x by using **Matsolve**:

```
> I := Idmat(2);
```
$$I := \begin{bmatrix} 1 & 0 \\ 0 & 1 \end{bmatrix}$$

```
> A := M-I;
```
$$A := \begin{bmatrix} -.06 & .09 \\ .06 & -.09 \end{bmatrix}$$

```
> x := Matsolve(A,zero);
```
$$x := \begin{bmatrix} \frac{3}{2} x_2 \\ x_2 \end{bmatrix}$$

The components of x must add to 1. Solving $\frac{3 x_2}{2} + x_2 = 1$ for x_2 yields $x_2 = \frac{2}{5}$, which we then substitute back into the expression for x:

$$x = \begin{bmatrix} \frac{3}{5} \\ \frac{2}{5} \end{bmatrix} = \begin{bmatrix} .6 \\ .4 \end{bmatrix}$$

========================

Maple note: In the `Matsolve` command above, the answer is an exact rational expression even though the matrix M has "floating-point" (i.e., decimal) entries. The command `Matsolve(A,b)` actually converts all entries in A and b to exact rationals before solving $A\,x = b$. To instruct `Matsolve` to skip this conversion and instead use an approximate method of solution, use the option method=numerical:

```
> Matsolve(A,zero,method=numerical);
```

$$\begin{bmatrix} 1.50000\,X_2 \\ X_2 \end{bmatrix}$$

Investigation of Question (b)

- (b) How does the long-term distribution of the two populations depend on the initial 45-55 distribution? For example, if the urban population had started out at 75%, how different would the long-term distribution be?

Exercise 2.4: (a) Find the long-term behavior of x_k for several different values of x_0. For example, use

$$x_0 = \begin{bmatrix} .50 \\ .50 \end{bmatrix}, \quad x_0 = \begin{bmatrix} .20 \\ .80 \end{bmatrix}, \quad x_0 = \begin{bmatrix} .75 \\ .25 \end{bmatrix}$$

(Suggestion: For each x_0, compute x_k for some large values of k.)
(b) Describe in words your answer to question (b).

+ Student Workspace

 Answer 2.4

```
> (M^150).Vector([.50,.50]);
  (M^150).Vector([.20,.80]);
  (M^150).Vector([.75,.25]);
```

$$\begin{bmatrix} .599999 \\ .400001 \end{bmatrix}, \begin{bmatrix} .599999 \\ .400001 \end{bmatrix}, \begin{bmatrix} .599998 \\ .400002 \end{bmatrix}$$

(b) The long-term distribution appears to be completely independent of the initial distribution! Why should that be?

One way to get some insight into the behavior of the long-term distribution of the populations is to recall that $x_k = M^k\,x_0$ and to therefore study M^k instead of x_k. Execute the following loop, which computes M^k for values of k from 10 to 150 in steps of 10:

```
> for k from 10 to 150 by 10
    do M^k end do;
```

So the powers M^k also approach a limit N, and both of the columns of the limit matrix appear to equal the steady-state vector $x = \langle .4, .6 \rangle$ found in Example 2A. If this is true, then no matter what initial vector x_0

we choose, the limit vector will always be Nx_0, where $N = [x, x]$.

```
> x := Vector([.6,.4]);
  N := Matrix([x,x]);
```

$$x := \begin{bmatrix} .6 \\ .4 \end{bmatrix}$$

$$N := \begin{bmatrix} .6 & .6 \\ .4 & .4 \end{bmatrix}$$

Exercise 2.5: (By hand) Calculate the product $N x_0$, where x_0 is any initial vector whose entries are non-negative numbers that add to 1, and show that the product is always the steady-state vector x. (Therefore the state vectors $M^k x_0$ always converge to the limit vector x, which depends only on M and not on x_0.)

 Student Workspace

 Answer 2.5

Write $x_0 = \begin{bmatrix} a \\ b \end{bmatrix}$, where $a + b = 1$ and a and b are non-negative. Then

$$N x_0 = [x, x] \begin{bmatrix} a \\ b \end{bmatrix} = a x + b x = (a + b) x = x.$$

The matrix M has to be of a rather special type for this surprising behavior to occur: the matrices M^k converge to a limit, and all the columns of the limit matrix are equal. In the remaining sections we will see what features of M lead to this behavior.

 ### Section 3: General Observations About Markov Chains

The type of mathematical model that we used for our migration problem can be applied to a large class of similar problems. The aspects of our model that generalize are: (1) the recursion relation $x_{k+1} = M x_k$ that links the matrix M with the state vectors x_k, and (2) the fact that the columns of M and the state vectors have the property that their components add to 1 and are non-negative. The second fact is important, since it suggests that the concept of probability is involved; probabilities are non-negative, and they add to 1 if they take into account all possible outcomes. The following definition therefore extracts some key parts of our set-up of the problem.

> **Definition**: A vector whose entries are non-negative numbers adding to 1 is called a ***probability vector***. A square matrix P whose columns are probability vectors is called a ***stochastic matrix*** (or a ***Markov matrix***). Suppose x_0 is a probability vector and P is a stochastic matrix, and suppose we then construct a sequence of ***state*** vectors $x_1, x_2, ...$ by using the recursion relation $x_{k+1} = P x_k$. This sequence of state vectors, linked to one another by the matrix P, is called a ***Markov chain***.

The following observations that we made about our migration matrix M and our population vectors x_k

can be deduced readily for every Markov chain:
- The state vectors x_k can be computed directly from P and x_0 by the formula $x_k = P^k x_0$.
- Every state vector x_k is a probability vector.

However, some of the conclusions we reached about the limit of the state vectors and the steady-state vector are not valid for every Markov chain, as the next example shows.

=====================

Example 3A (Random walks): For the matrix W below,
(a) find the limit of the state vectors $x_k = W^k x_0$ for each of the initial probability vectors u and v below;
(b) find all the probability vectors x that satisfy the equation $W x = x$ (i.e., the "steady-state vectors").

```
> W := Matrix([[1,.5,0,0],[0,0,.5,0],[0,.5,0,0],[0,0,.5,1]]);
```

$$W := \begin{bmatrix} 1 & .5 & 0 & 0 \\ 0 & 0 & .5 & 0 \\ 0 & .5 & 0 & 0 \\ 0 & 0 & .5 & 1 \end{bmatrix}$$

```
> u := Vector([0,1,0,0]);
  v := Vector([1,0,0,0]);
```

$$u := \begin{bmatrix} 0 \\ 1 \\ 0 \\ 0 \end{bmatrix}, \quad v := \begin{bmatrix} 1 \\ 0 \\ 0 \\ 0 \end{bmatrix}$$

Solution: (a)

```
> for k from 10 to 150 by 10
    do (W^k).u end do;
```

The limit of the state vectors x_k appears to be $\langle \frac{2}{3}, 0, 0, \frac{1}{3} \rangle$ when the initial vector is $\langle 0, 1, 0, 0 \rangle$.

```
> for k from 10 to 150 by 10
    do (W^k).v end do;
```

The limit of the state vectors x_k is $\langle 1, 0, 0, 0 \rangle$ when the initial vector is $\langle 1, 0, 0, 0 \rangle$. In fact, when $\langle 1, 0, 0, 0 \rangle$ is the initial vector, every x_k equals the initial vector!
(b)
```
> Matsolve(W-Idmat(4),zero);
```
For the above solution to be a probability vector, x_1 and x_4 must be non-negative and add to 1. Thus, the solutions of $W x = x$ that are probability vectors can be written $\langle a, 0, 0, 1-a \rangle$. Note that both limit vectors we found in part (a) have this form.

So, unlike the migration matrix M, the limiting state for the matrix W depends on the initial state, and there is more than one steady-state probability vector. In Exercise 3.1, you will see another difference in

Interpretation of the Matrix W

The matrix W in Example 3A is an example of a "random walk" matrix, which has the following interpretation. Suppose you are on one of the four blocks pictured below and you walk left or right at random, either moving to an adjacent block or standing still in each time increment. If you are on one of the middle two blocks, we assume that you will move either right or left with equal probability. If you are on one of the two end blocks, we assume that you remain standing there. The numbers in the picture indicate the probabilities governing your movements.

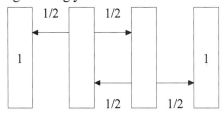

Let $x_k = \langle a_k, b_k, c_k, d_k \rangle$ be the probability vector that represents your position after k time increments. That is, a_k is the probability that you are on the first (i.e., leftmost) block after k time increments; b_k is the probability that you are on the second block after k time increments, etc. The vectors x_k are then state vectors in the Markov chain linked by the stochastic matrix W above.

The initial state $x_0 = \langle 0, 1, 0, 0 \rangle$ means that your initial position is on the second block, and the corresponding limit state, $\langle \frac{2}{3}, 0, 0, \frac{1}{3} \rangle$, means that two-thirds of the time you will eventually land on the leftmost block and one-third of the time you will eventually land on the rightmost block.

Exercise 3.1: (a) Find the limit of the state vectors x_k in Example 3A, if the initial vector is the vector x_0 below.

(b) Find the limit of the matrices W^k. Are all the columns of the limit matrix equal?

```
> x0 := Vector([0,0,1,0]);
```

$$x0 := \begin{bmatrix} 0 \\ 0 \\ 1 \\ 0 \end{bmatrix}$$

 Student Workspace

Answer 3.1

(a)
```
> (W^150).x0;
```

$$\begin{bmatrix} .333333 \\ 0. \\ .700629\ 10^{-45} \\ .666667 \end{bmatrix}$$

So the limit vector appears to be $\langle \frac{1}{3}, 0, 0, \frac{2}{3} \rangle$, which again has the form $\langle a, 0, 0, 1-a \rangle$ of the steady-state vectors.

(b)
```
> W^150;
```

$$\begin{bmatrix} 1. & .666667 & .333333 & 0. \\ 0. & .700629\ 10^{-45} & 0. & 0. \\ 0. & 0. & .700629\ 10^{-45} & 0. \\ 0. & .333333 & .666667 & 1. \end{bmatrix}$$

The limit of W^k, unlike the limit of M^k, does not have equal columns. Where have we seen the first three columns of this limit before?

In Problem 5, where we change only slightly our assumptions about the random walk, we will find yet another behavior for a Markov chain.

Section 4: Regular Stochastic Matrices

In this section we look at conditions which guarantee that a stochastic matrix behaves like the migration matrix M rather than the random walk matrix W. That is, these conditions will guarantee that the state vectors x_k approach a limit that is independent of the initial state x_0. For this, we introduce the concept of a "regular" stochastic matrix.

A stochastic matrix P is said to be *regular* if for some power k all the entries of P^k are positive (i.e, none of the entries are zero). Our migration matrix M is therefore regular. The following fact can be proven (using methods beyond this course) for all regular stochastic matrices P:

- The powers P^k converge to a limit matrix Q, and all the columns of Q are equal to the same probability vector x.

From this fact we can readily deduce the following properties, which we observed in our analysis of the migration problem:

- For every initial probability vector x_0, the state vectors $x_k = P^k x_0$ approach a limit vector, and the limit vector is the vector x that is in every column of Q and therefore does not depend on x_0.

- Furthermore, the limit vector x is a solution of the equation $Px = x$ and is the only probability vector in the solution set of this homogeneous system.

Exercise 4.1: Confirm that the stochastic matrix P below is regular and that P^k does converge to a matrix whose columns are all the same probability vector.

```
> P := Matrix([[0,.6,0],[1,0,.5],[0,.4,.5]]);
```

$$P := \begin{bmatrix} 0 & .6 & 0 \\ 1 & 0 & .5 \\ 0 & .4 & .5 \end{bmatrix}$$

+ Student Workspace

− Answer 4.1

Although P, P^2, and P^3 all have at least one zero entry, no entry of P^4 is zero, which shows that P is regular:

```
> P^2, P^3, P^4;
```

$$\begin{bmatrix} .6 & 0. & .30 \\ 0. & .80 & .25 \\ .4 & .20 & .45 \end{bmatrix}, \begin{bmatrix} 0. & .480 & .150 \\ .80 & .100 & .525 \\ .20 & .420 & .325 \end{bmatrix}, \begin{bmatrix} .480 & .0600 & .3150 \\ .100 & .6900 & .3125 \\ .420 & .2500 & .3725 \end{bmatrix}$$

```
> P^150;
```

$$\begin{bmatrix} .249997 & .249997 & .249998 \\ .416661 & .416661 & .416661 \\ .333329 & .333330 & .333330 \end{bmatrix}$$

The limit of P^k appears to be $[s,s,s]$, where $s = \langle \frac{1}{4}, \frac{5}{12}, \frac{1}{3} \rangle$.

Problems

− Problem 1: Compare two stochastic matrices

(a) For each of the two matrices below, find all probability vectors x such that $Px = x$.
(b) For each of the two matrices below, find the limit of the matrices P^k.
(c) Explain how the long-term behavior of the state vectors x_k differs for the two matrices. Base your explanation on the computations in parts (a) and (b).

```
> P1 := Matrix([[.5,0,1],[.1,1,0],[.4,0,0]]);
  P2 := Matrix([[.6,0,1],[0,1,0],[.4,0,0]]);
```

+ Student Workspace

Problem 2: What do the entries of M^2 mean?

(a) For our migration matrix M, compute M^2 and (by hand) use the entries of M^2 to express u_2 and r_2 in terms of u_0 and r_0: $u_2 = a\,u_0 + b\,r_0$, $r_2 = c\,u_0 + d\,r_0$. That is, find the constant coefficients a, b, c, d.

(b) By examining your answer to (a), describe the meaning of each of the four entries of M^2. (These meanings should be similar to the meanings of the four entries of M. For example, the (1,2) entry of M is the percentage of the rural population that moves into urban areas in one year.)

Student Workspace

Problem 3: Markov chain of healthy/sick workers

The personnel department of a large corporation is interested in constructing a mathematical model that will help them predict work force reduction due to worker illness. After researching employee attendence records over a five year period, they reached the following conclusions: Of the workers who are healthy and show up for work on a given day, 96% will be healthy and come to work the following (work) day. On the other hand, of the workers who call in sick on a given day, 60% will call in sick the next (work) day. (We'll assume that each worker is either healthy or sick on any given day and that healthy workers go to work.)

(a) Find the stochastic matrix that represents this situation. (Let the percent of healthy workers each day be the first component of the state vectors and the percent of sick workers be the second component.)

(b) Suppose that on one Monday in the middle of winter 15% of the workers call in sick. According to the assumptions of the model, what percent will be sick on Tuesday? on Wednesday? on Thursday? What percent will be sick ten weeks later?

(c) Find the steady-state vector by both of the methods we used in the Tutorial.

Student Workspace

Problem 4: Markov chain of planes at three hubs

The United Package Service has a fleet of 240 planes that shuttle between three national distribution hubs: Seattle, Chicago and Atlanta. Each day, most of the planes fly round trips returning to their starting location; however, some remain at the other two airports. Here are the percentages that move from each location:

- Each day 2% of the planes from Seattle remain at Chicago and 8% remain at Atlanta; the rest return to Seattle;

- Each day 3% of the planes from Chicago remain at Seattle and 6% remain at Atlanta; the rest return to Chicago;

- Each day 9% of the planes from Atlanta remain at Seattle and 4% remain at Chicago; the rest return to Atlanta.

(a) Find the stochastic matrix that represents this situation. (Let the three components of the state vectors represent the percent of the total number of planes that are in Seattle, Chicago, and Atlanta, respectively.)
(b) Assuming that the distribution of planes has reached a steady state, how many planes are at each location?

 Student Workspace

 Problem 5: Random walks -- a variation

In this problem we change slightly our assumptions about the random walk in Example 3A. Let's assume that, when we are on the leftmost block or the rightmost block, we always walk next to the adjacent middle block. Here is a picture of the probabilities:

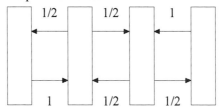

The corresponding Markov matrix is:
```
> W := Matrix([[0,.5,0,0],[1,0,.5,0],[0,.5,0,1],[0,0,.5,0]]);
```
(a) Find all the probability vectors that satisfy the equation $W\,x = x$. (There is only one such vector, even though W is not regular.)
(b) For the initial state vector x_0 below, compute $W\,x_0$, $W^2\,x_0$, $W^3\,x_0$. Interpret the results of these computations.
```
> x0 := Vector([0,1,0,0]):
```
(c) Check that the state vectors x_k do not have a limit by computing W^k for several large consecutive values of k.

(d) Discuss the meaning of W^k for large k. Specifically, your computation for (c) should indicate that it matters whether k is even or odd and it matters which block you start on. Discuss what these facts mean.

 Student Workspace

 Problem 6: Gambler's ruin

The stochastic matrix P below represents a situation that is sometimes called "Gambler's Ruin." Imagine a game of chance in which the bettor has a 60% chance of winning (and hence a 40% chance of losing). Suppose the bet for each game is $1 and that the bettor will quit if he goes broke (i.e. he reaches $0) or if he reaches his goal (which we assume is $6). The seven components of the state vector represent the following seven probabilities: the probability that the bettor has, respectively, $0, $1, $2, $3, $4, $5, $6.
(a) Suppose the bettor starts with $2. (So $x_0 = \langle 0, 0, 1, 0, 0, 0, 0 \rangle$). What is the probability that after three games the bettor has $0? $1? $2? $3? $4? $5? $6? (Do a matrix computation to get your answers.)

(b) Suppose the bettor starts with $2. What is the probability that he will eventually go broke and the probability that he will eventually reach his goal? Answer the same question for a bettor starting with $3.

(c) P is not a regular stochastic matrix. Use your answer for (b) to explain how we know this.

(d) Find all the solutions of $Px = x$ that are also probability vectors. Show that your answers to (b) are included among these solutions.

(e) Find the limit matrix of P^k. (That is, approximate the limit by calculating P^k for some large powers k.) Which entries of this matrix give you the answers you found in (b)? Why do they give you those answers? What other probabilities can you read off of this limit matrix?

(f) Suppose a more skillful player has a 70% chance of winning. Modify the matrix P to create the stochastic matrix representing this player's results. Find the probability that this player will eventually go broke and the probability that she will eventually reach her goal (assuming a starting bet of $2).

```
> P := Matrix([[1,.4,0,0,0,0,0],[0,0,.4,0,0,0,0],
  [0,.6,0,.4,0,0,0],[0,0,.6,0,.4,0,0],
  [0,0,0,.6,0,.4,0],[0,0,0,0,.6,0,0],[0,0,0,0,0,.6,1]]);
```

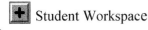 Student Workspace

Problem 7: Products of stochastic matrices

(a) Let P be a 2 by 2 stochastic matrix and y be a probability vector with 2 components. By a direct hand computation, show that Py is a probability vector.

(b) Let P and Q be 2 by 2 stochastic matrices. Use (a) to give a quick proof that PQ is a stochastic matrix.

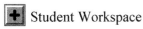 Student Workspace

Linear Algebra Modules Project
Chapter 3, Module 5

Inverse of a Matrix -- An Introduction

Purpose of this module

The purpose of this module is to introduce the concept of the inverse of a matrix and to explore its basic properties. In Section 2, we make connections between the concept of invertibility and earlier ideas from the course, such as linear independence and span of vectors.

Prerequisites

Matrix algebra, solutions of linear systems, and linear independence and span of vectors.

Commands used in this module

```
> restart: with(LinearAlgebra): with(Lamp): with(plots):
  UseHardwareFloats := false: Digits := 6:
```

Tutorial

Section 1: The Inverse of a Matrix

Throughout this module we will only consider square matrices, as these are the only matrices that have an inverse.

Definition: The **inverse** of an n by n matrix A is an n by n matrix B such that

$$AB = I \text{ and } BA = I$$

where I is the n by n identity matrix. The matrix A is said to be **invertible** if it has an inverse. The notation for the inverse of A is $A^{(-1)}$.

Warning: We never write $A^{(-1)}$ as $\frac{1}{A}$, since $\frac{C}{A}$ is ambiguous. (Is it $C\frac{1}{A}$ or $\frac{1}{A}C$?)

Exercise 1.1: For the matrices A, B, C below, form all products AB, AC, BC (and in the opposite order too). Which matrices are inverses of which?

```
> A := Matrix([[5,3],[3,2]]);
  B := Matrix([[-3,2],[5,-3]]);
  C := Matrix([[2,-3],[-3,5]]);
```

Module 3.5 Inverse of a Matrix -- An Introduction

$$A := \begin{bmatrix} 5 & 3 \\ 3 & 2 \end{bmatrix}, \quad B := \begin{bmatrix} -3 & 2 \\ 5 & -3 \end{bmatrix}, \quad C := \begin{bmatrix} 2 & -3 \\ -3 & 5 \end{bmatrix}$$

 Student Workspace

Answer 1.1

A and C are inverses of one another, since $A\,C = I$ and $C\,A = I$:
> A.C;

$$\begin{bmatrix} 1 & 0 \\ 0 & 1 \end{bmatrix}$$

> C.A;

$$\begin{bmatrix} 1 & 0 \\ 0 & 1 \end{bmatrix}$$

We can use Maple to find the inverse of a matrix:

==================

Example 1A: The matrix A below has an inverse (i.e., is invertible), and we can find it by using either of the commands `A^(-1)` or `Inverse(A)`.
> A := Matrix([[-3,2],[5,4]]);

$$A := \begin{bmatrix} -3 & 2 \\ 5 & 4 \end{bmatrix}$$

> B := A^(-1);
  B := Inverse(A);

$$B := \begin{bmatrix} -\dfrac{2}{11} & \dfrac{1}{11} \\ \dfrac{5}{22} & \dfrac{3}{22} \end{bmatrix}, \quad B := \begin{bmatrix} -\dfrac{2}{11} & \dfrac{1}{11} \\ \dfrac{5}{22} & \dfrac{3}{22} \end{bmatrix}$$

Let's check that B is the inverse of A by having Maple calculate the matrix products $A\,B$ and $B\,A$:
> A.B;

$$\begin{bmatrix} 1 & 0 \\ 0 & 1 \end{bmatrix}$$

> B.A;

$$\begin{bmatrix} 1 & 0 \\ 0 & 1 \end{bmatrix}$$

==================

Here are two basic properties of the inverse:

- If B is the inverse of A, then A is the inverse of B. (In other words, the inverse of the inverse of A is A.)
- If B and C are both inverses of A, then $B = C$. (In other words, a matrix can have only one inverse.)

It is not hard to see why these two facts are true. If B is the inverse of A, then $A\,B = I$ and $B\,A = I$ by the above definition. But these two equations also say that A is the inverse of B, by the same definition. To check this, we ask Maple to compute the inverse of B. Do we get A?

```
> Inverse(B), A;
```

$$\begin{bmatrix} -3 & 2 \\ 5 & 4 \end{bmatrix} \begin{bmatrix} -3 & 2 \\ 5 & 4 \end{bmatrix}$$

Here is why the second fact is true. We write the product BAC two ways by using the associative law of multiplication:

$$BAC = B(AC) = BI = B$$
$$BAC = (BA)C = IC = C$$

Therefore $B = C$.

You can find the inverse of a 2 by 2 matrix quite easily by hand. Here's how: The inverse of the matrix $A = \begin{bmatrix} a & b \\ c & d \end{bmatrix}$ is the matrix

$$\begin{bmatrix} \dfrac{d}{ad-bc} & -\dfrac{b}{ad-bc} \\ -\dfrac{c}{ad-bc} & \dfrac{a}{ad-bc} \end{bmatrix}$$

Maple confirms this fact:

```
> A := Matrix([[a,b],[c,d]]);
```

$$A := \begin{bmatrix} a & b \\ c & d \end{bmatrix}$$

```
> Inverse(A);
```

$$\begin{bmatrix} \dfrac{d}{ad-bc} & -\dfrac{b}{ad-bc} \\ -\dfrac{c}{ad-bc} & \dfrac{a}{ad-bc} \end{bmatrix}$$

Of course this formula is of no use if $ad - bc = 0$. In fact, a 2 by 2 matrix is invertible if and only if $ad - bc \neq 0$.

Aside on determinants: If you are familiar with determinants of 2 by 2 matrices, you will recognize the number $ad - bc$ as the determinant of the matrix A. So a 2 by 2 matrix A is invertible if and only if $\det(A) \neq 0$.

Exercise 1.2: (By hand) (a) Apply the above formula to calculate the inverse N of the matrix M defined below. Then check your work by calculating the products MN and NM.
(b) Check that the matrix S below does not have an inverse. (Apply the above determinant test, $ad - bc \neq 0$; also see what the command `Inverse(S)` does.)

```
> M := Matrix([[2,6],[3,6]]);
```
$$M := \begin{bmatrix} 2 & 6 \\ 3 & 6 \end{bmatrix}$$

```
> S := Matrix([[2,6],[1,3]]);
```
$$S := \begin{bmatrix} 2 & 6 \\ 1 & 3 \end{bmatrix}$$

Student Workspace

Answer 1.2

(a) Since $ad - bc = -6$, $N = M^{(-1)} = \begin{bmatrix} -1 & 1 \\ \frac{1}{2} & -\frac{1}{3} \end{bmatrix}$. Checking with Maple:

```
> N := Matrix([[-1,1],[1/2,-1/3]]);
```
$$N := \begin{bmatrix} -1 & 1 \\ \frac{1}{2} & \frac{-1}{3} \end{bmatrix}$$

```
> N.M;
```
$$\begin{bmatrix} 1 & 0 \\ 0 & 1 \end{bmatrix}$$

```
> M.N;
```
$$\begin{bmatrix} 1 & 0 \\ 0 & 1 \end{bmatrix}$$

(b) Since $ad - bc = 0$, S does not have an inverse.

```
> Inverse(S);
Error, (in Inverse) singular matrix
```

In Answer 1.2, note that Maple referred to the matrix S above as "singular". A matrix that does not have an inverse is called a *singular* matrix. Thus, "singular" is a synonym for "noninvertible". Likewise, "nonsingular" is a synonym for "invertible." Here it is one more time:

- If a matrix A <u>does not</u> have an inverse, we say that A is *noninvertible*, i.e., *singular*.

- If a matrix A <u>does</u> have an inverse, we say that A is *invertible*, i.e., *nonsingular*.

Exercise 1.3: The inverse of a diagonal matrix is especially easy to compute. Recall that the product of two diagonal matrices A and B is the diagonal matrix whose diagonal entries are simply the product of the corresponding diagonal entries of A and B.
(a) Use this idea to find (by hand) the inverse of the diagonal matrix K below. Check your answer using Maple.
(b) Which diagonal matrices are invertible? That is, state a necessary and sufficient condition for a diagonal matrix to be invertible.

```
> K := Diagmat([-1,3, 1/4]);
```

$$K := \begin{bmatrix} -1 & 0 & 0 \\ 0 & 3 & 0 \\ 0 & 0 & \dfrac{1}{4} \end{bmatrix}$$

 Student Workspace

Answer 1.3

(a) The inverse is found by taking the reciprocal of each diagonal entry:

```
> Inverse(K);
```

$$\begin{bmatrix} -1 & 0 & 0 \\ 0 & \dfrac{1}{3} & 0 \\ 0 & 0 & 4 \end{bmatrix}$$

(b) A diagonal matrix is invertible if and only if every diagonal entry is nonzero.

===================

Example 1B: If A and B are n by n invertible matrices, then it turns out that their product AB is also invertible. Let's look at an example. The matrices A and B below each have an inverse. Notice that the product matrix $C = AB$ also has an inverse.

```
> A := Matrix([[1,2,3],[2,1,4],[2,3,0]]);
  B := Matrix([[3,2,1],[5,-3,0],[2,3,3]]);
```

$$A := \begin{bmatrix} 1 & 2 & 3 \\ 2 & 1 & 4 \\ 2 & 3 & 0 \end{bmatrix}, \quad B := \begin{bmatrix} 3 & 2 & 1 \\ 5 & -3 & 0 \\ 2 & 3 & 3 \end{bmatrix}$$

```
> A^(-1), B^(-1);
```

$$A^{(-1)} = \begin{bmatrix} -\dfrac{3}{4} & \dfrac{9}{16} & \dfrac{5}{16} \\ \dfrac{1}{2} & -\dfrac{3}{8} & \dfrac{1}{8} \\ \dfrac{1}{4} & \dfrac{1}{16} & -\dfrac{3}{16} \end{bmatrix}, \quad B^{(-1)} = \begin{bmatrix} \dfrac{1}{4} & \dfrac{1}{12} & -\dfrac{1}{12} \\ \dfrac{5}{12} & -\dfrac{7}{36} & -\dfrac{5}{36} \\ -\dfrac{7}{12} & \dfrac{5}{36} & \dfrac{19}{36} \end{bmatrix}$$

```
> (A.B)^(-1);
```

$$\begin{bmatrix} \dfrac{-1}{6} & \dfrac{5}{48} & \dfrac{5}{48} \\ \dfrac{-4}{9} & \dfrac{43}{144} & \dfrac{19}{144} \\ \dfrac{23}{36} & \dfrac{-25}{72} & \dfrac{-19}{72} \end{bmatrix}$$

===================

Module 3.5 Inverse of a Matrix -- An Introduction

A natural question to ask at this point is whether $(AB)^{(-1)} = A^{(-1)} B^{(-1)}$?

> A^(-1).B^(-1);

$$\begin{bmatrix} \frac{-13}{96} & \frac{-37}{288} & \frac{43}{288} \\ \frac{-5}{48} & \frac{19}{144} & \frac{11}{144} \\ \frac{19}{96} & \frac{-5}{288} & \frac{-37}{288} \end{bmatrix}$$

No, $(AB)^{(-1)} \neq A^{(-1)} B^{(-1)}$. However, by another clever use of the associative law of multiplication, we can find a relationship among these three inverses (see Exercise 1.4).

Exercise 1.4: (By hand) Simplify the product below. From the result, state the relationship among the inverses $A^{(-1)}$, $B^{(-1)}$, and $(AB)^{(-1)}$.

$$(AB)(B^{(-1)} A^{(-1)})$$

 Student Workspace

 Answer 1.4

$(AB)(B^{(-1)} A^{(-1)}) = A(B B^{(-1)}) A^{(-1)} = A I A^{(-1)} = A A^{(-1)} = I$. Similarly, $(B^{(-1)} A^{(-1)})(AB) = I$. These two equations say that the inverse of the product AB is the product of the inverses in the reversed order, $B^{(-1)} A^{(-1)}$. In symbols, we have proved

- $(AB)^{(-1)} = B^{(-1)} A^{(-1)}$

Let's check with our example:

> (A.B)^(-1);

$$\begin{bmatrix} \frac{-1}{6} & \frac{5}{48} & \frac{5}{48} \\ \frac{-4}{9} & \frac{43}{144} & \frac{19}{144} \\ \frac{23}{36} & \frac{-25}{72} & \frac{-19}{72} \end{bmatrix}$$

> B^(-1).A^(-1);

$$\begin{bmatrix} \frac{-1}{6} & \frac{5}{48} & \frac{5}{48} \\ \frac{-4}{9} & \frac{43}{144} & \frac{19}{144} \\ \frac{23}{36} & \frac{-25}{72} & \frac{-19}{72} \end{bmatrix}$$

Exercise 1.5: (By hand) Assume that R, S and T are three invertible matrices. Find a formula for the inverse of the product $R\,S\,T$. (Your formula should be in terms of $R^{(-1)}$, $S^{(-1)}$, and $T^{(-1)}$.) Then test your conclusion on some random matrices.

 Student Workspace

 Answer 1.5

$(R\,S\,T)^{(-1)} = T^{(-1)}\,S^{(-1)}\,R^{(-1)}$. Note that the inverses are again multiplied in reverse order. Checking with Maple:

```
> R := Randmat(3,3);
  S := Randmat(3,3);
  T := Randmat(3,3);
```

$$R := \begin{bmatrix} 8 & 4 & -5 \\ -5 & 3 & -5 \\ 8 & 5 & -1 \end{bmatrix},\quad S := \begin{bmatrix} 0 & 3 & -4 \\ -5 & -2 & -1 \\ -9 & 5 & 8 \end{bmatrix},\quad T := \begin{bmatrix} 4 & -3 & 8 \\ -9 & 7 & 5 \\ 0 & 4 & 4 \end{bmatrix}$$

```
> (R.S.T)^(-1);
```

$$\begin{bmatrix} \dfrac{-63447}{13991978} & \dfrac{105965}{27983956} & \dfrac{883}{635999} \\ \dfrac{-38083}{6995989} & \dfrac{17302}{6995989} & \dfrac{1471}{635999} \\ \dfrac{54963}{13991978} & \dfrac{-76579}{27983956} & \dfrac{-3291}{635999} \end{bmatrix}$$

```
> T^(-1).S^(-1).R^(-1);
```

$$\begin{bmatrix} \dfrac{-63447}{13991978} & \dfrac{105965}{27983956} & \dfrac{883}{635999} \\ \dfrac{-38083}{6995989} & \dfrac{17302}{6995989} & \dfrac{1471}{635999} \\ \dfrac{54963}{13991978} & \dfrac{-76579}{27983956} & \dfrac{-3291}{635999} \end{bmatrix}$$

We summarize the main facts we have established for invertible matrices:

Theorem 5: Suppose A and B are n by n invertible matrices. Then:
(a) A has only one inverse matrix;
(b) $A^{(-1)}$ is invertible and $(A^{(-1)})^{(-1)} = A$;
(c) $A\,B$ is invertible and $(A\,B)^{(-1)} = B^{(-1)}\,A^{(-1)}$.

==================

Example 1C (Matrix equations revisited): The inverse of a matrix can be used to solve certain matrix equations, as follows. Let's assume we have a matrix equation $A\,x = b$, and suppose the matrix A

happens to be invertible. Then we can solve this equation by the following steps:

Multiply the equation $Ax = b$ on the left by $A^{(-1)}$:

$$A^{(-1)} A x = A^{(-1)} b$$

Simplify the left side using the properties $A^{(-1)} A = I$ and $I x = x$:

$$x = A^{(-1)} b$$

Notice that when the coefficient matrix A is invertible, the solution set of $A x = b$ is a single vector, namely $A^{(-1)} b$. Here is an example:

```
> A := Matrix([[2,3,5],[3,2,7],[3,2,4]]);
> b := Vector([3,4,1]);
```

$$A := \begin{bmatrix} 2 & 3 & 5 \\ 3 & 2 & 7 \\ 3 & 2 & 4 \end{bmatrix}$$

$$b := \begin{bmatrix} 3 \\ 4 \\ 1 \end{bmatrix}$$

```
> A^(-1).b;
```

$$\begin{bmatrix} -1 \\ 0 \\ 1 \end{bmatrix}$$

And here, for sake of comparison, is the solution we get using **Matsolve**:

```
> Matsolve(A,b);
```

$$\begin{bmatrix} -1 \\ 0 \\ 1 \end{bmatrix}$$

===================

Caution: The method illustrated above has two significant limitations. First, it only works if A is invertible. Second, as elegant at this method seems to be, in practice it is not an efficient way to solve a linear system, since calculating an inverse requires many more additions and multiplications than row reduction.

Exercise 1.6: Solve the linear system $A x = b$, where A and b are defined below. Use the method of Example 1C.

```
> A := Matrix([[3,2,4],[1,3,5],[8,6,7]]);
> b := Vector([1,4,8]);
```

$$A := \begin{bmatrix} 3 & 2 & 4 \\ 1 & 3 & 5 \\ 8 & 6 & 7 \end{bmatrix}, \quad b := \begin{bmatrix} 1 \\ 4 \\ 8 \end{bmatrix}$$

 Student Workspace

 Answer 1.6

```
> A^(-1).b;
```

$$\begin{bmatrix} -5/11 \\ 3 \\ -10/11 \end{bmatrix}$$

Section 2: The Invertibility Theorem

This section is devoted entirely to describing and understanding the Invertibility Theorem, which provides important connections between the concept of the inverse of a matrix, linear independence and span of vectors, and existence of solutions of linear systems:

> **Theorem 6 (Invertibility Theorem):** Suppose A is a n by n matrix. Then the statement that A is invertible is equivalent to each of the following statements:
> (a) the only solution to the homogeneous linear system $Ax = 0$ is the trivial solution $x = 0$;
> (b) the columns of A are linearly independent;
> (c) the linear system $Ax = b$ is consistent for every vector b in R^n;
> (d) the columns of A span R^n.

This theorem gives us a lot of information. In part, it says that an invertible matrix A has all four of the properties (a), (b), (c), (d) above; furthermore, it says that if A has any one of these four properties, then A is invertible and thus A has the other three properties as well.

Before we discuss the proof of this theorem, let's practice using it:

=====================

Example 2A: Use part (a) of Theorem 6 to determine whether the matrix A below is invertible or not. Does $Ax = b$ have a solution for every vector b in R^3?

```
> A := Matrix([[4,3,6],[5,8,7],[4,5,3]]);
```

$$A := \begin{bmatrix} 4 & 3 & 6 \\ 5 & 8 & 7 \\ 4 & 5 & 3 \end{bmatrix}$$

Solution: We use **Matsolve** to solve the equation $Ax = 0$:

```
> Matsolve(A,zero);
```
$$\begin{bmatrix} 0 \\ 0 \\ 0 \end{bmatrix}$$

Since $A x = 0$ has only the trivial solution $x = 0$, the matrix A is invertible by the Invertibility Theorem. We confirm by observing that Maple can find the inverse of A:

```
> Inverse(A);
```
$$\begin{bmatrix} \frac{11}{47} & \frac{-21}{47} & \frac{27}{47} \\ \frac{-13}{47} & \frac{12}{47} & \frac{-2}{47} \\ \frac{7}{47} & \frac{8}{47} & \frac{-17}{47} \end{bmatrix}$$

Since A is invertible, part (c) of the Invertibility Theorem tells us that $A x = b$ does have a solution for every vector b in R^3.

========================

Exercise 2.1: Use part (b) of Theorem 6 to determine whether the matrices B and C below are invertible or not. Try to determine the linear independence or dependence of the columns of B simply by inspection. Do the columns of B span R^3? Do the columns of C span R^4?

```
> B := Matrix([[1,-1,0],[-2,3,-1],[0,2,-2]]);
```
$$B := \begin{bmatrix} 1 & -1 & 0 \\ -2 & 3 & -1 \\ 0 & 2 & -2 \end{bmatrix}$$

```
> C := Matrix([[-1,-1,-1,-4],[2,1,0,3],[2,1,-2,-1],[1,1,0,2]]);
```
$$C := \begin{bmatrix} -1 & -1 & -1 & -4 \\ 2 & 1 & 0 & 3 \\ 2 & 1 & -2 & -1 \\ 1 & 1 & 0 & 2 \end{bmatrix}$$

 Student Workspace

Answer 2.1

The sum of the columns of B is zero; that is, we have the linear dependence relation $B_1 + B_2 + B_3 = 0$ for the columns of B. Therefore B is not invertible.

To determine whether the columns of C, namely C_1, C_2, C_3, C_4, are linearly dependent, we solve the equation $x_1 C_1 + x_2 C_2 + x_3 C_3 + x_4 C_4 = 0$. This is the same as the matrix-vector equation $C x = 0$, which we solve using **Matsolve(C,zero)** as in Example 2A:

```
> Matsolve(C,zero);
```

$$\begin{bmatrix} -X_4 \\ -X_4 \\ -2X_4 \\ X_4 \end{bmatrix}$$

Since this equation has a nontrivial solution, the columns of C are linearly dependent and therefore C is not invertible.

Since B is not invertible, the columns of B do not span R^3 by part (d) of the Invertibility Theorem. Likewise, the columns of C do not span R^4.

About the Proof of the Invertibility Theorem

Since the complete proof of the Invertibility Theorem is long, we will not include all the details here. However, some parts of the proof are quite instructive. First we recall the following facts from Module 1:

- The columns of a matrix A are linearly independent if and only if the matrix-vector equation $Ax = 0$ has only the trivial solution $x = 0$.

- The vector b is in the span of the columns of a matrix A if and only if the matrix-vector equation $Ax = b$ has a solution.

Thus parts (a) and (b) of the Invertibility Theorem are equivalent to one another, and (c) and (d) are equivalent to one another. Hence, we can complete the proof of Theorem 6 just by showing that if A is invertible then (a) and (c) hold, and by showing that A is invertible if either (a) or (c) holds.

======================

Example 2B: If A is an invertible matrix, show that the only solution to $Ax = 0$ is the trivial solution $x = 0$; i.e., deduce property (a).

Solution: Multiply both sides of the equation $Ax = 0$ on the left by $A^{(-1)}$ and simplify:

$$A^{(-1)} A x = A^{(-1)} 0$$
$$I x = 0$$

Thus the only solution to $Ax = 0$ is $x = 0$.

======================

Exercise 2.2: If A is an invertible n by n matrix, show that the linear system $Ax = b$ is consistent for every vector b in R^n; i.e., deduce property (c). Hint: Example 1C shows what the solution x must be.

➕ Student Workspace

Answer 2.2

Check that the vector $x = A^{(-1)} b$ is a solution of $A x = b$:

$$A A^{(-1)} b = I b = b$$

The following two theorems are used in proving the converse of what we have shown above, namely that A is invertible if either (a) or (c) holds. Furthermore, they are also useful in their own right. For example, Theorem 8 says that we do not need to confirm both $A B = I$ and $B A = I$ when showing that A and B are inverses of one another. Either one of these equations implies the other. Theorem 7 is proved in Module 6; for Theorem 8, see Problem 12.

Theorem 7: A square matrix A is invertible if and only if it is row equivalent to the identity matrix.

Theorem 8: Suppose A and B are n by n matrices such that $A B = I$. Then A and B are both invertible, and $A^{(-1)} = B$.

Here is yet another property that is equivalent to invertibility (see Problem 13):

- An n by n matrix A is invertible if and only if the rank of A is equal to n.

Problems

 Problem 1: Solve a matrix equation

Let B and C be the invertible matrices shown below. Find the matrix A that satisfies the equation

$$B A C - C = B^4 + 5 B C$$

First, use the rules of matrix algebra to solve this equation for A by hand. Then use your result to find A with the help of Maple. Also, test your answer by substituting A into the above equation.

```
> B := Matrix([[3,2,5],[4,6,3],[2,4,3]]);
> C := Matrix([[8,3,2],[4,3,5],[0,8,0]]);
```

 Student Workspace

 Problem 2: Cancellation is not always valid

(a) The following statement is FALSE. Give a counter-example using 2 by 2 matrices.
"If A and B are n by n matrices and A is not the zero matrix and $A B = A$, then B must be the identity matrix."

(b) Modify the above statement so that it is true. That is, modify one of the hypotheses so the conclusion is true.

 Student Workspace

Problem 3: Polynomial equations and inverses

Let A be the 3 by 3 matrix defined below.
```
> A := Matrix([[-1,4,-4],[1,-3,1],[1,-2,0]]);
```
Execute the next input region to check that A satisfies the polynomial equation $x^3 + 4x^2 + 5x + 2 = 0$. That is,
$$A^3 + 4A^2 + 5A + 2I = O$$
(Note that the number x is replaced by the matrix A, the number 2 by the matrix $2I$, and the number 0 by the zero matrix.)
```
> I := Idmat(3):
  A^3+4*A^2+5*A+2*I;
```
(a) (By hand) Rewrite the matrix equation $A^3 + 4A^2 + 5A + 2I = O$ to find a formula for the inverse of A as a polynomial in A.
(b) Use Maple to calculate the inverse of A from your formula in (a).

+ Student Workspace

Problem 4: Simplify an expression involving inverses

(By hand) Simplify the following expression, where A, B, C are n by n invertible matrices. (It simplifies drastically.)

$$(AB^{(-1)})^{(-1)} (CA^{(-1)})^{(-1)} (BC^{(-1)})^{(-1)}$$

+ Student Workspace

Problem 5: Hand method for finding inverses

Execute the four commands below, which illustrate a common method for computing the inverse of a matrix by hand.
```
> A := Randmat(4,4);
  I := Idmat(4):
> B := Matrix([A,I]);
> Reduce(B,form=rref);
> Inverse(A);
```
The steps of this method for finding $A^{(-1)}$ are: (1) form the augmented matrix $[A, I]$, where I is the identity matrix; (2) reduce $[A, I]$ to reduced row echelon form. Then $A^{(-1)}$ appears where I had been.
(a) Apply this method <u>by hand</u> to the matrix B below. (You may use Maple to check your answer.)

$$B = \begin{bmatrix} 2 & 2 & 1 \\ -2 & -2 & 0 \\ -1 & 0 & 0 \end{bmatrix}$$

Explanation of why this method works: If you were to solve the equation $B x_1 = e_1$, where e_1 is the first column of I, the solution x_1 would be the first column of $B^{(-1)}$. The other columns of $B^{(-1)}$ are found similarly. By reducing the matrix $[B, I]$ to reduced row echelon form, we are simply solving all of the equations $B x_1 = e_1$, $B x_2 = e_2$, ... simultaneously.

(b) (By hand or using Maple) Using the matrix B above, solve the equations $B x_1 = e_1$, $B x_2 = e_2$, $B x_3 = e_3$, and check that the solutions are the columns of $B^{(-1)}$.

+ Student Workspace

Problem 6: Inverse of a triangular matrix

A square matrix is "upper triangular" if all its entries below the main diagonal are zero. The matrix T below is an example.

(a) Explain why the product $T_1 T_2$ of any two n by n upper triangular matrices is also upper triangular. (For example, what entries of T_1 and T_2 are used to compute the (2, 1) entry of $T_1 T_2$?)

(b) By similar reasoning, the inverse of any upper triangular matrix is also upper triangular. (See T and $T^{(-1)}$ below for an example.) How are the diagonal entries of the inverse related to the diagonal entries of the matrix? Explain.

(c) Which triangular matrices are invertible? That is, state a necessary and sufficient condition for a triangular matrix to be invertible.

```
> T := Matrix([[2,-1,3],[0,-3,5],[0,0,1/2]]);
> T^(-1);
```

+ Student Workspace

Problem 7: Inverse of A^T

If a matrix A has an inverse, then so will its transpose. In fact, these inverses are closely related. You can discover this relationship quickly by looking at a couple of examples below.

(a) Once you have found the relationship, state it in words.

(b) Prove that the relationship you found in (a) is true. Hint: Apply the transpose operation to the identity $A A^{(-1)} = I$, and simplify by using the rule for the transpose of a product.

```
> A := Randmat(3,3);
> B := Transpose(A);
> A^(-1), B^(-1);
```

+ Student Workspace

Problem 8: Rows and columns

(a) For each of the matrices below, check that the rows are linearly independent if and only if the columns are.

(b) Explain why, for every square matrix A, the rows of A will be linearly independent if and only if the columns are. Hint: Consider the transpose of A.

```
> A := Matrix([[1,3],[4,-1]]);
> B := Matrix([[1,3],[-2,-6]]);
> C := Matrix([[0, 3, -4], [-5, -2, -1], [-9, 5, 8]]);
> E := Matrix([[-2,0,-2,-1],[2,1,1,2],[6,4,2,5],[-3,-5,2,0]]);
```

Student Workspace

Problem 9: Invertible or noninvertible at a glance

For each of the matrices below, determine whether or not the matrix is invertible by inspection (i.e., doing no calculation). For each matrix, base your determination on one of the four properties of the Invertibility Theorem, and explain how you know that the matrix does or does not have that property.

(a) $A = \begin{bmatrix} 1 & a & b \\ 0 & 4 & c \\ 0 & 0 & 6 \end{bmatrix}$ a, b, c real numbers (Do the values of a, b, c matter?)

(b) $A = \begin{bmatrix} 1 & 2 & 3 \\ 0 & k & 4 \\ 0 & 0 & 9 \end{bmatrix}$ k a real number (Does the value of k matter?)

(c) $A = \begin{bmatrix} 1 & 2 & 7 \\ 2 & 4 & 9 \\ 3 & 6 & 11 \end{bmatrix}$ (d) $A = \begin{bmatrix} 1 & -3 & 2 \\ 2 & 0 & 1 \\ 0 & 4 & -2 \end{bmatrix}$ (e) $A = \begin{bmatrix} 0 & 1 & 0 & 0 \\ 0 & 0 & 1 & 0 \\ 0 & 0 & 0 & 1 \\ 1 & 0 & 0 & 0 \end{bmatrix}$

Student Workspace

Problem 10: Invertibility and linear independence

For each of the following matrices, determine whether the matrix is invertible or not by determining whether its columns satisfy a linear dependence relation. If the columns do satisfy a linear dependence relation, find one. If you can make your determination by inspection (i.e., without any calculation), please do so.

```
> A := Matrix([[1,3],[4,-1]]);
> B := Matrix([[1,3],[-2,-6]]);
> C := Matrix([[0, 3, -4], [-5, -2, -1], [-9, 5, 8]]);
> E := Matrix([[-2,0,-2,-1],[2,1,1,2],[6,4,2,5],[-3,-5,2,0]]);
```

Student Workspace

Problem 11: More criteria for invertibility

If A is a square matrix, each of the statements (a) and (b) below is equivalent to the statement that A is

not invertible. Prove this equivalence by using part of the Invertibility Theorem or other theorems.
(a) There are two vectors $u \neq v$ such that $A u = A v$.
(b) There is a nonzero vector x such that $A^T x = 0$.

+ Student Workspace

Problem 12: If $A B = I$ then $B A = I$

Prove Theorem 8:

> **Theorem 8:** Suppose A and B are n by n matrices such that $A B = I$. Then A and B are both invertible, and $A^{(-1)} = B$.

Hint: First prove that B is invertible by using the equation $A B = I$ to show that B satisfies property (a) of the Invertibility Theorem.

+ Student Workspace

Problem 13: Rank and invertibility

Use Theorem 3 of Chapter 1 (stated below) to prove the following property:

- An n by n matrix A is invertible if and only if the rank of A is equal to n.

Hint: Use property (a) of the Invertibility Theorem.

> **Theorem 3:** Suppose we are given a consistent linear system with m equations and n unknowns. If a row echelon form of the augmented matrix of this system has exactly r nonzero rows (i.e., rows that are not made up entirely of zeros), then the solution set of the system has exactly $n - r$ free variables.

+ Student Workspace

Problem 14: Inverses of permutation matrices

Explain how to find the inverse of any permutation matrix and use this explanation to find (by hand) the inverse of the permutation matrix P below. (See Example 2B in Module 2.)

```
> P := Matrix([[0,0,1,0],[0,0,0,1],[0,1,0,0],[1,0,0,0]]);
```

Hint: A permutation matrix is obtained by permuting the columns of the identity matrix I. Furthermore, any product $A P$, where P is a permutation matrix, can be obtained by permuting the columns of A in the same way as P was obtained from I. Since the inverse Q of P must satisfy $Q P = I$, we want to choose Q so that P permutes its columns back to I.

+ Student Workspace

Linear Algebra Modules Project
Chapter 3, Module 6

Inverse of a Matrix and Elementary Matrices

Purpose of this module

The purpose of this module is to introduce elementary matrices and use them to further our understanding of inverses.

Prerequisites

Matrix algebra and matrix inverses, solutions of linear systems, and elementary row operations.

Commands used in this module

```
> restart: with(LinearAlgebra): with(Lamp): with(plots):
  UseHardwareFloats := false: Digits := 6:
```

Tutorial

Section 1: Elementary Matrices

Definition: An ***elementary matrix*** is a matrix that created by performing a single elementary row operation on an identity matrix.

====================

Example 1A: Below are three 2 by 2 elementary matrices and the elementary row operations that created them:

$\begin{bmatrix} \frac{1}{2} & 0 \\ 0 & 1 \end{bmatrix}$: Multiply row 1 of I by 1/2.

$\begin{bmatrix} 1 & 0 \\ -4 & 1 \end{bmatrix}$: Add -4 times row 1 of I to row 2 of I.

$\begin{bmatrix} 0 & 1 \\ 1 & 0 \end{bmatrix}$: Interchange row 1 and row 2 of I.

=====================

Exercise 1.1: (By hand) For each of the three elementary matrices A, B, C below, describe in words the elementary row operation on the identity matrix that created it:

$$A = \begin{bmatrix} 1 & 0 & 0 \\ 0 & 1 & 0 \\ 0 & 5 & 1 \end{bmatrix} \quad B = \begin{bmatrix} 0 & 0 & 1 & 0 \\ 0 & 1 & 0 & 0 \\ 1 & 0 & 0 & 0 \\ 0 & 0 & 0 & 1 \end{bmatrix} \quad C = \begin{bmatrix} 1 & 0 & 0 \\ 0 & -6 & 0 \\ 0 & 0 & 1 \end{bmatrix}$$

Student Workspace

Answer 1.1

A: Add 5 times row 2 of I to row 3 of I.
B: Interchange row 1 and row 3 of I.
C: Multiply row 2 of I by -6.

Inverses of Elementary Matrices

==================

Example 1B: We now calculate the inverses of the above elementary matrices A, B and C. Notice that in each case the inverse matrix is also an elementary matrix. In fact, the inverse corresponds to the "inverse" of the original row operation. By the inverse of a row operation we mean the row operation of the same type which undoes the effect of the original row operation, returning the elementary matrix to the identity matrix.

> `A := Matrix([[1,0,0],[0,1,0],[0,5,1]]);`

$$A := \begin{bmatrix} 1 & 0 & 0 \\ 0 & 1 & 0 \\ 0 & 5 & 1 \end{bmatrix}$$

> `Inverse(A);`

$$\begin{bmatrix} 1 & 0 & 0 \\ 0 & 1 & 0 \\ 0 & -5 & 1 \end{bmatrix}$$

A: add 5 times row 2 to row 3; $A^{(-1)}$: add -5 times row 2 to row 3.

> `B := Matrix([[0,0,1,0],[0,1,0,0],[1,0,0,0],[0,0,0,1]]);`

$$B := \begin{bmatrix} 0 & 0 & 1 & 0 \\ 0 & 1 & 0 & 0 \\ 1 & 0 & 0 & 0 \\ 0 & 0 & 0 & 1 \end{bmatrix}$$

> `Inverse(B);`

$$\begin{bmatrix} 0 & 0 & 1 & 0 \\ 0 & 1 & 0 & 0 \\ 1 & 0 & 0 & 0 \\ 0 & 0 & 0 & 1 \end{bmatrix}$$

B: interchange row 1 and row 3; $B^{(-1)}$: interchange row 1 and row 3. This is an example of a matrix that is its own inverse.

```
> C := Matrix([[1,0,0],[0,-6,0],[0,0,1]]);
```

$$C := \begin{bmatrix} 1 & 0 & 0 \\ 0 & -6 & 0 \\ 0 & 0 & 1 \end{bmatrix}$$

```
> Inverse(C);
```

$$\begin{bmatrix} 1 & 0 & 0 \\ 0 & \frac{-1}{6} & 0 \\ 0 & 0 & 1 \end{bmatrix}$$

C: multiply row 2 by -6; $C^{(-1)}$: multiply row 2 by -1/6.

========================

Exercise 1.2: (By hand) For each of the elementary matrices P and Q defined below, describe in words the elementary row operation that created it and describe the inverse of this elementary row operation. You can check your answers by having Maple calculate the inverse.

```
> P := Matrix([[1,0,-1/4,0],[0,1,0,0],[0,0,1,0],[0,0,0,1]]);
> Q := Matrix([[1,0,0,0,0],[0,0,0,1,0],[0,0,1,0,0],
  [0,1,0,0,0],[0,0,0,0,1]]);
```

$$P := \begin{bmatrix} 1 & 0 & -\frac{1}{4} & 0 \\ 0 & 1 & 0 & 0 \\ 0 & 0 & 1 & 0 \\ 0 & 0 & 0 & 1 \end{bmatrix}, \quad Q := \begin{bmatrix} 1 & 0 & 0 & 0 & 0 \\ 0 & 0 & 0 & 1 & 0 \\ 0 & 0 & 1 & 0 & 0 \\ 0 & 1 & 0 & 0 & 0 \\ 0 & 0 & 0 & 0 & 1 \end{bmatrix}$$

+ Student Workspace

− Answer 1.2

P: Add -1/4 times row 3 of I to row 1 of I. $P^{(-1)}$: Add 1/4 times row 3 of I to row 1 of I.

```
> Inverse(P);
```

$$\begin{bmatrix} 1 & 0 & \frac{1}{4} & 0 \\ 0 & 1 & 0 & 0 \\ 0 & 0 & 1 & 0 \\ 0 & 0 & 0 & 1 \end{bmatrix}$$

Q: Interchange row 2 and row 4 of I. $Q^{(-1)}$: Interchange row 2 and row 4 of I.

```
> Inverse(Q);
```

$$\begin{bmatrix} 1 & 0 & 0 & 0 & 0 \\ 0 & 0 & 0 & 1 & 0 \\ 0 & 0 & 1 & 0 & 0 \\ 0 & 1 & 0 & 0 & 0 \\ 0 & 0 & 0 & 0 & 1 \end{bmatrix}$$

Elementary Matrices Perform Row Operations

Example 1C: Let A be a general 2 by 2 matrix, and let E be the elementary matrix formed by adding 5 times row 1 of I to row 2 of I:

```
> A := Matrix([[a,b],[c,d]]);
```
$$A := \begin{bmatrix} a & b \\ c & d \end{bmatrix}$$

```
> E := Matrix([[1,0],[5,1]]);
```
$$E := \begin{bmatrix} 1 & 0 \\ 5 & 1 \end{bmatrix}$$

Now we multiply A on the left by E:
```
> E.A;
```
$$\begin{bmatrix} a & b \\ 5a+c & 5b+d \end{bmatrix}$$

Note that multiplying A on the left by E had the effect of performing the same row operation on A as the row operation that created E.

Exercise 1.3: Below is a general 3 by 2 matrix B and two elementary matrices F and G.
(a) Without doing any calculation, predict what the product FB will equal.
(b) Similarly, predict what the product GFB will equal.

```
> B := Matrix([[a,b],[c,d],[e,f]]);
  F := Matrix([[1,0,0],[0,1,0],[0,0,-2]]);
  G := Matrix([[0,0,1],[0,1,0],[1,0,0]]);
```
$$B := \begin{bmatrix} a & b \\ c & d \\ e & f \end{bmatrix}, \quad F := \begin{bmatrix} 1 & 0 & 0 \\ 0 & 1 & 0 \\ 0 & 0 & -2 \end{bmatrix}, \quad G := \begin{bmatrix} 0 & 0 & 1 \\ 0 & 1 & 0 \\ 1 & 0 & 0 \end{bmatrix}$$

 Student Workspace

Answer 1.3

(a) FB is the result of multiplying the third row of B by -2:
```
> F.B;
```
$$\begin{bmatrix} a & b \\ c & d \\ -2e & -2f \end{bmatrix}$$

(b) GFB is the result of interchanging the first and third rows of FB:
```
> G.F.B;
```

$$\begin{bmatrix} -2e & -2f \\ c & d \\ a & b \end{bmatrix}$$

The following statement describes this key property of elementary matrices:

Theorem 9: Suppose A and B are matrices where B is obtained from A by applying a single elementary row operation to A. Then $B = EA$, where E is the elementary matrix obtained by applying the same elementary row operation to the identity matrix.

In other words, we can carry out any elementary row operation simply by multiplying by the corresponding elementary matrix.

Section 2: Row Reduction by Elementary Matrices

We can now prove Theorem 7 of Module 5 (see below), with the help of Theorem 9 and the fact that elementary matrices are invertible.

Theorem 7: A square matrix A is invertible if and only if it is row equivalent to the identity matrix.

Specifically, to prove this theorem, we reduce a square matrix A to reduced row echelon form, and we represent each row operation by an elementary matrix. If A reduces to the identity matrix, the elementary matrices give us the inverse of A. The following example illustrates the process.

===================

Example 2A: We will reduce the matrix A defined below to reduced row echelon form, and we will carry out the reduction by multiplying A on the left by the elementary matrices corresponding to the row operations we wish to perform on A.

```
> A := Matrix([[2,4],[1,0]]);
```
$$A := \begin{bmatrix} 2 & 4 \\ 1 & 0 \end{bmatrix}$$

Step 1: Interchange row 1 and row 2. We multiply by the corresponding elementary matrix E_1:

```
> E1 := Matrix([[0,1],[1,0]]);
```
$$E1 := \begin{bmatrix} 0 & 1 \\ 1 & 0 \end{bmatrix}$$

```
> A1 := E1.A;
```
$$A1 := \begin{bmatrix} 1 & 0 \\ 2 & 4 \end{bmatrix}$$

Step 2: Add -2 times row 1 to row 2. We multiply by the corresponding elementary matrix E_2:

```
> E2 := Matrix([[1,0],[-2,1]]);
```
$$E2 := \begin{bmatrix} 1 & 0 \\ -2 & 1 \end{bmatrix}$$

```
> A2 := E2.A1;
```
$$A2 := \begin{bmatrix} 1 & 0 \\ 0 & 4 \end{bmatrix}$$

Step 3: Multiply row 2 by 1/4. We multiply by the corresponding elementary matrix E_3:
```
> E3 := Matrix([[1,0],[0,1/4]]);
```
$$E3 := \begin{bmatrix} 1 & 0 \\ 0 & \frac{1}{4} \end{bmatrix}$$

```
> A3 := E3.A2;
```
$$A3 := \begin{bmatrix} 1 & 0 \\ 0 & 1 \end{bmatrix}$$

We have found the reduced row echelon form of the matrix A by multiplying by three elementary matrices, and we have discovered that the reduced row echelon form is the identity matrix. In fact, we have done much more. We have found the inverse of matrix A! Here's why: Let B denote the product $E_3\, E_2\, E_1$. Then the computation above shows that $B\,A = I$. We must also show that $A\,B = I$. But we know that B is an invertible matrix, since B is the product of invertible matrices. Therefore we can multiply the equation $B\,A = I$ on the left by $B^{(-1)}$ and simultaneously on the right by B and then simplify:

$$B\,A = I$$
$$B^{(-1)}\,B\,A\,B = B^{(-1)}\,I\,B$$
$$A\,B = I$$

The following Maple calculations confirm that B is indeed the inverse of A:
```
> B := E3.E2.E1;
```
$$B := \begin{bmatrix} 0 & 1 \\ \frac{1}{4} & \frac{-1}{2} \end{bmatrix}$$

```
> B.A;
```
$$\begin{bmatrix} 1 & 0 \\ 0 & 1 \end{bmatrix}$$

```
> A.B;
```
$$\begin{bmatrix} 1 & 0 \\ 0 & 1 \end{bmatrix}$$

Summary: If we can reduce A to the identity matrix via a sequence of elementary row operations, then A has an inverse. The inverse is simply the product of the elementary matrices that correspond to the elementary row operations we perform on A.

Chapter 3 Matrix Algebra

Note: We have illustrated the process by which we can show that if A is row equivalent to the identity matrix, then A is invertible. The converse is also true: If A is invertible, then A is row equivalent to the identity matrix (see Problem 6).

Exercise 2.1: If we reduce A to the identity matrix using a different sequence of row operations, would we get a different inverse matrix? No, we proved in Module 5 that a matrix has only one inverse. Check this assertion by reducing A to the identity matrix again, but this time start with the elementary matrices F_1 and F_2 below. Then use this new collection of elementary matrices to compute $A^{(-1)}$.

```
> A := Matrix([[2,4],[1,0]]);
```
$$A := \begin{bmatrix} 2 & 4 \\ 1 & 0 \end{bmatrix}$$

```
> F1 := Matrix([[1/2,0],[0,1]]);
```
$$F1 := \begin{bmatrix} \frac{1}{2} & 0 \\ 0 & 1 \end{bmatrix}$$

```
> A1 := F1.A;
```
$$A1 := \begin{bmatrix} 1 & 2 \\ 1 & 0 \end{bmatrix}$$

```
> F2 := Matrix([[1,0],[-1,1]]);
```
$$F2 := \begin{bmatrix} 1 & 0 \\ -1 & 1 \end{bmatrix}$$

```
> A2 := F2.A1;
```
$$A2 := \begin{bmatrix} 1 & 2 \\ 0 & -2 \end{bmatrix}$$

Student Workspace

Answer 2.1

```
> F3 := Matrix([[1,0],[0,-1/2]]);
```
$$F3 := \begin{bmatrix} 1 & 0 \\ 0 & \frac{-1}{2} \end{bmatrix}$$

```
> A3 := F3.A2;
```
$$A3 := \begin{bmatrix} 1 & 2 \\ 0 & 1 \end{bmatrix}$$

```
> F4 := Matrix([[1,-2],[0,1]]);
```
$$F4 := \begin{bmatrix} 1 & -2 \\ 0 & 1 \end{bmatrix}$$

```
> A4 := F4.A3;
```

$$A4 := \begin{bmatrix} 1 & 0 \\ 0 & 1 \end{bmatrix}$$

So the inverse of A should be the product $F_4 F_3 F_2 F_1$. Let's check it:

```
> C := F4.F3.F2.F1;
```

$$C := \begin{bmatrix} 0 & 1 \\ 1 & -1 \\ 4 & 2 \end{bmatrix}$$

```
> C.A, A.C;
```

$$\begin{bmatrix} 1 & 0 \\ 0 & 1 \end{bmatrix}, \begin{bmatrix} 1 & 0 \\ 0 & 1 \end{bmatrix}$$

Exercise 2.2: Use Theorem 7 to determine whether the matrices A and B below are invertible.

```
> A := Matrix([[2,-1,-4,-1],[-3,0,5,1],[0,2,0,2],[1,1,-1,0]]);
> B := Matrix([[2,-1,-4,-1],[-3,0,5,3],[0,2,0,2],[1,1,-1,0]]);
```

$$A := \begin{bmatrix} 2 & -1 & -4 & -1 \\ -3 & 0 & 5 & 1 \\ 0 & 2 & 0 & 2 \\ 1 & 1 & -1 & 0 \end{bmatrix}, \quad B := \begin{bmatrix} 2 & -1 & -4 & -1 \\ -3 & 0 & 5 & 3 \\ 0 & 2 & 0 & 2 \\ 1 & 1 & -1 & 0 \end{bmatrix}$$

 Student Workspace

Answer 2.2

```
> Reduce(A,form=rref);
```

$$\begin{bmatrix} 1 & 0 & 0 & -2 \\ 0 & 1 & 0 & 1 \\ 0 & 0 & 1 & -1 \\ 0 & 0 & 0 & 0 \end{bmatrix}$$

```
> Reduce(B,form=rref);
```

$$\begin{bmatrix} 1 & 0 & 0 & 0 \\ 0 & 1 & 0 & 0 \\ 0 & 0 & 1 & 0 \\ 0 & 0 & 0 & 1 \end{bmatrix}$$

Since the reduced row echelon form of A is not I, A is not invertible; by the same reasoning, B is invertible. Let's check:

```
> Inverse(A);
Error, (in Inverse) singular matrix
```

```
> Inverse(B);
```

$$\begin{bmatrix} \frac{1}{2} & 1 & \frac{-5}{4} & 3 \\ \frac{-1}{2} & -1 & \frac{1}{2} & \frac{-1}{2} \\ 0 & \frac{1}{2} & \frac{-3}{4} & \frac{3}{2} \\ \frac{1}{2} & \frac{1}{2} & 0 & \frac{1}{2} \end{bmatrix}$$

Problems

Problem 1: Inverses of elementary matrices

Define the matrix A to be the product $A = FE$ of the elementary matrices E and F below.

(a) Describe in words the elementary row operations that create E, $E^{(-1)}$, F, and $F^{(-1)}$.

(b) Find $E^{(-1)}$ and $F^{(-1)}$ by using the row operations you described in (a).

(c) Compute $A^{(-1)}$ from your answers in (b).

```
> E := Matrix([[1,0,0],[0,1,0],[-2,0,1]]);
  F := Matrix([[1,0,0],[0,0,1],[0,1,0]]);
```

Student Workspace

Problem 2: Invertible matrices and products of elementary matrices

(a) Reduce the matrix A below to the identity matrix, and find elementary matrices E_1, E_2, E_3, E_4 such that $E_4 E_3 E_2 E_1 A = I$.

(b) (By hand) Express $A^{(-1)}$ as a product of elementary matrices. Use Maple to check your answer.

(c) (By hand) Express A as a product of elementary matrices. Hint: Solve for A in the equation $E_4 E_3 E_2 E_1 A = I$. Use Maple to check your answer.

(d) Based on your work above, describe a general method for writing every invertible matrix as a product of elementary matrices.

```
> A := Matrix([[-2,4],[6,-8]]);
```

Student Workspace

Problem 3: Reduced row echelon form and invertible matrices

For each of the matrices A and B below,

(a) determine whether the matrix is invertible or not by computing its reduced row echelon form;

(b) state all the conclusions you can about the matrix from Theorem 6 (see Module 5).

```
> A := Matrix([[-5,4,3],[-1,0,3],[0,2,4]]);
> B := Matrix([[2,0,0,-2],[1,4,0,-3],[1,1,-1,-1],[1,5,-1,-3]]);
```

Problem 4: Elementary matrices perform row operations

Recall that the vector-matrix product wB can be written as a linear combination of the rows of B, where the weights are the components of w. Also, each row of the matrix-matrix product AB is the product of the corresponding row of A multiplied by all of B. Thus, combining these two ideas, we see that each row of AB is a linear combination of the rows of B where the weights are the entries of the corresponding row of A. For example, if A and B are 2 by 2 matrices,

$$A = \begin{bmatrix} r & s \\ t & u \end{bmatrix} \quad B = \begin{bmatrix} a & b \\ c & d \end{bmatrix}$$

then

$$AB = \begin{bmatrix} r[a,b] + s[c,d] \\ t[a,b] + u[c,d] \end{bmatrix}$$

(a) (By hand) For the matrices E and C below, write the product EC in the above form (i.e., where each row of the product is a linear combination of the rows of C).

(b) Suppose A is an n by m matrix and E is an elementary n by n matrix produced by a row operation of the type $r_j \Leftarrow r_j + c\, r_k$. Explain why the product EA is exactly the matrix we would get by applying the same row operation to A as the row operation that created E. Hint: Your explanation should be based on the results you observed in part (a).

$$E = \begin{bmatrix} 1 & 0 & 0 \\ 0 & 1 & 0 \\ -4 & 0 & 1 \end{bmatrix} \quad C = \begin{bmatrix} a & b \\ c & d \\ e & f \end{bmatrix}$$

Problem 5: If A is row equivalent to I, then A is invertible (an essay)

Explain clearly and carefully how we justify the following central fact developed in this module: If a matrix A is row equivalent to the identity matrix, then the matrix A is invertible. Your explanation should be an extrapolation of Example 2A. That is, your explanation should apply to <u>every</u> square matrix that is row equivalent to the identity matrix, not just the specific matrix in Example 2A.

Problem 6: Completing the proof of Theorem 7

Theorem 7 also includes the converse of the statement you discussed in Problem 5: If A is invertible, then A is row equivalent to the identity matrix. Prove this direction of Theorem 7. Hint: If the reduced row echelon form of A is not the identity matrix, explain why A is not invertible.

Linear Algebra Modules Project
Chapter 3, Module 7

Determinants

Purpose of this module

The purpose of this module is to present the most important algebraic properties of determinants.

Prerequisites

Evaluation of determinants via minors and cofactors. This module assumes assumes that you have practiced finding determinants by hand using cofactor expansions along rows and columns. (For example, first work through the Appendix "Pencil and Paper Tutorial: Determinants and Cofactors" in the accompanying book.) Also, matrix algebra, the inverse of a matrix, and elementary row operations.

Commands used in this module

```
> restart: with(plots): with(LinearAlgebra): with(Lamp):
  UseHardwareFloats := false: Digits := 6:
```

Tutorial

Section 1: Quick Review of Cofactor Expansions of Determinants

This module assumes that you are familiar with some basic facts about determinants. In this section, we quickly review these facts.

The determinants of 1 by 1 and 2 by 2 matrices are given by the formulas

$$\text{Det}([a]) = a \quad \text{and} \quad \text{Det}\left(\begin{bmatrix} a & b \\ c & d \end{bmatrix}\right) = ad - bc$$

The determinant of an n by n matrix when $n > 2$ is defined in terms of determinants of certain submatrices of size $n-1$ by $n-1$. Specifically, the **(i, j) minor** of an n by n matrix A, denoted $A_{i,j}$, is the $n-1$ by $n-1$ submatrix that remains when we delete row i and column j of A. Also:

> **Definition:** If A is a square matrix and $A_{i,j}$ is the (i,j) minor of A, then $C_{i,j}$, the **(i, j) cofactor** of A, is defined as:
>
> $$C_{i,j} = \text{Det}(A_{i,j}) \quad \text{if } i+j \text{ is an even number}$$
>
> $$C_{i,j} = (-1)\,\text{Det}(A_{i,j}) \quad \text{if } i+j \text{ is an odd number}$$

We can also write this in a single formula:

$$C_{i,j} = (-1)^{(i+j)} \text{Det}(A_{i,j})$$

The following theorem describes how to compute determinants by using cofactors.

Theorem 10: The determinant of an *n* by *n* matrix *A* can be calculated using the cofactor expansion across any row or down any column:

The cofactor expansion across row *i* is: $\quad \text{Det}(A) = \sum_{j=1}^{n} a_{i,j} C_{i,j}$

The cofactor expansion down column *j* is: $\quad \text{Det}(A) = \sum_{i=1}^{n} a_{i,j} C_{i,j}$

where $C_{i,j} = (-1)^{(i+j)} \text{Det}(A_{i,j})$ and $A_{i,j}$ denotes the (i, j) minor of *A*.

==================

Example 1A: Use Theorem 10 to calculate the determinant of the matrix *A* below by using the cofactor expansion down column 3.

$$A = \begin{bmatrix} 3 & 4 & 6 \\ 5 & 6 & 7 \\ -3 & 9 & 2 \end{bmatrix}$$

Solution:

$$\text{Det}\left(\begin{bmatrix} 3 & 4 & 6 \\ 5 & 6 & 7 \\ -3 & 9 & 2 \end{bmatrix}\right) = 6 \, \text{Det}\left(\begin{bmatrix} 5 & 6 \\ -3 & 9 \end{bmatrix}\right) - 7 \, \text{Det}\left(\begin{bmatrix} 3 & 4 \\ -3 & 9 \end{bmatrix}\right) + 2 \, \text{Det}\left(\begin{bmatrix} 3 & 4 \\ 5 & 6 \end{bmatrix}\right)$$

$$= 6\,(45 + 18) - 7\,(27 + 12) + 2\,(18 - 20) = 378 - 273 - 4 = 101$$

==================

Determinant of an Upper Triangular Matrix

The determinant of an upper triangular matrix is very easy to calculate. For example, the determinant of the 2 by 2 upper triangular matrix $U = \begin{bmatrix} 2 & 5 \\ 0 & 3 \end{bmatrix}$ is simply the product of its diagonal entries, $(2)(3) = 6$. In a similar way, the determinant of any upper triangular matrix is simply the product of its diagonal entries. We record this fact as Theorem 11:

Theorem 11: The determinant of an *n* by *n* upper triangular matrix is the product of its diagonal entries.

Exercise 1.1: (By hand) (a) Find the determinant of the matrix *T* below by using the cofactor expansion down column 1.

178   Chapter 3 Matrix Algebra

$$T = \begin{bmatrix} 4 & 5 & 6 \\ 0 & -2 & 7 \\ 0 & 0 & 3 \end{bmatrix}$$

(b) How would you use this technique to prove Theorem 11?

+ Student Workspace

− Answer 1.1

(a) $\text{Det}(T) = 4\,\text{Det}\left(\begin{bmatrix} -2 & 7 \\ 0 & 3 \end{bmatrix}\right) = (4)(-2)(3) = -24$.

(b) Expand along the first column repeatedly.

− Section 2: Algebraic Properties of Determinants

We can calculate the determinant of a matrix A by using the command **Det(A)**:

```
> A := Matrix([[3,4,6],[5,6,7],[-3,9,2]]);
```

$$A := \begin{bmatrix} 3 & 4 & 6 \\ 5 & 6 & 7 \\ -3 & 9 & 2 \end{bmatrix}$$

```
> Det(A);
```

$$101$$

In this section you will use the **Det** command to explore the algebraic properties of determinants. Suppose A and B are square matrices of the same size and k is any real number. If we know the values of Det(A) and Det(B), what can we say about the values of the following determinants?

$$\text{Det}(kA),\ \text{Det}(A+B),\ \text{Det}(AB),\ \text{Det}(A^T)$$

For example, let's experiment with values of Det(kA):

```
> A := Randmat(2,2);
  k := 10;
```

$$A := \begin{bmatrix} 8 & 4 \\ -5 & -5 \end{bmatrix}$$

$$k := 10$$

```
> Det(A);
```

$$-20$$

```
> Det(k*A);
```

$$-2000$$

After changing the values of A and k and repeating the above computation a few times, you will soon be convinced that

- If A is an n by n matrix and k is a scalar, then $\text{Det}(kA) = k^n\,\text{Det}(A)$.

Module 3.7 Determinants

In the three mini-explorations that follow, you will seek similar algebraic properties (if they exist!) for $\text{Det}(A + B)$, $\text{Det}(A\,B)$, and $\text{Det}(A^T)$. In each case you will generate one or two random matrices and then use Maple to calculate the appropriate determinants. Try executing the commands more than once to generate several random matrices. You may also want to look at matrices of size 3 by 3 or 4 by 4.

Exploration I ~~ Is it true that $\text{Det}(A + B) = \text{Det}(A) + \text{Det}(B)$?

```
> A := Randmat(2,2);
  B := Randmat(2,2);
```

$$A := \begin{bmatrix} 3 & -5 \\ 8 & 5 \end{bmatrix}, \quad B := \begin{bmatrix} -1 & 0 \\ 3 & -4 \end{bmatrix}$$

```
> Det(A);
  Det(B);
```

$$55$$
$$4$$

```
> Det(A)+Det(B);
```

$$59$$

```
> Det(A+B);
```

$$57$$

Answer to Exploration I

The conjecture is false.

Exploration II ~~ Is it true that $\text{Det}(A\,B) = \text{Det}(A)\,\text{Det}(B)$?

```
> A := Randmat(2,2);
  B := Randmat(2,2);
```

$$A := \begin{bmatrix} 8 & -9 \\ 7 & 5 \end{bmatrix}, \quad B := \begin{bmatrix} 0 & 4 \\ 4 & 2 \end{bmatrix}$$

```
> Det(A);
  Det(B);
```

$$103$$
$$-16$$

```
> Det(A)*Det(B);
```

$$-1648$$

```
> Det(A.B);
```

$$-1648$$

Answer to Exploration II

Yes! The conjecture is true. While this is a simple fact to state, proving it takes a bit of work. A proof of this important fact about determinants is developed in Problem 5.

Exploration III ~~ Is it true that $\text{Det}(A^T) = \text{Det}(A)$?

```
> A := Randmat(2,2);
```
$$A := \begin{bmatrix} 4 & 0 \\ -8 & -5 \end{bmatrix}$$

```
> Det(A);
```
$$-20$$

```
> Det(Transpose(A));
```
$$-20$$

 Answer to Exploration III

Yes, the conjecture is true.

For future reference we state the two algebraic properties that you discovered in the explorations:

Theorem 12: If A and B are n by n matrices, then $\text{Det}(AB) = \text{Det}(A)\,\text{Det}(B)$.

Theorem 13: If A is an n by n matrix, then $\text{Det}(A^T) = \text{Det}(A)$.

Exercise 2.1: (By hand) Suppose that A and B are 4 by 4 matrices with $\text{Det}(A) = 5$ and $\text{Det}(B) = -4$. Calculate (a) $\text{Det}(AA^T)$ and (b) $\text{Det}(AB^2 A)$.

 Student Workspace

 Answer 2.1

(a) $\text{Det}(AA^T) = \text{Det}(A)\,\text{Det}(A^T) = \text{Det}(A)\,\text{Det}(A) = 25$.

(b) $\text{Det}(AB^2 A) = \text{Det}(A)\,\text{Det}(B)^2\,\text{Det}(A) = 5\,(-4)^2\,5 = 400$.

 Section 3: Determinants and Invertibility

One of the more important properties of a determinant of a square matrix A is that it tells us immediately whether or not A is invertible. To see how the two concepts are related, we will look at the 2 by 2 and 3 by 3 cases.

====================

Example 3A: To investigate the relationship between inverses and determinants, we have Maple calculate each for a general 2 by 2 matrix A.

```
> A := Matrix([[a,b],[c,d]]);
```
$$A := \begin{bmatrix} a & b \\ c & d \end{bmatrix}$$

```
> Inverse(A);
```
$$\begin{bmatrix} \dfrac{d}{ad-bc} & -\dfrac{b}{ad-bc} \\ -\dfrac{c}{ad-bc} & \dfrac{a}{ad-bc} \end{bmatrix}$$

```
> Det(A);
```
$$ad - bc$$

Observe that the determinant of A shows up as the denominator for each of the entries of the inverse. So if $\text{Det}(A) \neq 0$, the inverse of A exists. In fact, the inverse of A exists if and only if $\text{Det}(A) \neq 0$. Before we read too much into this single example, let's look at the 3 by 3 case.

```
> A := Matrix([[a,b,c],[d,e,f],[g,h,j]]);
```
$$A := \begin{bmatrix} a & b & c \\ d & e & f \\ g & h & j \end{bmatrix}$$

```
> Inverse(A);
```
$$\begin{bmatrix} \dfrac{ej-fh}{gbf-gce-dbj+dch+aej-afh} & -\dfrac{bj-ch}{gbf-gce-dbj+dch+aej-afh} & \dfrac{-bf+ce}{gbf-gce-dbj+dch+aej-afh} \\ -\dfrac{gf-dj}{gbf-gce-dbj+dch+aej-afh} & \dfrac{-gc+aj}{gbf-gce-dbj+dch+aej-afh} & -\dfrac{-dc+af}{gbf-gce-dbj+dch+aej-afh} \\ \dfrac{ge-dh}{gbf-gce-dbj+dch+aej-afh} & -\dfrac{-gb+ah}{gbf-gce-dbj+dch+aej-afh} & \dfrac{-db+ae}{gbf-gce-dbj+dch+aej-afh} \end{bmatrix}$$

```
> Det(A);
```
$$gbf - gce - dbj + dch + aej - afh$$

Both the inverse and the determinant are more complicated this time, but again the computation suggests that the inverse of A exists if and only if $\text{Det}(A) \neq 0$.

========================

The relationship between the inverse and the determinant that we observed in the case of 2 by 2 and 3 by 3 matrices is true in general. We record this important fact as Theorem 14:

> **Theorem 14:** A square matrix A is invertible if and only if $\text{Det}(A) \neq 0$. Equivalently, A is singular if and only if $\text{Det}(A) = 0$.

How might we prove this theorem? While calculating inverses and determinants for the 2 by 2 and 3 by 3 cases was helpful in spotting the pattern, continuing in this fashion is not very illuminating. Instead we will pursue a completely different approach, one that does not require us to actually calculate determinants. In Section 4 you will see that the key to understanding why Theorem 14 is true is to examine how determinants are affected by elementary row operations.

 Section 4: Row Operations and Determinants

Applying a row operation to a matrix has an easily described effect on the determinant. This relationship between row operations and determinants will let us develop some deeper results about determinants.

Explore ~~ The three basic row operations are applied to the matrix A below. In each case we calculate the determinant of the resulting matrix. Try changing these row operations. Can you predict how each type of row operation changes the value of the determinant?

```
> A := Matrix([[1,3,5],[4,2,3],[1,3,2]]);
```

$$A := \begin{bmatrix} 1 & 3 & 5 \\ 4 & 2 & 3 \\ 1 & 3 & 2 \end{bmatrix}$$

```
> Det(A);
```
$$30$$

Case 1: Interchange two rows of A. Does it matter which two rows we interchange?

```
> M := Rowop(A,r1 <> r2);
```

$$M := \begin{bmatrix} 4 & 2 & 3 \\ 1 & 3 & 5 \\ 1 & 3 & 2 \end{bmatrix}$$

```
> Det(M);
```
$$-30$$

Case 2: Multiply a row by a scalar. Try different scalars.

```
> N := Rowop(A,r1 <= 10*r1);
```

$$N := \begin{bmatrix} 10 & 30 & 50 \\ 4 & 2 & 3 \\ 1 & 3 & 2 \end{bmatrix}$$

```
> Det(N);
```
$$300$$

Case 3: Now for a surprise, add a multiple of one row to another row.

```
> P := Rowop(A,r1 <= r1+10*r2);
```

$$P := \begin{bmatrix} 41 & 23 & 35 \\ 4 & 2 & 3 \\ 1 & 3 & 2 \end{bmatrix}$$

```
> Det(P);
```
$$30$$

Compare your conclusions from the above explorations to the following summary.

Theorem 15: If a matrix B is obtained from an n by n matrix A by:
(a) interchanging two rows of A, then $\text{Det}(B) = -\text{Det}(A)$;
(b) multiplying one row of A by a scalar k, then $\text{Det}(B) = k\,\text{Det}(A)$;
(c) adding a multiple of one row of A to another row of A, then $\text{Det}(B) = \text{Det}(A)$.

==================

Example 4A: Prove part (b) of Theorem 15 by carrying out a cofactor expansion along the row that is changed.

Solution: Suppose the matrix B is obtained by multiplying row i of matrix A by k. Then A and B are identical except for the entries of row i, and therefore the cofactors $C_{i,j}$ associated with row i are identical for A and B. The cofactor expansion of B along row i is then:

$$\text{Det}(B) = \sum_{j=1}^{n} b_{i,j} C_{i,j} = \sum_{j=1}^{n} k\, a_{i,j} C_{i,j} = k\left(\sum_{j=1}^{n} a_{i,j} C_{i,j}\right) = k\,\text{Det}(A)$$

Note: Proofs for parts (a) and (c) of Theorem 15 are addressed in Problem 8.

Exercise 4.1: (By hand) Use Theorem 15 to prove the following for an n by n matrix A:
(a) If two rows of A are identical, then $\text{Det}(A) = 0$.
(b) If one row of A is a multiple of another row of A, then $\text{Det}(A) = 0$.
(c) If k is any real number, then $\text{Det}(kA) = k^n \text{Det}(A)$.

 Student Workspace

Answer 4.1

(a) If we interchange two identical rows then we still have the matrix A. But by Theorem 15(a) we must have $\text{Det}(A) = -\text{Det}(A)$. This implies $\text{Det}(A) = 0$.
(b) Suppose row j of A is k times row i. Factor out the constant k from row j of A, and call the new matrix B. Since rows i and j of B are identical, it follows from part(a) that $\text{Det}(B) = 0$. But by Theorem 15(b), we have $\text{Det}(A) = k\,\text{Det}(B)$. Therefore $\text{Det}(A) = 0$.
(c) The matrix kA is obtained by multiplying each of the n rows of A by k. By Theorem 15(b), each such row multiplication multiplies the determinant by k. So if we multiply each of the n rows by k we have $\text{Det}(kA) = k^n \text{Det}(A)$.

Calculating Determinants by Reducing to Echelon Form

Recall that we can use row operations to reduce a matrix A to an echelon form U. In fact, we can do so using only row interchanges and adding multiples of one row to another (that is, without scaling and hence requiring only parts (a) and (c) of Theorem 15). Therefore, if we carry out such a row reduction and use an even number of row interchanges, then by Theorem 15 $\text{Det}(A) = \text{Det}(U)$; if we use an odd number of row interchanges, then by Theorem 15 $\text{Det}(A) = -\text{Det}(U)$. But $\text{Det}(U)$ is easy to calculate since U is upper triangular; by Theorem 11, $\text{Det}(U)$ is the product of its diagonal entries. In summary:

- Suppose A is a square matrix and U is an echelon form of A obtained by row reduction without scaling, and suppose p is the product of the diagonal entries of U. Then $\text{Det}(A) = p$ if an <u>even</u> number of row interchanges are used in reducing A to U, and $\text{Det}(A) = -p$ if an <u>odd</u> number of row interchanges are used in reducing A to U.

Example 4B: Use the above method of Gaussian elimination and Theorem 15 to find $\text{Det}(A)$.
```
> A := Matrix([[0,1,2],[1,-1,3],[2,0,1]]);
```

$$A := \begin{bmatrix} 0 & 1 & 2 \\ 1 & -1 & 3 \\ 2 & 0 & 1 \end{bmatrix}$$

Solution: We begin by interchanging the first and second rows:

```
> A1 := Rowop(A,r1 <> r2);
```

$$A1 := \begin{bmatrix} 1 & -1 & 3 \\ 0 & 1 & 2 \\ 2 & 0 & 1 \end{bmatrix}$$

Next we add -2 times row 1 to row 3:

```
> A2 := Rowop(A1,r3 <= r3-2*r1);
```

$$A2 := \begin{bmatrix} 1 & -1 & 3 \\ 0 & 1 & 2 \\ 0 & 2 & -5 \end{bmatrix}$$

Adding -2 times row 2 to row 3 produces an echelon form:

```
> A3 := Rowop(A2,r3 <= r3-2*r2);
```

$$A3 := \begin{bmatrix} 1 & -1 & 3 \\ 0 & 1 & 2 \\ 0 & 0 & -9 \end{bmatrix}$$

Since we used one row interchange, we have $\text{Det}(A) = -\text{Det}(A_3)$. Since, by inspection, $\text{Det}(A_3) = -9$, we conclude that $\text{Det}(A) = 9$. Let's check this with Maple:

```
> Det(A);
```

$$9$$

============================

Exercise 4.2: (By hand) Two matrices M and N are shown below.

$$M = \begin{bmatrix} 12 & 24 & 38 \\ -15 & -26 & -52 \\ 3 & 6 & 9 \end{bmatrix} \quad N = \begin{bmatrix} 3 & 6 & 9 \\ 0 & 4 & -7 \\ 0 & 0 & 2 \end{bmatrix}$$

N is an echelon form for the matrix M. In fact, you can get from M to N by applying the following sequence of three row operations:

- (1) interchange row 1 and row 3;
- (2) add 5 times row 1 to row 2;
- (3) add -4 times row 1 to row 3.

Use this information to determine $\text{Det}(M)$. (You can do this computation in your head.)

 Student Workspace

Answer 4.2

The first row operation changes the sign of the determinant. The remaining two row operations have no effect. So Det(M) = −Det(N). Since Det(N) = 24, we conclude that Det(M) = −24. Checking with Maple:

```
> M := Matrix([[12,24,38],[-15,-26,-52],[3,6,9]]);
```

$$M := \begin{bmatrix} 12 & 24 & 38 \\ -15 & -26 & -52 \\ 3 & 6 & 9 \end{bmatrix}$$

```
> Det(M);
```

$$-24$$

Determinants and Invertibility - A Second Look

In Section 3 we stated the following important connection between determinants and invertibility. We are now in a position to prove this theorem.

Theorem 14: A square matrix A is invertible if and only if Det(A) ≠ 0. Equivalently, A is singular if and only if Det(A) = 0.

Proof: According to Theorem 15, the only change that a row operation can cause in the determinant of a matrix is to multiply it by a nonzero scalar. Therefore two matrices that are row equivalent either both have zero determinant or both have nonzero determinant. If A is invertible, then, by Theorem 7, A is row equivalent to I, the identity matrix. Since Det(I) = 1, Det(A) must also be nonzero. On the other hand, if A is not invertible, then its reduced row echelon form has at least one zero row and therefore has determinant = 0; hence Det(A) = 0.

 Section 5: Geometric Properties of Determinants (optional)

We begin by recalling a simple fact from elementary geometry. In the figure below, lines l and m are parallel. Let h be the constant distance between these lines. In the figure we also have line segments AC parallel to BD and AE parallel to BF. These lines create two parallelograms: ACDB and AEFB.

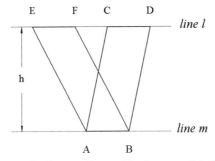

Recall that the area of a parallelogram is the product of its base and height. Since both parallelograms have base AB and height h, they have equal areas. So if we think of the side AB of the parallelogram as fixed and the opposite side as sliding along the line l, we see that all such parallelograms have equal area. We record this fact:

- Parallelograms with a common base and with their side opposite the base lying on the same line have equal area.

We now apply this fact to determinants.

===================

Example 5A: Let A be the 2 by 2 matrix defined below. We will consider each row of A as a vector in R^2; we denote the rows u and v.

```
> A := Matrix([[4,2],[2,6]]);
```

$$A := \begin{bmatrix} 4 & 2 \\ 2 & 6 \end{bmatrix}$$

```
> u := Row(A,1);
> v := Row(A,2);
```

$$u := [4, 2], \quad v := [2, 6]$$

```
> Det(A);
```

$$20$$

We form the unit span of u and v (i.e., the parallelogram with u and v as two of its sides):

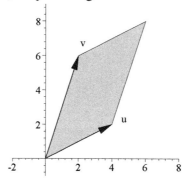

The area of this parallelogram is 20 square units, and it is also true that $\text{Det}(A) = 20$. Why are these numbers the same? This example will answer that question.

The key to our understanding will be a geometric visualization of the familiar row reduction process. We will reduce the given 2 by 2 matrix A to a diagonal matrix and use the above geometric fact to show that the area of the unit span of the rows is unchanged. For a diagonal matrix, the area of the unit span of the rows is easy to compute.

Suppose we add some multiple k of row 1 (vector u) to row 2 (vector v). Then the resulting matrix B has the form:

$$B = \begin{bmatrix} u \\ w \end{bmatrix} = \begin{bmatrix} u \\ v + k u \end{bmatrix}$$

In particular, note that the new second row (vector w) will lie somewhere on the line $v + t u$ (thick, red).

See the figure below.

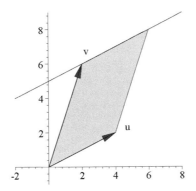

By our simple fact about areas of parallelograms, the area of the unit span of u and w equals the area of the unit span of u and v. Geometrically, we can say that our row operation slides one side of the unit span along a line parallel to the vector u, and therefore the area of the unit span is unchanged.
If in particular we take $k = -1/2$, then B is in echelon form.

```
> B := Rowop(A,r2 <= r2-(1/2)*r1);
```

$$B := \begin{bmatrix} 4 & 2 \\ 0 & 5 \end{bmatrix}$$

The figure below shows the unit span of u and this particular vector w. Note that this vector w lies at the point where the line $v + t\,u$ intersects the y axis since its first component is 0.

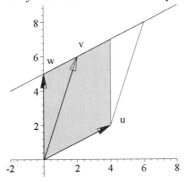

To further row reduce matrix B, we next add -2/5 times row 2 to row 1.

```
> C := Rowop(B,r1 <= r1-(2/5)*r2);
```

$$C := \begin{bmatrix} 4 & 0 \\ 0 & 5 \end{bmatrix}$$

Now we are adding a multiple of w (row 2) to u (row 1). So this time we slide one side of the parallelogram along a line parallel to vector w until it touches the x axis, as shown below. Again the area of the parallelogram does not change, and the parallelogram becomes a rectangle. So the area of the rectangle is equal to the area of our original parallelogram.

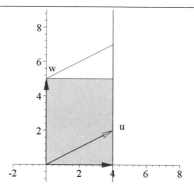

Finally, the area of the rectangle is (4)(5), which is also the determinant of C.

================================

Let's summarize what we have seen. Whenever we add a multiple of one row to another, the area of the unit span of the resulting rows is unchanged. Furthermore, if we reduce the original matrix A until we have a diagonal matrix C, we will have transformed the unit span into a rectangle, whose area is easy to calculate and which clearly equals the determinant of C. But we also know that adding a multiple of one row to another has no effect on the value of the determinant. So we also have $\text{Det}(A) = \text{Det}(C)$. Therefore $\text{Det}(A)$ equals the area of the unit span of its rows.

If we carry out this process on any 2 by 2 matrix, we might also need to use a row interchange. That would only introduce a sign change. So the general result is as follows:

- If A is a 2 by 2 matrix with row vectors u and v, then the area of the unit span of u and v equals the absolute value of $\text{Det}(A)$.

In the event that u and v are parallel, then $\text{Det}(A) = 0$ since A is singular. This is consistent with the unit span having zero area since geometrically it is a line segment.

The result above can be extended to 3 by 3 matrices as illustrated in the following example.
=====================

Example 5B: If A is a 3 by 3 matrix with row vectors u, v and w then $\left|\text{Det}(A)\right|$ (the absolute value of the determinant of A) is equal to the volume of the unit span of u, v and w (i.e., the parallelepiped with u, v and w as three of its sides). Here is an example:

```
> A := Matrix([[4,0,3],[-1,5,1],[-2,3,7]]);
```

$$A := \begin{bmatrix} 4 & 0 & 3 \\ -1 & 5 & 1 \\ -2 & 3 & 7 \end{bmatrix}$$

```
> u := Row(A,1);
  v := Row(A,2);
  w := Row(A,3);
```

$$u := [4, 0, 3], \quad v := [-1, 5, 1], \quad w := [-2, 3, 7]$$

```
> Det(A);
```
$$149$$

So the volume of the unit span below is 149 (see Problem 9):

===================

Section 6: Cramer's Rule (optional)

In this section we describe how determinants can be used to solve certain matrix equations. Before stating the main result, we make a preliminary definition.

Definition: If A is an n by n matrix and b is a vector in R^n, define Ai_b to be the n by n matrix created by replacing column i of A by the vector b.

For example, for the matrix A and vector b shown below, we display $A1\_b$, $A2\_b$ and $A3\_b$.

$$A := \begin{bmatrix} 3 & 4 & 5 \\ 6 & 7 & 8 \\ 4 & 2 & 8 \end{bmatrix} \qquad b := \begin{bmatrix} 4 \\ 4 \\ 4 \end{bmatrix}$$

$$A1\_b := \begin{bmatrix} 4 & 4 & 5 \\ 4 & 7 & 8 \\ 4 & 2 & 8 \end{bmatrix} \qquad A2\_b := \begin{bmatrix} 3 & 4 & 5 \\ 6 & 4 & 8 \\ 4 & 4 & 8 \end{bmatrix} \qquad A3\_b := \begin{bmatrix} 3 & 4 & 4 \\ 6 & 7 & 4 \\ 4 & 2 & 4 \end{bmatrix}$$

Recall that if A is an invertible matrix, the matrix-vector equation $A\,x = b$ has the unique solution $x = A^{(-1)}\,b$. One way to compute the solution x is by a formula known as "Cramer's Rule". In this formula, the components of the solution vector x are each described in terms of determinants. (While this formula is attractively simple to state, it is rarely used in practice, since determinants are so cumbersome to calculate.)

Theorem 16: (**Cramer's Rule**) If A is an n by n invertible matrix and b is a vector in R^n, then the components of the unique solution x of the matrix-vector equation $A\,x = b$ are given by:

$$x_i = \frac{\mathrm{Det}(Ai\_b)}{\mathrm{Det}(A)} \quad \text{for } i = 1 \,..\, n$$

Example 6A: Use Cramer's Rule to find the solution of the matrix-vector equation $A x = b$.

```
> A := Matrix([[3,4,5],[6,7,8],[4,2,8]]);
> b := Vector([4,4,4]);
```

$$A := \begin{bmatrix} 3 & 4 & 5 \\ 6 & 7 & 8 \\ 4 & 2 & 8 \end{bmatrix}, \quad b := \begin{bmatrix} 4 \\ 4 \\ 4 \end{bmatrix}$$

```
> A1_b := Matrix([[4, 4, 5], [4, 7, 8], [4, 2, 8]]);
> A2_b := Matrix([[3, 4, 5], [6, 4, 8], [4, 4, 8]]);
> A3_b := Matrix([[3, 4, 4], [6, 7, 4], [4, 2, 4]]);
```

$$A1\_b := \begin{bmatrix} 4 & 4 & 5 \\ 4 & 7 & 8 \\ 4 & 2 & 8 \end{bmatrix}, \quad A2\_b := \begin{bmatrix} 3 & 4 & 5 \\ 6 & 4 & 8 \\ 4 & 4 & 8 \end{bmatrix}, \quad A3\_b := \begin{bmatrix} 3 & 4 & 4 \\ 6 & 7 & 4 \\ 4 & 2 & 4 \end{bmatrix}$$

```
> x1 := Det(A1_b)/Det(A);
  x2 := Det(A2_b)/Det(A);
  x3 := Det(A3_b)/Det(A);
```

$$x1 := -\frac{5}{2}, \quad x2 := 1, \quad x3 := \frac{3}{2}$$

Let's compare our answer to what we get from **Matsolve**.

```
> Matsolve(A,b);
```

$$\begin{bmatrix} \frac{-5}{2} \\ 1 \\ \frac{3}{2} \end{bmatrix}$$

Maple note: To produce the matrices Ai_b using Maple, you can use the **Copyinto** command:

```
> A1_b := Copyinto(b,A,1,1);
  A2_b := Copyinto(b,A,1,2);
  A3_b := Copyinto(b,A,1,3);
```

$$A1\_b := \begin{bmatrix} 4 & 4 & 5 \\ 4 & 7 & 8 \\ 4 & 2 & 8 \end{bmatrix}, \quad A2\_b := \begin{bmatrix} 3 & 4 & 5 \\ 6 & 4 & 8 \\ 4 & 4 & 8 \end{bmatrix}, \quad A3\_b := \begin{bmatrix} 3 & 4 & 4 \\ 6 & 7 & 4 \\ 4 & 2 & 4 \end{bmatrix}$$

Exercise 6.1 Use Cramer's Rule to find the solution to $A x = b$ for the matrix A and vector b below.

```
> A := Matrix([[3,4,6,2],[5,3,6,4],[7,3,6,3],[1,0,4,5]]);
> b := Vector([4,5,9,1]);
```

$$A := \begin{bmatrix} 3 & 4 & 6 & 2 \\ 5 & 3 & 6 & 4 \\ 7 & 3 & 6 & 3 \\ 1 & 0 & 4 & 5 \end{bmatrix}, \quad b := \begin{bmatrix} 4 \\ 5 \\ 9 \\ 1 \end{bmatrix}$$

+ Student Workspace

− Answer 6.1

```
> A1_b := Copyinto(b,A,1,1);
  A2_b := Copyinto(b,A,1,2);
  A3_b := Copyinto(b,A,1,3);
  A4_b := Copyinto(b,A,1,4);
```

$$A1\_b := \begin{bmatrix} 4 & 4 & 6 & 2 \\ 5 & 3 & 6 & 4 \\ 9 & 3 & 6 & 3 \\ 1 & 0 & 4 & 5 \end{bmatrix}, \quad A2\_b := \begin{bmatrix} 3 & 4 & 6 & 2 \\ 5 & 5 & 6 & 4 \\ 7 & 9 & 6 & 3 \\ 1 & 1 & 4 & 5 \end{bmatrix}, \quad A3\_b := \begin{bmatrix} 3 & 4 & 4 & 2 \\ 5 & 3 & 5 & 4 \\ 7 & 3 & 9 & 3 \\ 1 & 0 & 1 & 5 \end{bmatrix}, \quad A4\_b := \begin{bmatrix} 3 & 4 & 6 & 4 \\ 5 & 3 & 6 & 5 \\ 7 & 3 & 6 & 9 \\ 1 & 0 & 4 & 1 \end{bmatrix}$$

```
> x1 := Det(A1_b)/Det(A);
  x2 := Det(A2_b)/Det(A);
  x3 := Det(A3_b)/Det(A);
  x4 := Det(A4_b)/Det(A);
```

$$x1 := \frac{33}{29}, \quad x2 := -\frac{63}{29}, \quad x3 := \frac{123}{58}, \quad x4 := -\frac{50}{29}$$

Check with **Matsolve**:

```
> Matsolve(A,b);
```

$$\begin{bmatrix} \dfrac{33}{29} \\ \dfrac{-63}{29} \\ \dfrac{123}{58} \\ \dfrac{-50}{29} \end{bmatrix}$$

Problems

Problem 1: Evaluating determinants by using row operations

Evaluate the determinants of the two matrices below by reducing the matrices to echelon form, as in Example 4B. You may use the **Rowop** command, but do not use scaling.

```
> A := Matrix([[-4,-4,3],[-4,-1,4],[1,-3,-1]]);
> B := Matrix([[0,-2,-2,2],[0,1,-1,-3],
    [-1,-1,1,2],[-1,-2,-3,-1]]);
```
[+] Student Workspace

Problem 2: Lower triangular matrices

(a) Evaluate the lower triangular matrix below by expanding by cofactors along the first row.
(b) State a general theorem about the determinant of an n by n lower triangular matrix.
(c) Using cofactor expansions, explain why this theorem is true.

```
> C := Matrix([[3,0,0,0],[2,1,0,0],[-1,-4,-4,0],[2,-3,2,-1]]);
```
[+] Student Workspace

Problem 3: Algebra and determinants

If A and B are n by n matrices and c is a scalar, use the algebraic properties of determinants in Section 2 to prove each of the following identities:

(a) $\text{Det}(A\,B) = \text{Det}(B\,A)$.
(b) $\text{Det}(A^{(-1)}\,B\,A) = \text{Det}(B)$, if A is invertible.
(c) $\text{Det}(A^{(-1)}\,B\,A - c\,I) = \text{Det}(B - c\,I)$ if A is invertible.
(d) $\text{Det}(B^T - c\,I) = \text{Det}(B - c\,I)$.

[+] Student Workspace

Problem 4: Invertibility of A^T

Use Theorems 13 and 14 (see below) to prove that A is invertible if and only if A^T is invertible.

Theorem 13: If A is an n by n matrix, then $\text{Det}(A^T) = \text{Det}(A)$.

Theorem 14: A square matrix A is invertible if and only if $\text{Det}(A) \neq 0$. Equivalently, A is singular if and only if $\text{Det}(A) = 0$.

[+] Student Workspace

Problem 5: Proof of $\text{Det}(A\,B) = \text{Det}(A)\,\text{Det}(B)$

In this problem we guide you through a proof of the fact that $\text{Det}(A\,B) = \text{Det}(A)\,\text{Det}(B)$. We begin the proof by considering the determinants of elementary matrices (see Module 6).

Determinants of Elementary Matrices

Recall that, for each of the three basic row operations, there is a corresponding elementary matrix E

obtained by applying the row operation to the identity matrix I. If we multiply a matrix A on the left by the elementary matrix E, then the product EA is the same as if the row operation that created E were applied to A (Theorem 9 in Module 6).

Theorem 15 shows us how to find the determinant of an elementary matrix. For example, if E is obtained by interchanging two rows of I, then by Theorem 15(a), $\text{Det}(E) = -\text{Det}(I) = -1$. Here is the complete list of determinants of elementary matrices:

- If E is an elementary matrix obtained by interchanging two rows of I, then $\text{Det}(E) = -1$.
- If E is an elementary matrix obtained by multiplying a row of I by a nonzero scalar k, then $\text{Det}(E) = k$.
- If E is an elementary matrix obtained by adding a multiple of one row of I to another, then $\text{Det}(E) = 1$.

From these facts and Theorem 15, we can prove the following Lemma.

Lemma: If A is an n by n matrix and E is an n by n elementary matrix, then

$$\text{Det}(EA) = \text{Det}(E)\,\text{Det}(A)$$

 Proof of Lemma

If E interchanges two rows, then by Theorem 15(a), $\text{Det}(E) = -1$ and $\text{Det}(EA) = -\text{Det}(A)$. So $\text{Det}(EA) = \text{Det}(E)\,\text{Det}(A)$.

If E multiplies a row by a nonzero scalar k, then by Theorem 15(b), $\text{Det}(E) = k$ and $\text{Det}(EA) = k\,\text{Det}(A)$. So $\text{Det}(EA) = \text{Det}(E)\,\text{Det}(A)$.

If E adds a multiple of one row to another row, then by Theorem 15(c), $\text{Det}(E) = 1$ and $\text{Det}(EA) = \text{Det}(A)$. So $\text{Det}(EA) = \text{Det}(E)\,\text{Det}(A)$.

Use the Lemma to prove that $\text{Det}(AB) = \text{Det}(A)\,\text{Det}(B)$. Hints: If A is invertible, we can write A as a product of finitely many elementary matrices (Problem 2 of Module 6). If A is not invertible, show that both sides of the equation $\text{Det}(AB) = \text{Det}(A)\,\text{Det}(B)$ are zero and hence equal.

 Student Workspace

 Problem 6: Analyzing symbolic matrices with determinants

Use determinants to answer the following questions.
(a) Check that the matrix A below is invertible. If we add a constant to each of its entries, it appears that the matrix remains invertible. For example, matrix B is formed by adding 1 to each entry of A and matrix C is formed by adding 2 to each entry. Check that these matrices are invertible. If you try other values of q at random, chances are that the resulting matrix will still be invertible. Is there any number that we could add to each entry of A to make the resulting matrix singular?

```
> A := Matrix([[4,3],[2,5]]);
> M := Matrix([[4+q,3+q],[2+q,5+q]]);
```

```
> B := subs(q=1,M);
> C := subs(q=2,M);
```

(b) Consider the matrix *N* shown below. For each value that we choose for *h* we get a new matrix *B*. If you try values at random for *h* you will find that the resulting matrix seems to always be invertible. Are there any values of *h* that will result in a singular matrix? Support your answer by either finding the exact value(s) of *h* or demonstrating algebraically that no such value of *h* exists.

```
> N := Matrix([[3+h,-2],[15,5+h]]);
> B := subs(h=1,N);
```

(c) For which values of *g* will the matrix *P* below be singular ?

```
> P := Matrix([[1-g,-2,4,2],[-2,1-g,4,2],
    [0,-2,5-g,2],[-2,-2,4,5-g]]);
```

 Student Workspace

 Problem 7: Vandermonde matrices

(a) Consider the symbolic matrix *V* below. Show that the columns of this matrix form a linearly independent set, provided that *a*, *b* and *c* are all different. Hint: Compute the determinant of *V* and apply the `factor` command.

```
> a := 'a': b := 'b': c := 'c':
  V := Matrix([[1, a, a^2], [1, b, b^2], [1, c, c^2]]);
```

(b) The matrix *V* is an example of a Vandermonde matrix. Use the command below to generate a 4 by 4 Vandermonde matrix. Form a conjecture about the columns of this matrix and prove it.

```
> W := Vandermat([a,b,c,d]);
```

 Student Workspace

 Problem 8: Proof of Theorem 15

[Note: Experience with proof by mathematical induction is required for this problem.]
Recall Theorem 15 from Section 4:

> **Theorem 15:** If a matrix *B* is obtained from an *n* by *n* matrix *A* by:
> (a) interchanging two rows of *A*, then $\text{Det}(B) = -\text{Det}(A)$;
> (b) multiplying one row of *A* by a scalar *k*, then $\text{Det}(B) = k\,\text{Det}(A)$;
> (c) adding a multiple of one row of *A* to another row of *A*, then $\text{Det}(B) = \text{Det}(A)$.

Recall that we proved part (b) in that section. Your task in this problem is to write proofs for parts (a) and (c) of the theorem. Specifically, you are to write proofs using mathematical induction. Some preliminary ideas relating to a proof of part (a) are provided below to help you get started.

Suppose we interchange two rows of a 3 by 3 matrix *A*. Let's call the new matrix *B*. If we carry out a cofactor expansion along a different row of *B* (that is, not one of the interchanged rows), then note that all the 2 by 2 minors associated with this row will also have two rows interchanged. The corresponding cofactors will each therefore have a sign change (since we know that Theorem 15(a) is true for 2 by 2 matrices). Hence Det(B) = –Det(A). Thus Theorem 15(a) is true for $n = 3$.

(a) Write a proof by mathematical induction for Theorem 15(a). You can use the 3 by 3 case above as a model to write a proof that if Theorem 15(a) is true for $n = k$ then it will also be true for $n = k + 1$.
(b) Write a similar proof for Theorem 15(c).

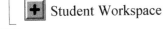 Student Workspace

 Problem 9: Geometry and reduction of 3 by 3 matrices

(a) Use the **Drawvec3d** command to draw the three row vectors of the matrix A below. Then partly reduce A so that the (2, 1) and (3, 1) entries are zero, and again plot the three row vectors. (Use no scaling here or in the later parts of this problem.) What plane do the second and third row vectors lie in? Confirm your answer visually.

(b) Reduce A further so that the (3, 2) entry is also zero, and again plot the three row vectors. Confirm visually that the second and third row vectors still lie in the plane you found in (a). What else is special about the third row vector? What is the area of the parallelogram defined by the unit span of the second and third row vectors?

(c) Reduce A further so that the reduced matrix is diagonal. What is special about the three row vectors now? What is the volume of the parallelepiped defined by the unit span of the three row vectors?

```
> A := Matrix([[-3,0,3],[4,-1,-3],[3,-2,-2]]);
```

 Student Workspace

Chapter 4: Matrix Transformations

Module 1. Geometry of Matrix Transformations of the Plane

Module 2. Geometry of Matrix Transformations of 3-Space

Module 3. Computer Graphics

 Commands used in this chapter

diff(f(x),x); calculates the derivative of f(x).
display([pict1, pict2]); displays together a group of previously defined pictures.
Matrix([[a,b],[c,d]]); defines the matrix with rows [a, b] and [c, d].
plot([exprx,expry,t=a..b]); plots a parametrized curve, where *exprx* and *expry* describe x and y, respectively, as functions of t.
seq(f,i=m..n); constructs the sequence of expressions f(m), ..., f(n), where f is a funtion of i.
subs({a=3,b=13},expr); substitutes the values for a and b in the expression *expr*.
Vector([a,b,c]); defines the vector $\langle a, b, c \rangle$.

LAMP commands:
Clock(A); an animation which simultaneously displays a rotating unit vector v and its image $A\,v$, where A is a 2 by 2 matrix.
Diagmat([a,b,c]); produces a diagonal matrix with diagonal entries a, b, c.
Drawmatrix(F); draws the 2d or 3d figure whose vertices are the columns of the matrix F.
Drawvec2d(u,[v,w]); draws the vector u with tail at the origin and the vector with tail at v and head at w. (Vectors are in 2-space.)
evalf(expr); evaluates the expression *expr* as a decimal approximation.
Inverse(M); produces the inverse of the matrix M.
Movie(M,F,frames=n); a 2d or 3d animation whose successive frames are plots of the figures $M^k F$, for k from 0 to n.
Projectmat(theta); produces the 2 by 2 matrix that projects vectors onto the line making the angle θ with the x axis.
Projectmat3d(n); produces the 3 by 3 matrix that projects vectors onto the plane with normal vector n.
Reflectmat(theta); produces the 2 by 2 matrix that reflects vectors across the line making the angle θ with the x axis.

Reflectmat3d(n); produces the 3 by 3 matrix that reflects vectors across the plane with normal vector *n*.
Rotatemat(theta); produces the 2 by 2 matrix that rotates vectors through the angle θ.
Rotatemat3d(v,theta); produces the 3 by 3 matrix that rotates vectors about the axis with direction vector *v* through the angle θ.
Transform(M,F); draws the 2d or 3d figure whose vertices are the columns of the matrix *F* and also draws the transformed figure *MF*.
Translatemat(v); produces the 3 by 3 matrix that translates vectors in 2-space by the vector *v*.
Translatemat3d(v); produces the 4 by 4 matrix that translates vectors in 3-space by the vector *v*.
Xshearmat(k); produces the 2 by 2 horizontal shear matrix with parameter *k*.
Yshearmat(k); produces the 2 by 2 vertical shear matrix with parameter *k*.

Linear Algebra Modules Project
Chapter 4, Module 1

Geometry of Matrix Transformations of the Plane

 Purpose of this module

The purpose of this module is to introduce the basic geometric matrix transformations of the plane and to study their effects on figures in the plane. From this study of the geometric behavior of matrix transformations, you will gain insights into topics studied earlier (such as matrix multiplication) and topics to be studied later (such as eigenvectors).

 Prerequisites

Matrix multiplication; inverse of a matrix.

 Commands used in this module

```
> restart: with(LinearAlgebra): with(plots): with(Lamp):
  UseHardwareFloats := false: Digits := 6:
```

Tutorial

 Section 1: Introduction to Matrix Functions

One way of understanding the meaning of multiplying a vector x by a matrix A to obtain a vector $y = A\,x$ is to think of the vector y as a function of the vector x:

$$y = f(x) = A\,x$$

We might describe this matrix function in words as "multiply by A"; that is, to find the "output" vector y from the "input" vector x, we multiply the input by A.

It is common practice to refer to these matrix functions as *matrix transformations*. Rather than use the letter f as the name of such a function, we will normally use the capital letter T (for transformation). Thus, we will write $y = T(x)$ rather than $y = f(x)$. We will also say that the transformation T "maps" an input vector x to the output vector $y = T(x)$ or simply "T maps x to y."

In this module you will investigate five types of geometric transformations: dilation, shear, rotation, reflection and projection. Each is defined by a particular type of 2 by 2 matrix.

Section 2: Dilation Transformations

The standard form for the dilation matrix is $\begin{bmatrix} r & 0 \\ 0 & r \end{bmatrix}$. We use the command **Diagmat([r,r])** to construct a dilation matrix with dilation factor r. For example:

```
> A := Diagmat([3/4,3/4]);
```

$$A := \begin{bmatrix} \frac{3}{4} & 0 \\ 0 & \frac{3}{4} \end{bmatrix}$$

We apply this dilation transformation to a vector by multiplying the vector on the left by A. For example, here is the calculation of $A\,u$ when $u = \begin{bmatrix} 1 \\ 2 \end{bmatrix}$:

```
> u := Vector([1,2]);
  Au := A.u;
```

$$u := \begin{bmatrix} 1 \\ 2 \end{bmatrix}$$

$$Au := \begin{bmatrix} \frac{3}{4} \\ \frac{3}{2} \end{bmatrix}$$

Alternatively, we can use function notation. Let T be the function that takes as input the vector x and returns as output the vector $A\,x$. Note that the domain for this function is the set of all vectors in R^2. Symbolically, we write $T(x) = A\,x$. We can define this function in Maple as shown on the next line.

```
> T := x -> A.x;
```

$$T := x \rightarrow A\,.\,x$$

Now we find $A\,u$ again, this time by applying the function T to u:

```
> T(u);
```

$$\begin{bmatrix} \frac{3}{4} \\ \frac{3}{2} \end{bmatrix}$$

So T maps $\begin{bmatrix} 1 \\ 2 \end{bmatrix}$ to $\begin{bmatrix} .75 \\ 1.5 \end{bmatrix}$.

The geometric effect of applying this transformation is very simple. T simply scales a vector by the factor r. When $r > 1$ the vector is stretched (dilated), and when $0 < r < 1$ the vector is contracted. What does T do when $r < 0$?

We can see the effect of the dilation transformation by looking at the relationship between its input and output vectors. Execute the next input region to see the effect of the transformation $T(x) = Ax$ on two vectors u and v.

Note: The input vectors u and v are colored red and their corresponding output vectors Au and Av are blue. Throughout, we will follow this color scheme: inputs red and outputs blue.

```
> u := Vector([1,2]);
  v := Vector([3,1]);
```

$$u := \begin{bmatrix} 1 \\ 2 \end{bmatrix}, \quad v := \begin{bmatrix} 3 \\ 1 \end{bmatrix}$$

```
> pict1 := Drawvec(u,v):
  pict2 := Drawvec(T(u),T(v),headcolor=blue):
  display([pict1,pict2],view=[-2..4,-2..4]);
```

Exercise 2.1: (By hand) Identify the vector v in the plot above. Calculate the coordinates of the vector $T(v)$ without using Maple. Check that your answer is consistent with what you see in the plot.

+ Student Workspace

 Answer 2.1

$$T(v) = \begin{bmatrix} 2.25 \\ .75 \end{bmatrix}$$

Applying the same transformation to a larger collection of input vectors gives us a picture that helps us visualize how this transformation contracts vectors towards the origin:

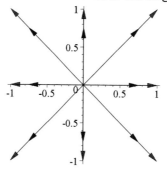

Transforming Figures

Often we want to think of a matrix transformation as acting on points rather than vectors. From this perspective, the transformation $T(x) = A\,x$ moves the point at the head of vector x to the point at the head of vector $A\,x$. Now we can imagine the transformation being applied to arbitrarily complicated figures, since any figure can be thought of as a collection of points. Furthermore, a matrix transformation T will map the line segment from point P to point Q to the line segment from $T(P)$ to $T(Q)$ (unless $T(P) = T(Q)$, in which case the line segment maps to a single point). (See Problem 3.) Therefore it will be convenient to describe figures by a sequence of points connected by straight line segments.

Here is an example involving a figure described by a predefined matrix called "House." Execute the next input region to see the entries in this matrix and the figure it describes (first with axes, then without):

```
> H := House;
```

$$H := \begin{bmatrix} 0 & 0 & 2 & 4 & 4 & 0 \\ 0 & 3 & 4 & 3 & 0 & 0 \end{bmatrix}$$

```
> Drawmatrix(H);
```

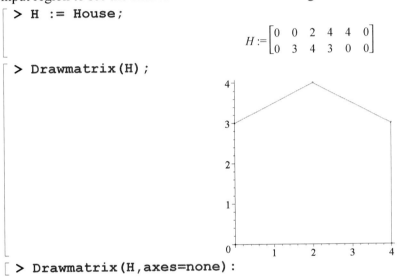

```
> Drawmatrix(H,axes=none):
```

Maple note: The command **Drawmatrix** works as follows. Suppose you have in mind a figure made up of n points connected by line segments. If you store these points as the columns of a 2 by n matrix M, then **Drawmatrix(M)** will draw your figure by connecting consecutive points with line segments. To connect the last point back to the first point, we add a final column equal to the first column. You can also change the color of the figure from the default color red to any other color by using **Drawmatrix(M,figcolor=colorname)**.

==================

Example 2A: In the input region below, we define a dilation matrix A with $r = 3$ and then compute the matrix product $A\,H$. Compare the result to the original House matrix H. Notice that each column of $A\,H$, the output matrix, is the result of multiplying the corresponding column of the input matrix H on the left by A.

```
> A := Diagmat([3,3]);
```
$$A := \begin{bmatrix} 3 & 0 \\ 0 & 3 \end{bmatrix}$$

```
> H;
```
$$\begin{bmatrix} 0 & 0 & 2 & 4 & 4 & 0 \\ 0 & 3 & 4 & 3 & 0 & 0 \end{bmatrix}$$

```
> AH := A.H;
```
$$AH := \begin{bmatrix} 0 & 0 & 6 & 12 & 12 & 0 \\ 0 & 9 & 12 & 9 & 0 & 0 \end{bmatrix}$$

Now to see the input (in red) and output (in blue) together, we use the display command:

```
> part1 := Drawmatrix(H):
> part2 := Drawmatrix(AH,figcolor=blue):
> display([part1,part2]):
```

==================

Maple note: We can accomplish all of this in one step by using the command **Transform(M,F)**. Here M is the matrix for the transformation and F is the matrix of points that defines the figure. The result is a plot of both the original figure (thin red lines) and the transformed figure (thick blue lines). Execute the next line to try it out. Both **Drawmatrix** and **Transform** will be used throughout the rest of the module.

```
> Transform(A,H);
```

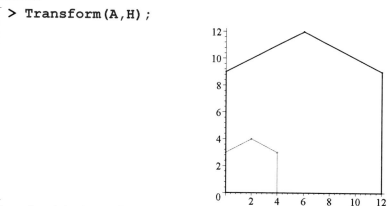

Exercise 2.2: (a) Define your own figure as follows: Pick three points from quadrants I and II, and form the 2 by 4 matrix that has these points as its columns (the final column being a repeat of the first). Call this matrix "triangle".

(b) Apply the **Drawmatrix** command to your figure to see what it looks like.

(c) Apply the dilation matrix A to your figure by using the **Transform** command. What are the coordinates of the vertices of the transformed triangle?

+ Student Workspace

Answer 2.2

We will use the three points (6, 3), (-2,9), and (-3,4). Answers to (a), (b), and (c):

```
> A := Diagmat([3,3]):
  triangle := Matrix([[6,-2,-3,6],[3,9,4,3]]);
```

$$triangle := \begin{bmatrix} 6 & -2 & -3 & 6 \\ 3 & 9 & 4 & 3 \end{bmatrix}$$

```
> Drawmatrix(triangle);
```

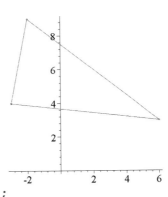

```
> Transform(A,triangle);
```

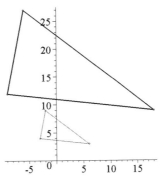

```
> A.triangle;
```

$$\begin{bmatrix} 18 & -6 & -9 & 18 \\ 9 & 27 & 12 & 9 \end{bmatrix}$$

Therefore the new vertices are (18,9), (-6,27), and (-9,-12).

Scaling Transformations

As you have observed, a dilation matrix expands (or contracts) vectors by an equal factor in the x and y directions. It is therefore a special case of a more general scaling transformation which scales in the x and y directions by different factors. This takes the familiar form of a diagonal matrix:

$$\begin{bmatrix} r & 0 \\ 0 & s \end{bmatrix}$$

Example 2B: The command `Diagmat([r,s])` constructs a scaling matrix for given values of *r* and *s*. Here we define a scaling matrix *A* and apply it to the House matrix:

```
> A := Diagmat([1/4,2]);
```

$$A := \begin{bmatrix} \dfrac{1}{4} & 0 \\ 0 & 2 \end{bmatrix}$$

```
> Transform(A,House);
```

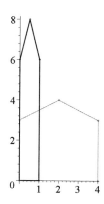

We have been concentrating on the effect of matrix transformations on various figures, but of course a transformation maps <u>every</u> point in the plane to an image point. To visualize this movement of all the points of the plane, think of the transformation as distorting the coordinate grid of the plane. We have predefined a "Grid" figure that represents a portion of the first quadrant:

```
> Drawmatrix(Grid);
```

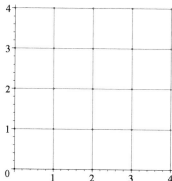

Execute the next command to see how the scaling matrix *A* transforms this part of the plane. Note that the Grid is compressed horizontally by a factor of 4 and stretched vertically by a factor of 2; these factors correspond to the diagonal entries of *A* (1/4 and 2).

```
> Transform(A,Grid);
```

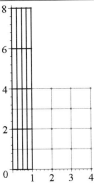

Try to predict how the Grid will be transformed by the following scaling matrix.

```
> A := Diagmat([3,-1/2]);
```

$$A := \begin{bmatrix} 3 & 0 \\ 0 & \frac{-1}{2} \end{bmatrix}$$

```
> Transform(A,Grid);
```

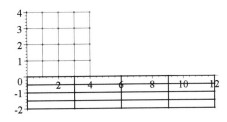

Now that we have established a quick way of visualizing the effect of transformations, we can move on to some more interesting transformations and figures.

Section 3: Shear Transformations

The standard matrices for the horizontal and vertical shear transformations are shown below.

$$\text{horizontal shear} = \begin{bmatrix} 1 & k \\ 0 & 1 \end{bmatrix} \qquad \text{vertical shear} = \begin{bmatrix} 1 & 0 \\ k & 1 \end{bmatrix}$$

The commands **Xshearmat(k)** and **Yshearmat(k)**, respectively, construct these matrices. Execute the next input region to see how the vertical shear matrix A transforms the Grid figure.

```
> A := Yshearmat(2);
```

$$A := \begin{bmatrix} 1 & 0 \\ 2 & 1 \end{bmatrix}$$

```
> Transform(A,Grid);
```

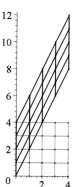

Observe the following about this picture. The points on the *y* axis are mapped to themselves. (We say these are "fixed points" of the mapping.) Each of the other vertical lines in the Grid is mapped to itself, although the individual points of the line are moved up. Each horizontal line is transformed to a line with slope 2, where 2 is the value of the parameter *k* in the shear matrix. In the next example, we verify one of these observations for all vertical shear transformations.

======================

Example 3A: Show that a vertical shear matrix with parameter *k* maps each horizontal line $y = b$ to the line with slope *k* and *y* intercept *b*.

Solution: We can represent a horizontal line $y = b$ parametrically by $u = \begin{bmatrix} t \\ b \end{bmatrix}$, where *t* is a parameter. Multiplying this vector by a general vertical shear matrix gives us a parametric description of the corresponding image vectors *A u*:

```
> u := Vector([t,b]);
```

$$u := \begin{bmatrix} t \\ b \end{bmatrix}$$

```
> A := Yshearmat(k);
```

$$A := \begin{bmatrix} 1 & 0 \\ k & 1 \end{bmatrix}$$

```
> Au := A.u;
```

$$Au := \begin{bmatrix} t \\ kt+b \end{bmatrix}$$

Note that *A u* describes a line with slope *k* and *y* intercept *b*.

======================

Similarly, a horizontal shear matrix with parameter *k* maps each vertical line $x = a$ to the line with slope $1/k$ and *x* intercept *a* (see Problem 5):

```
> Transform(Xshearmat(2),Grid);
```

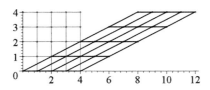

Section 4: Composite Transformations

If A and B are two matrix transformations of the plane, their product $A\,B$ is another matrix transformation. The product $A\,B$ is referred to as a "composite" transformation for the following reason. If T_1 and T_2 are the matrix transformations defined by $T_1(x) = A\,x$ and $T_2(x) = B\,x$, then the composite $T_1(T_2(x))$ equals $A\,B\,x$:

$$T_1(T_2(x)) = A\,(B\,x) = (A\,B)\,x \quad \text{and likewise} \quad T_2(T_1(x)) = B\,(A\,x) = (B\,A)\,x$$

====================

Example 4A: Let A be the horizontal shear matrix with $k=1$ and B be the vertical shear matrix with $k=1$. The **Drawvec** command below shows the composite $T_1(T_2(x)) = A\,(B\,x)$ for a typical vector x. Note that B, the vertical shear, is applied to x first; then A, the horizontal shear, is applied to the output vector $B\,x$.

```
> A := Xshearmat(1);
  B := Yshearmat(1);
> x := Vector([2,1]):
```

$$A := \begin{bmatrix} 1 & 1 \\ 0 & 1 \end{bmatrix}, \quad B := \begin{bmatrix} 1 & 0 \\ 1 & 1 \end{bmatrix}$$

```
> Drawvec([x,label="x"],[B.x,label="Bx",headcolor=blue],
    [A.B.x,label="ABx",headcolor=black],headlength=0.5);
```

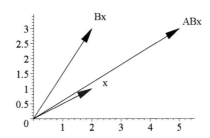

The next command draws this picture for the composite in the opposite order, $T_2(T_1(x)) = B\,(A\,x)$.

```
> Drawvec([x,label="x"],[A.x,label="Ax",headcolor=blue],
    [B.A.x,label="BAx",headcolor=black],headlength=0.5);
```

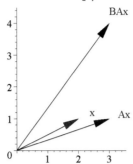

Note that the two composites have quite different effects on the vector x, which reflects the fact that $AB \neq BA$. So we can see geometrically why these two matrices do not commute: Shearing vertically and then horizontally is not the same as shearing horizontally and then vertically.

====================

Exercise 4.1: From what you know about the geometric behavior of shears, would you predict that horizontal shears commute? Let A and B be horizontal shears with $k = -2$ and $k = 3$, respectively:

```
> A := Xshearmat(-2);
  B := Xshearmat(3);
```

$$A := \begin{bmatrix} 1 & -2 \\ 0 & 1 \end{bmatrix}, \quad B := \begin{bmatrix} 1 & 3 \\ 0 & 1 \end{bmatrix}$$

(a) Apply the matrix transformations AB and BA to the Grid figure. Verify that they yield the same result.

(b) (By hand) Show that for <u>every</u> pair of shear matrices, A and B, we have $AB = BA$.

➕ Student Workspace

➖ Answer 4.1

(a)
```
> Transform(A.B,Grid);
```

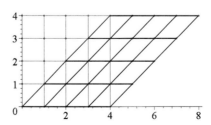

> Transform(B.A,Grid);

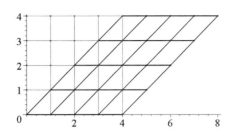

(b) $\begin{bmatrix} 1 & j \\ 0 & 1 \end{bmatrix}\begin{bmatrix} 1 & k \\ 0 & 1 \end{bmatrix} = \begin{bmatrix} 1 & j+k \\ 0 & 1 \end{bmatrix}$ and $\begin{bmatrix} 1 & k \\ 0 & 1 \end{bmatrix}\begin{bmatrix} 1 & j \\ 0 & 1 \end{bmatrix} = \begin{bmatrix} 1 & j+k \\ 0 & 1 \end{bmatrix}$.

So the product in either order is a horizontal shear with parameter $j + k$.

Section 5: Rotation Transformations

========================

Example 5A: Find the matrix transformation R that rotates every vector $\begin{bmatrix} a \\ b \end{bmatrix}$ counter-clockwise 90 degrees. For example, the matrix R transforms the vector $\begin{bmatrix} 3 \\ 2 \end{bmatrix}$ to the vector $\begin{bmatrix} -2 \\ 3 \end{bmatrix}$ as shown below:

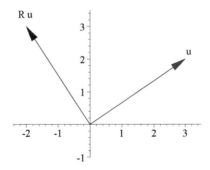

Solution: Just as in the example above, R transforms a general vector $\begin{bmatrix} a \\ b \end{bmatrix}$ to the vector $\begin{bmatrix} -b \\ a \end{bmatrix}$:

$$R\begin{bmatrix} a \\ b \end{bmatrix} = \begin{bmatrix} -b \\ a \end{bmatrix}$$

We can now find R by using the method of decomposition:

$$\begin{bmatrix} -b \\ a \end{bmatrix} = a \begin{bmatrix} 0 \\ 1 \end{bmatrix} + b \begin{bmatrix} -1 \\ 0 \end{bmatrix} = \begin{bmatrix} 0 & -1 \\ 1 & 0 \end{bmatrix} \begin{bmatrix} a \\ b \end{bmatrix}. \text{ Hence } R = \begin{bmatrix} 0 & -1 \\ 1 & 0 \end{bmatrix}.$$

General Rotation Formula

More generally, the matrix transformation that rotates every vector in the plane by θ radians in the counter-clockwise direction is:

$$\begin{bmatrix} \cos(\theta) & -\sin(\theta) \\ \sin(\theta) & \cos(\theta) \end{bmatrix}$$

The **Rotatemat(θ)** command constructs a rotation matrix for any angle θ. For example, we construct a rotation matrix A corresponding to a rotation angle of π/6 radians (30 degrees) and apply it to the Grid figure:

```
> A := Rotatemat(Pi/6);
```

$$A := \begin{bmatrix} \frac{1}{2}\sqrt{3} & \frac{-1}{2} \\ \frac{1}{2} & \frac{1}{2}\sqrt{3} \end{bmatrix}$$

```
> Transform(A,Grid);
```

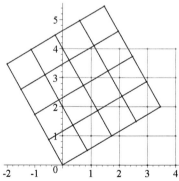

Next let B be the rotation matrix corresponding to an angle of π/3 radians (60 degrees):

```
> B := Rotatemat(Pi/3);
```

$$B := \begin{bmatrix} \frac{1}{2} & -\frac{1}{2}\sqrt{3} \\ \frac{1}{2}\sqrt{3} & \frac{1}{2} \end{bmatrix}$$

Exercise 5.1: (a) Check that the matrices A and B above are related by the equation $A^2 = B$, and give a geometric explanation for this relationship.

(b) Find a positive integer k such that $A^k = I$.

Student Workspace

 Answer 5.1

```
> A^2;
```

$$\begin{bmatrix} \frac{1}{2} & -\frac{1}{2}\sqrt{3} \\ \frac{1}{2}\sqrt{3} & \frac{1}{2} \end{bmatrix}$$

```
> B;
```

$$\begin{bmatrix} \frac{1}{2} & -\frac{1}{2}\sqrt{3} \\ \frac{1}{2}\sqrt{3} & \frac{1}{2} \end{bmatrix}$$

(a) If applying the matrix A once rotates the plane by 30 degrees, then applying it twice in succession will rotate the plane by 60 degrees; thus $A^2 = A\,A = B$.

(b) To rotate every vector 360 degrees, we apply A twelve times (12*30 = 360); therefore $A^{12} = I$.

```
> A^12;
```

$$\begin{bmatrix} 1 & 0 \\ 0 & 1 \end{bmatrix}$$

Next we construct a new rotation matrix U and apply it to a figure called "Bug":

```
> U := Rotatemat(Pi/4);
```

$$U := \begin{bmatrix} \frac{1}{2}\sqrt{2} & -\frac{1}{2}\sqrt{2} \\ \frac{1}{2}\sqrt{2} & \frac{1}{2}\sqrt{2} \end{bmatrix}$$

```
> Transform(U,Bug);
```

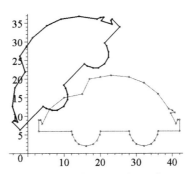

Exercise 5.2: Finding an inverse of the above rotation matrix U is equivalent to finding a second rotation transformation V that returns the Bug to its original position (since $V\,U = I$, and I maps all points to themselves).

(a) Find the inverse V of U two ways, one where V rotates vectors clockwise and another where V rotates vectors counterclockwise. Check that both ways of finding the inverse of U give the same result as the command **Inverse(U)**.

(b) Express the inverse of U as a positive power of U; i.e., find a positive integer k such that $V = U^k$.

+ Student Workspace

− Answer 5.2

(a) Since U rotates vectors by $\dfrac{\pi}{4}$ radians, we need an additional rotation of $\dfrac{7\pi}{4}$ radians to return a vector to its original position. Alternatively, we could rotate clockwise, in which case we use the angle $\theta = -\dfrac{\pi}{4}$.

```
> V := Rotatemat(7*Pi/4);
> V := Rotatemat(-Pi/4);
> V := Inverse(U);
```

$$V := \begin{bmatrix} \dfrac{1\sqrt{2}}{2} & \dfrac{1\sqrt{2}}{2} \\ -\dfrac{1\sqrt{2}}{2} & \dfrac{1\sqrt{2}}{2} \end{bmatrix},\ V := \begin{bmatrix} \dfrac{1\sqrt{2}}{2} & \dfrac{1\sqrt{2}}{2} \\ -\dfrac{1\sqrt{2}}{2} & \dfrac{1\sqrt{2}}{2} \end{bmatrix},\ V := \begin{bmatrix} \dfrac{1\sqrt{2}}{2} & \dfrac{1\sqrt{2}}{2} \\ -\dfrac{1\sqrt{2}}{2} & \dfrac{1\sqrt{2}}{2} \end{bmatrix}$$

(b) A rotation of $\dfrac{7\pi}{4}$ radians is equivalent to a sequence of seven rotations of $\dfrac{\pi}{4}$ radians. We can therefore express the inverse as U^7. We confirm that U^7 equals the inverse of U:

```
> U^7;
```

$$\begin{bmatrix} \dfrac{1}{2}\sqrt{2} & \dfrac{1}{2}\sqrt{2} \\ -\dfrac{1}{2}\sqrt{2} & \dfrac{1}{2}\sqrt{2} \end{bmatrix}$$

− Section 6: Using Transformations to Animate Figures

If we take a rotation matrix A with a small angle such as $\pi/18$ and apply it to the House figure H, we will get a slightly turned house $A\,H$. If we apply the matrix A again to the slightly turned house, we will turn it a bit more. Now we have $A\,A\,H$ or $A^2\,H$. If we keep doing this, we will produce a succession of still pictures that can be used to create an animation of a rotating house.

=====================

Example 6A: The **Movie** command displays just such a sequence of still pictures that constitute an animation. The syntax is **Movie(A,H,frames=n)**, where A is the matrix that you want to apply repeatedly, H is the matrix describing the figure, and n is the number of applications (i.e., iterations) of the matrix A. Keep in mind that the frames consist of H, $A\,H$, $A^2\,H$, ..., $A^n\,H$. Execute the next input region for an example. (This may take some time to compute.) Then click on the picture to bring up the animation buttons, and click on the "play" button.

```
> H := House;
  A := Rotatemat(Pi/18);
```

$$H := \begin{bmatrix} 0 & 0 & 2 & 4 & 4 & 0 \\ 0 & 3 & 4 & 3 & 0 & 0 \end{bmatrix}$$

$$A := \begin{bmatrix} \cos\left(\dfrac{1}{18}\pi\right) & -\sin\left(\dfrac{1}{18}\pi\right) \\ \sin\left(\dfrac{1}{18}\pi\right) & \cos\left(\dfrac{1}{18}\pi\right) \end{bmatrix}$$

```
> Movie(A,H,frames=36);
```

To see the succession of frames that make up the animation, add the option **trace=on**:

```
> Movie(A,H,frames=36,trace=on);
```

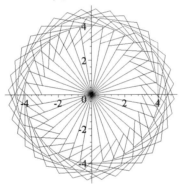

Example 6B: If we form a composite transformation made up of a contraction and a rotation, we can produce a more dynamic movie. We will use a contraction followed by the rotation above.

```
> C := Diagmat([.95,.95]);
```

$$C := \begin{bmatrix} .95 & 0 \\ 0 & .95 \end{bmatrix}$$

```
> A;
```

$$\begin{bmatrix} \cos\left(\dfrac{1}{18}\pi\right) & -\sin\left(\dfrac{1}{18}\pi\right) \\ \sin\left(\dfrac{1}{18}\pi\right) & \cos\left(\dfrac{1}{18}\pi\right) \end{bmatrix}$$

```
> M := A.C;
```

$$M := \begin{bmatrix} .95\cos\left(\frac{1}{18}\pi\right) & -.95\sin\left(\frac{1}{18}\pi\right) \\ .95\sin\left(\frac{1}{18}\pi\right) & .95\cos\left(\frac{1}{18}\pi\right) \end{bmatrix}$$

```
> Movie(M,H,frames=36);
```

Since animations cannot be printed, we provide the option snapshot=on, which produces a printable but nonanimated plot of all the frames:

```
> Movie(M,H,frames=36,snapshot=on);
```

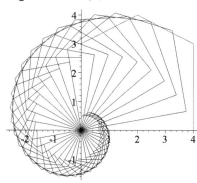

Feel free to experiment. Just keep in mind that computing time will increase with the complexity of the figure and the number of iterations. In Module 3, "Computer Graphics," we will come back to this topic.

Section 7: Reflection Transformations

Example 7A: Find the matrix transformation R that reflects every vector $\begin{bmatrix} a \\ b \end{bmatrix}$ across the x axis. For example, the matrix R transforms the vector $\begin{bmatrix} 3 \\ 2 \end{bmatrix}$ to the vector $\begin{bmatrix} 3 \\ -2 \end{bmatrix}$ as shown below:

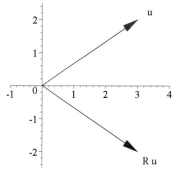

Solution: R simply changes the sign of the second coordinate, transforming $\begin{bmatrix} a \\ b \end{bmatrix}$ to $\begin{bmatrix} a \\ -b \end{bmatrix}$. We use the method of decomposition to find the matrix transformation R:

$$R\begin{bmatrix} a \\ b \end{bmatrix} = \begin{bmatrix} a \\ -b \end{bmatrix} = a\begin{bmatrix} 1 \\ 0 \end{bmatrix} + b\begin{bmatrix} 0 \\ -1 \end{bmatrix} = \begin{bmatrix} 1 & 0 \\ 0 & -1 \end{bmatrix}\begin{bmatrix} a \\ b \end{bmatrix}$$

Therefore R is given by:

```
> R := Matrix([[1,0],[0,-1]]);
```

$$R := \begin{bmatrix} 1 & 0 \\ 0 & -1 \end{bmatrix}$$

Let's check this matrix transformation by applying it to the Bug figure:

```
> Transform(R,Bug);
```

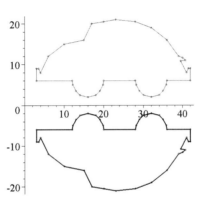

====================

Exercise 7.1: Use the method of decomposition to find the following matrices. Hint: Start by considering how each transformation maps a particular vector, such as $\begin{bmatrix} 3 \\ 2 \end{bmatrix}$.

(a) Find the matrix that reflects vectors across the y axis. Check your answer as we did above.
(b) Find the matrix that reflects vectors across the 45 degree line, $y = x$. Check your answer as above.

+ Student Workspace
− Answer 7.1

(a) Reflection across the y axis: The vector $\begin{bmatrix} a \\ b \end{bmatrix}$ is transformed to

$$\begin{bmatrix} -a \\ b \end{bmatrix} = a\begin{bmatrix} -1 \\ 0 \end{bmatrix} + b\begin{bmatrix} 0 \\ 1 \end{bmatrix} = \begin{bmatrix} -1 & 0 \\ 0 & 1 \end{bmatrix}\begin{bmatrix} a \\ b \end{bmatrix}$$

Therefore the reflection matrix is $\begin{bmatrix} -1 & 0 \\ 0 & 1 \end{bmatrix}$.

(b) Reflection across the line $y = x$: The vector $\begin{bmatrix} a \\ b \end{bmatrix}$ is transformed to

$$\begin{bmatrix} b \\ a \end{bmatrix} = a \begin{bmatrix} 0 \\ 1 \end{bmatrix} + b \begin{bmatrix} 1 \\ 0 \end{bmatrix} = \begin{bmatrix} 0 & 1 \\ 1 & 0 \end{bmatrix} \begin{bmatrix} a \\ b \end{bmatrix}$$

Therefore the reflection matrix is $\begin{bmatrix} 0 & 1 \\ 1 & 0 \end{bmatrix}$.

General Reflection Formula

More generally, the matrix transformation below reflects vectors across the line through the origin making an angle of θ radians with the *x* axis.

$$\begin{bmatrix} \cos(2\theta) & \sin(2\theta) \\ \sin(2\theta) & -\cos(2\theta) \end{bmatrix}$$

The command **Reflectmat(θ)** constructs a reflection matrix for an angle of θ radians. We check that this formula gives the three reflection matrices we have already considered, namely, reflection about the *x* axis, reflection about the *y* axis, and reflection about the line *y* = *x*:

```
> Reflectmat(0);
```
$$\begin{bmatrix} 1 & 0 \\ 0 & -1 \end{bmatrix}$$

```
> Reflectmat(Pi/2);
```
$$\begin{bmatrix} -1 & 0 \\ 0 & 1 \end{bmatrix}$$

```
> Reflectmat(Pi/4);
```
$$\begin{bmatrix} 0 & 1 \\ 1 & 0 \end{bmatrix}$$

When we apply a reflection transformation to a figure, we get its mirror image across the line of reflection. The input region below displays the result of applying a reflection matrix for the angle $\theta = \frac{\pi}{3}$ to the Bug. It also displays the line of reflection.

Explore~~~ Change the value of θ below to view the effect of other reflection matrices. Find a value of θ for which the image of the Bug lies entirely within the first quadrant.

```
> theta := Pi/3;
  R := Reflectmat(theta):
  Bugplot := Transform(R,Bug):
  lineplot :=
  plot([t*cos(theta),t*sin(theta),t=-5..40],color=black):
  display([Bugplot,lineplot]);
```
$$\theta := \frac{1}{3}\pi$$

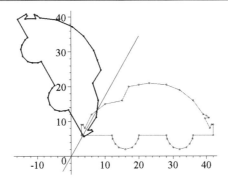

Exercise 7.2: (By hand) What are the fixed points for a reflection transformation T? (A "fixed point" is one that is mapped to itself by the transformation.) The following animation of a typical reflection may help you see the answers; this **Clock** animation shows a rotating unit input vector (in red) and its corresponding output vector (in blue).

```
> Clock(Reflectmat(Pi/3));
```

+ Student Workspace

− Answer 7.2

The fixed points for a reflection transformation are all of the points on the line of reflection.

− Section 8: Projection Transformations

Exercise 8.1: Use the method of decomposition to find the matrix transformation P that projects vectors perpendicularly onto the x axis. That is, P maps $\begin{bmatrix} a \\ b \end{bmatrix}$ to $\begin{bmatrix} a \\ 0 \end{bmatrix}$.

+ Student Workspace

− Answer 8.1

$$\begin{bmatrix} a \\ 0 \end{bmatrix} = a \begin{bmatrix} 1 \\ 0 \end{bmatrix} + b \begin{bmatrix} 0 \\ 0 \end{bmatrix} = \begin{bmatrix} 1 & 0 \\ 0 & 0 \end{bmatrix} \begin{bmatrix} a \\ b \end{bmatrix}$$

Therefore $P = \begin{bmatrix} 1 & 0 \\ 0 & 0 \end{bmatrix}$.

General Projection Formula

The matrix below is the transformation that projects vectors onto the line through the origin making an angle of θ radians with the x axis.

$$\begin{bmatrix} \cos(\theta)^2 & \cos(\theta)\sin(\theta) \\ \cos(\theta)\sin(\theta) & \sin(\theta)^2 \end{bmatrix}$$

Alternatively, if we apply double-angle identities, we get:

$$\begin{bmatrix} \dfrac{\cos(2\theta)+1}{2} & \dfrac{\sin(2\theta)}{2} \\ \dfrac{\sin(2\theta)}{2} & \dfrac{1-\cos(2\theta)}{2} \end{bmatrix}$$

The command **Projectmat(θ)** constructs this projection matrix for an angle of θ radians:

```
> theta := 'theta':
> Projectmat(theta);
```

$$\begin{bmatrix} \dfrac{1}{2}\cos(2\theta)+\dfrac{1}{2} & \dfrac{1}{2}\sin(2\theta) \\ \dfrac{1}{2}\sin(2\theta) & \dfrac{1}{2}-\dfrac{1}{2}\cos(2\theta) \end{bmatrix}$$

======================

Example 8A: The plot below displays the projections of three vectors u, v, w onto the line making an angle of 30 degrees with the x axis. If we were to drop a perpendicular from the head of the vector u to the line of projection, the foot of the perpendicular would land at the head of $A\,u$; the same is true for v and $A\,v$, and for w and $A\,w$. Note especially that the direction in which we project is <u>perpendicular to the line of projection</u>.

On a paper copy of the plot, draw the three perpendiculars, and label the vectors $A\,u$, $A\,v$, $A\,w$.

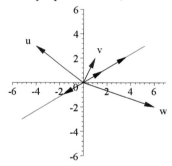

================================

Explore~~~ In the next input region, we project the Bug figure onto a 45 degree line. To change the line of projection, change the value of θ. Find a value of θ for which the image of the bug lies entirely within the fourth quadrant.

```
> theta := Pi/4;
  A := Projectmat(theta):
  Bugplot := Transform(A,Bug):
  lineplot :=
  plot([t*cos(theta),t*sin(theta),t=-5..40],color=black):
  display([Bugplot,lineplot]);
```

$$\theta := \frac{1}{4}\pi$$

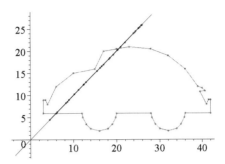

Exercise 8.2: (By hand) (a) What are the fixed points for a projection transformation T?
(b) Which points are mapped to the zero vector; that is, for which x is $T(x) = 0$?
The following animation of a typical projection may help you see the answers:
```
> Clock(Projectmat(Pi/3));
```
 Student Workspace

 Answer 8.2

(a) The fixed points for a projection transformation are all of the points on the line of projection.
(b) All of the points on the line through the origin perpendicular to the line of projection map to the zero vector.

Problems

 Problem 1: Guess the transformation

This problem consists of three questions. In each, you are presented with a picture that shows the effect of a transformation on the Bug. (The input is red, the output black.) Your task is to identify the specific transformation (or composite transformation) that is being applied. There may be more than one correct answer.

Execute the next input region to see Question (a).
```
> display(Question(a));
```
The next input region provides a quick way for you to check your answers. For example, suppose your answer to Question (a) is the rotation matrix A below. The **display** command lets you see the output from A (in blue) superimposed on the question plot.
```
> A := Rotatemat(Pi/2);
> your_answer := Drawmatrix(A.Bug,figcolor=blue):
> display([your_answer,Question(a)]);
```

Execute the next input region to see Question (b).
```
> display(Question(b));
```
Execute the next input region to see Question (c).
```
> display(Question(c));
```

Problem 2: Easy inverses

In Exercise 5.2 you found an easy way to state the inverse of a rotation matrix. Is there an easy way to state the inverse for each of the other types of transformations (i.e., dilations, shears, reflections, and projections)? If so, explain, from a geometric viewpoint, how the inverse of each type of transformation can be found. You may use Maple to check your answers.

 Student Workspace

Problem 3: Line segments transform to line segments

The line segment from a point p to another point q is expressed parametrically (i.e., in vector form) as $p + t(q - p)$, with the parameter t ranging from $t = 0$ to $t = 1$. Check that when $t = 0$ we are at p and when $t = 1$ we are at q. If we apply a matrix transformation A to the points on this line segment, then, using the laws of matrix algebra, we get:

$$A(p + t(q - p)) = Ap + t(Aq - Ap) \qquad [1]$$

Note that this is just the line segment between the points Ap and Aq. If $Ap = Aq$, then the image segment collapses to the single point Ap.

(a) Let A be the rotation matrix below and let p and q be the vectors below. Find the vector form of the line segment from p to q, and use formula [1] above to find the vector form of the image of this line segment when the rotation matrix is applied to it. Use **Drawvec** to display both the vector from p to q and the rotated vector from Ap to Aq.

```
> A := Rotatemat(Pi/3);
  p := Vector([3,4]);
  q := Vector([4,3]);
```

(b) Give an example of a nonzero matrix B that maps the line segment from p to q to a single point. Hint: Think geometrically.

 Student Workspace

Problem 4: Parallel lines transform to parallel lines

Show that a matrix transformation always takes a pair of parallel lines to parallel lines (unless it maps each line to a point). Hint: Express the two lines in vector form, $p + tv$.

 Student Workspace

Problem 5: Horizontal shears

Show that a horizontal shear matrix has the following properties:
(a) The points on the x axis are fixed points (i.e., are mapped to themselves).
(b) The image of each horizontal line is the same horizontal line.
(c) The image of each vertical line $x = a$ is the line with slope $1/k$ and x intercept a.

+ Student Workspace

Problem 6: Composition of transformations

As we saw in Section 4, composition of matrix transformations corresponds precisely to multiplication of matrices. Illustrate each of the following facts about matrix multiplication or matrix inverses by giving an example that uses geometric transformations. Explain what the composition does geometrically, and check your example by computing the corresponding matrix product or inverse.

For example, we could illustrate (a) by using a horizontal shear T_1 and a vertical shear T_2, as in Example 4A. The composites $T_1(T_2(x))$ and $T_2(T_1(x))$ are different since shearing a nonzero vector x horizontally and then vertically maps x to a different location than does shearing x vertically and then horizontally. (In your answer to (a), find an example that does not use shears.)

(a) Matrix multiplication is not commutative; that is, there exist matrices A and B such that $A B \neq B A$.
(b) The cancellation property does not hold for matrix multiplication; that is, there exist nonzero matrices A and B such that $A B = O$.
(c) There are nonzero square matrices that have no inverse; that is, there exists a nonzero square matrix A such that $B A \neq I$ for every square matrix B.

+ Student Workspace

Problem 7: Determinant of a matrix transformation

The matrices that you have worked with in this module are all "well-behaved" in the sense that each has an identifiable geometric effect. But since any matrix will suffice to define a matrix transformation, let's consider some random 2 by 2 matrix transformations and apply them to the Grid figure (see below). To keep things simple, we consider only matrices that have integer entries ranging between -2 and 2. Execute this region many times to see the variety of possibilities. As you do so, you may see some familiar geometric transformations from this module or composites of two (or more) of these basic geometric transformations.

Each time we generate a new matrix we also calculate the absolute value of its determinant. This number tells us something about the relationship between the original (red) Grid and its image (blue) Grid.
(a) Find the relationship and illustrate it by referring to several examples. If you find an example you want to save for purposes of illustration, just copy and paste it into your discussion.

(b) (Challenge) Give an algebraic proof of the relationship you found in part (a) using properties of determinants.

```
> A := Randmat(2,2,bound=2);
  abs(Det(A));
> Transform(A,Grid);
```
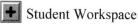 Student Workspace

Problem 8: True/False questions about reflections

(By hand) These are true/false questions. Give a geometric reason that supports your answer (by drawing a hand sketch, for example).

(a) If A is the reflection matrix for an angle θ, then A^2 is the reflection matrix for 2θ.

(b) If A is the reflection matrix for an angle θ and B is the reflection matrix for $-\theta$, then A and B must be inverse matrices.

(c) If A is the reflection matrix for an angle θ, then $A^n = A$, whenever n is a positive odd integer.

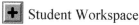 Student Workspace

Problem 9: Square roots, cube roots, ...

In the real numbers, equations such as $x^2 = 1$ and $x^3 = 1$ and $x^4 = 1$ have only one or two solutions. On the other hand, matrix equations of the same form have many more solutions. Using only our five special types of matrix transformations, find as many solutions as you can to each of the equations $A^2 = I$ and $A^3 = I$ and $A^4 = I$. Explain geometrically why your solutions work.

 Student Workspace

Problem 10: Products of projections I

Let A and B be the projection matrices below.

(a) Explain geometrically why $A B = 0$ and $B A = 0$. Include, as part of your explanation, a diagram that displays v, $B v$, and $A B v$ for one or more vectors v.

(b) Calculate A^2 and B^2. What do you observe? Is this true for all projection matrices? Explain geometrically.

```
> A := Projectmat(Pi/4);
  B := Projectmat(3*Pi/4);
```
 Student Workspace

Problem 11: Products of projections II

This is a continuation of Problem 10.

(a) Define a third projection matrix C (see below) and a vector $v = \begin{bmatrix} 3 \\ 1 \end{bmatrix}$. Make a hand sketch (or use **Drawvec**) that shows the following four vectors together in a single plot: $A\,v$, $C\,v$, $A\,C\,v$ and $C\,A\,v$. Use your sketch to explain why the matrix products $A\,C$ and $C\,A$ are not equal. Are there any vectors x where $A\,C\,x$ does equal $C\,A\,x$? Hint: Solve the homogeneous equation $(A\,C - C\,A)\,x = 0$ and thus find all vectors x such that $A\,C\,x = C\,A\,x$.

(b) Find another pair of projection matrices that, like A and B in Problem 10, commute (i.e., $A\,B = B\,A$). Under what conditions do two projection matrices commute?

(c) Is it possible to find two nonzero matrices from one of the other types of transformations (i.e., dilations, shears, rotations or reflections) whose product is the zero matrix? Explain why your answers are correct.

```
> A := Projectmat(Pi/4);
  B := Projectmat(3*Pi/4);
  C := Projectmat(Pi/3);
```

 Student Workspace

Problem 12: Reflections that commute

(a) If P is the matrix that projects onto the line through the origin making an angle of θ radians with the x axis, and R is the matrix that reflects across this line, show that $R = 2\,P - I$.

(b) Which reflection matrices commute? Hint: In Problem 11, you found the conditions under which projection matrices commute. Use your result from that problem and part (a) of this problem.

 Student Workspace

Problem 13: Derivation of rotation, projection, reflection matrices

(a) Let M be the matrix transformation that rotates vectors counter-clockwise by θ radians. It's not hard to see that M sends the vector $\begin{bmatrix} 1 \\ 0 \end{bmatrix}$ to $\begin{bmatrix} \cos(\theta) \\ \sin(\theta) \end{bmatrix}$, and the vector $\begin{bmatrix} 0 \\ 1 \end{bmatrix}$ to $\begin{bmatrix} -\sin(\theta) \\ \cos(\theta) \end{bmatrix}$. Use this information to derive the formula for M that was shown in Section 5.

(b) Let P be the matrix transformation that projects vectors onto the line through the origin making an angle of θ radians with the x axis. Let u be a unit vector lying along this line and v be a unit vector perpendicular to this line:

```
> theta := 'theta':
  u := Vector([cos(theta),sin(theta)]);
  v := Vector([-sin(theta),cos(theta)]);
```

Then $P\,u = u$ and $P\,v = 0$. Use this information to derive the formula for P that was shown in Section 8. (Hint: Write the equations $P\,u = u$ and $P\,v = 0$ as a single equation, $P\,A = B$, where A and B are 2 by 2 matrices; then solve this equation for P.)

(c) Let R be the matrix transformation that reflects vectors across the line through the origin making an angle of θ radians with the x axis. Use the method in part (b) to derive the formula for R that was shown in Section 7.

+ Student Workspace

- Problem 14: Transformations of lines

Let A be the projection matrix below and B be the rotation matrix below:
```
> A := Projectmat(Pi/2);
  B := Rotatemat(Pi/4);
```

(a) Find the image of the horizontal line $y = 1$ under each of the matrix transformations A, B, $A\,B$, and $B\,A$. (Try to determine your answers by geometric reasoning alone.)

(b) Find a matrix transformation that maps the horizontal line $y = 1$ onto the vertical line $x = 2$. (This can be done using only transformations of the type studied in this module.) Check your matrix by applying it to the vector $\begin{bmatrix} x \\ 1 \end{bmatrix}$, which represents the line $y = 1$.

(c) Can you find a matrix transformation that maps the x axis to the line $y = 1$?

+ Student Workspace

- Problem 15: Rotations from shears or reflections

(a) Check that the product $A\,B\,A$, where A and B are the shears below, is the rotation through the angle ϕ. (Hint: Apply `map(simplify,P)` to the product P.)
```
> A := Xshearmat(-tan(phi/2));
  B := Yshearmat(sin(phi));
```

(b) Apply the matrices A, B, A, in succession, to a figure such as Bug or Grid for a particular value of ϕ. For example, if G is the Grid matrix, apply A to G, then B to $A\,G$, and then A to $B\,A\,G$. Use these pictures to describe how the product of shears results in a rotation. (A rigorous geometric argument is not expected here.)

(c) Now we switch from using shears to using reflections. Every rotation can also be written as a product of two reflections in many different ways. Express the rotation through the angle $\dfrac{\pi}{2}$ as a product of two reflections in two different ways.

(d) (Challenge) Show that the product of two reflections is always a rotation, and find a formula for the angle of rotation in terms of the two angles of reflections. In how many ways can a rotation be written as a product of two reflections?

+ Student Workspace

Linear Algebra Modules Project
Chapter 4, Module 2

Geometry of Matrix Transformations of 3-Space

Purpose of this module

The purpose of this module is to study the basic geometric matrix transformations of R^3 and the concepts of kernel and range of a transformation.

Prerequisites

Chapter 4, Module 1: "Geometry of Matrix Transformations of the Plane."

Commands used in this module

```
> restart: with(LinearAlgebra): with(plots): with(Lamp):
  UseHardwareFloats := false: Digits := 6:
```

Tutorial

Section 1: Geometric Transformations of 3-Space

As in Module 1, we will study dilations, rotations, reflections, and projections. Now, however, these will be 3 by 3 matrix transformations that map vectors in R^3 to vectors in R^3.

Scaling Transformations

====================

Example 1A: The 3 by 3 diagonal matrix A defined below scales the x, y, and z components of vectors in R^3 by different factors. That is, if we multiply any vector in R^3 by A, the components of the vector will be scaled by the corresponding diagonal entries of A.

```
> A := Diagmat([2,3,5]);
```

$$A := \begin{bmatrix} 2 & 0 & 0 \\ 0 & 3 & 0 \\ 0 & 0 & 5 \end{bmatrix}$$

```
> v := Vector([a,b,c]);
  Av := A.v;
```

$$v := \begin{bmatrix} a \\ b \\ c \end{bmatrix}, \quad Av := \begin{bmatrix} 2a \\ 3b \\ 5c \end{bmatrix}$$

226   Chapter 4 Matrix Transformations

We use the **Drawmatrix** command to draw a figure in R^3, and the **Transform** command to draw both the original figure and the transformed figure. As in Module 1, we have several predefined figures available. Execute the next region to see the "Hotel3d" figure.

> `Drawmatrix(Hotel3d);`

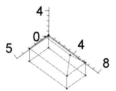

In the command below, we apply the scaling transformation A to Hotel3d. The dimensions of the original figure (in red) have not changed, but the scale of the plot has changed to accommodate the larger transformed figure (in blue).

> `Transform(A,Hotel3d);`

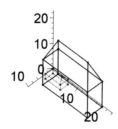

============================

The vectors e_1, e_2, e_3 below are called the "standard unit vectors" of R^3. They are vectors of length 1 along the positive x, y, and z axes, respectively. They are also the columns of the 3 by 3 identity matrix.

> `e1 := Vector([1,0,0]);`
  `e2 := Vector([0,1,0]);`
  `e3 := Vector([0,0,1]);`

$$e1 := \begin{bmatrix} 1 \\ 0 \\ 0 \end{bmatrix}, \quad e2 := \begin{bmatrix} 0 \\ 1 \\ 0 \end{bmatrix}, \quad e3 := \begin{bmatrix} 0 \\ 0 \\ 1 \end{bmatrix}$$

Since $A I = A$ for any 3 by 3 matrix A, the standard unit vectors have the following useful property:

- If A is a 3 by 3 matrix and T is the matrix transformation defined by $T(x) = A x$ for all vectors x in R^3, then the columns of A are $T(e_1)$, $T(e_2)$, $T(e_3)$.

Exercise 1.1: Let B be the matrix below, and let T be the transformation $T(x) = Bx$. Check that the bulleted property above is true in this case.

```
> B := Matrix([[2,-4,1],[3,0,-2],[-1,1,5]]);
```

$$B := \begin{bmatrix} 2 & -4 & 1 \\ 3 & 0 & -2 \\ -1 & 1 & 5 \end{bmatrix}$$

```
> T := x -> B.x;
```

$$T := x \to B \cdot x$$

+ Student Workspace

 Answer 1.1

```
> T(e1), T(e2), T(e3);
```

$$\begin{bmatrix} 2 \\ 3 \\ -1 \end{bmatrix}, \begin{bmatrix} -4 \\ 0 \\ 1 \end{bmatrix}, \begin{bmatrix} 1 \\ -2 \\ 5 \end{bmatrix}$$

Note that $T(e_1) = B_1$, $T(e_2) = B_2$, and $T(e_3) = B_3$, where B_1, B_2, B_3 are the columns of B.

The main use we will make of the above property is to construct matrix transformations. If we know the images of the standard unit vectors under a given matrix transformation, the above property tells us that these images are the columns of the matrix; hence we know the matrix.

======================

Example 1B: If T is the matrix transformation that projects vectors onto the xy plane, use the above property to find the matrix P such that $T(x) = Px$.

Solution: Since T projects vectors onto the xy plane, T maps e_1 onto e_1, e_2 onto e_2, and e_3 onto 0 (the zero vector in R^3). Thus,

$$T(e_1) = \begin{bmatrix} 1 \\ 0 \\ 0 \end{bmatrix}, \quad T(e_2) = \begin{bmatrix} 0 \\ 1 \\ 0 \end{bmatrix}, \quad T(e_3) = \begin{bmatrix} 0 \\ 0 \\ 0 \end{bmatrix}$$

Therefore the projection matrix is

$$P = \begin{bmatrix} 1 & 0 & 0 \\ 0 & 1 & 0 \\ 0 & 0 & 0 \end{bmatrix}$$

We check that the matrix P has the required effect on a general vector $\langle a, b, c \rangle$:

```
> P := Matrix([[1,0,0],[0,1,0],[0,0,0]]):
> P.Vector([a,b,c]);
```

$$\begin{bmatrix} a \\ b \\ 0 \end{bmatrix}$$

======================

Rotation Transformations

Example 1C: Let M be the 3 by 3 matrix that rotates vectors 90 degrees counter-clockwise about the z axis. Use the technique in Example 1B to find the matrix M.

Solution: Since the rotation is about the z axis, the transformation T defined by $T(x) = Mx$ maps e_1 to e_2, e_2 to $-e_1$, and e_3 to itself. Thus:

$$T(e_1) = \begin{bmatrix} 0 \\ 1 \\ 0 \end{bmatrix} \quad T(e_2) = \begin{bmatrix} -1 \\ 0 \\ 0 \end{bmatrix} \quad T(e_3) = \begin{bmatrix} 0 \\ 0 \\ 1 \end{bmatrix}$$

Therefore the rotation matrix is

$$M = \begin{bmatrix} 0 & -1 & 0 \\ 1 & 0 & 0 \\ 0 & 0 & 1 \end{bmatrix}$$

We check this by applying the matrix M to the Hotel3d figure:

```
> M := Matrix([[0,-1,0],[1,0,0],[0,0,1]]);
```

$$M := \begin{bmatrix} 0 & -1 & 0 \\ 1 & 0 & 0 \\ 0 & 0 & 1 \end{bmatrix}$$

```
> Transform(M,Hotel3d);
```

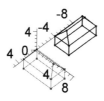

Exercise 1.2: Use the technique in Example 1B to find the 3 by 3 matrix that rotates vectors 90 degrees counter-clockwise about the x axis. Check your matrix by applying it to Hotel3d.

+ Student Workspace

− Answer 1.2

Since e_1 is fixed, $T(e_1) = e_1$; since e_2 rotates to e_3 and e_3 rotates to $-e_2$, we also have $T(e_2) = e_3$ and $T(e_3) = -e_2$. Therefore

$$M = \begin{bmatrix} 1 & 0 & 0 \\ 0 & 0 & -1 \\ 0 & 1 & 0 \end{bmatrix}$$

```
> M := Matrix([[1,0,0],[0,0,-1],[0,1,0]]):
> Transform(M,Hotel3d);
```

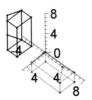

The command **Rotatemat3d(v,θ)** constructs the 3 by 3 rotation matrix that rotates vectors θ radians counter-clockwise about the axis with direction vector v. The rotation matrices we considered in Examples 1C and Exercise 1.2 are therefore:

```
> Rotatemat3d(<0,0,1>,Pi/2);
> Rotatemat3d(<1,0,0>,Pi/2);
```

$$\begin{bmatrix} 0 & -1 & 0 \\ 1 & 0 & 0 \\ 0 & 0 & 1 \end{bmatrix}, \begin{bmatrix} 1 & 0 & 0 \\ 0 & 0 & -1 \\ 0 & 1 & 0 \end{bmatrix}$$

Reflection Transformations

==========================

Example 1D: Let R be the 3 by 3 matrix that reflects vectors in R^3 across the yz plane. Use the method of decomposition from Module 1 to find R.

Solution: Note that $R \begin{bmatrix} a \\ b \\ c \end{bmatrix} = \begin{bmatrix} -a \\ b \\ c \end{bmatrix}$, and then decompose the vector on the right to find R:

$$\begin{bmatrix} -a \\ b \\ c \end{bmatrix} = a \begin{bmatrix} -1 \\ 0 \\ 0 \end{bmatrix} + b \begin{bmatrix} 0 \\ 1 \\ 0 \end{bmatrix} + c \begin{bmatrix} 0 \\ 0 \\ 1 \end{bmatrix} = \begin{bmatrix} -1 & 0 & 0 \\ 0 & 1 & 0 \\ 0 & 0 & 1 \end{bmatrix} \begin{bmatrix} a \\ b \\ c \end{bmatrix}$$

(Alternatively, by the technique in Example 1B, we have $T(e_1) = -e_1$, $T(e_2) = e_2$ and $T(e_3) = e_3$, which yields the same matrix.)

Chapter 4 Matrix Transformations

```
> R := Matrix([[-1,0,0],[0,1,0],[0,0,1]]);
```

$$R := \begin{bmatrix} -1 & 0 & 0 \\ 0 & 1 & 0 \\ 0 & 0 & 1 \end{bmatrix}$$

Here is the result of the **Transform** command applied to the Hotel3d figure:

```
> Transform(R,Hotel3d);
```

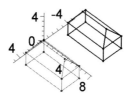

And here is the same transformation applied to a figure called "Jet3d."

```
> Drawmatrix(Jet3d):
> Transform(R,Jet3d);
```

======================

Exercise 1.3: (By hand) Find each of the following matrices. Use either the method of decomposition or the technique in Example 1B.

(a) The 3 by 3 matrix A that reflects vectors across the xy plane.

(b) The 3 by 3 matrix B that reflects vectors across the vertical plane $y = x$.

+ Student Workspace

− Answer 1.3

(a) The matrix A reflects the vector $\begin{bmatrix} a \\ b \\ c \end{bmatrix}$ to the vector $\begin{bmatrix} a \\ b \\ -c \end{bmatrix}$. So, by the method of decomposition,

$$A = \begin{bmatrix} 1 & 0 & 0 \\ 0 & 1 & 0 \\ 0 & 0 & -1 \end{bmatrix}$$

(b) The matrix B interchanges the vectors e_1 and e_2 and leaves e_3 fixed. So, by the technique in Example 1B,

$$B = \begin{bmatrix} 0 & 1 & 0 \\ 1 & 0 & 0 \\ 0 & 0 & 1 \end{bmatrix}$$

The command **Reflectmat3d(n)** constructs the 3 by 3 reflection matrix that reflects vectors across the plane through the origin with normal vector n. For example, if we want to reflect across the yz plane (as in Example 1D), we use $n = \langle 1, 0, 0 \rangle$:

```
> Reflectmat3d(<1,0,0>);
```

$$\begin{bmatrix} -1 & 0 & 0 \\ 0 & 1 & 0 \\ 0 & 0 & 1 \end{bmatrix}$$

Projection Transformations

==================

Example 1E: Use the technique in Example 1B to find the 3 by 3 matrix P that projects vectors onto the yz plane.

Solution: The vector e_1 is projected to the zero vector, and e_2 and e_3 are mapped to themselves. So

$$P = \begin{bmatrix} 0 & 0 & 0 \\ 0 & 1 & 0 \\ 0 & 0 & 1 \end{bmatrix}$$

```
> P := Matrix([[0,0,0],[0,1,0],[0,0,1]]):
```

Here is the transformation applied to Hotel3d. If you rotate the figure so you are looking directly at the yz plane, you will see the relationship between the 3d figure and its projection.

```
> Transform(P,Hotel3d);
```

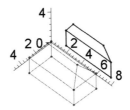

The projection of Jet3d is also interesting.
> `Transform(P,Jet3d,axes=none);`

Exercise 1.4: Find the 3 by 3 matrix P that projects vectors in R^3 onto the vertical plane $y = x$. (Hint: the vectors e_1 and e_2 are projected onto the same vector on the line $y = x$ in the xy plane.)

 Student Workspace

 Answer 1.4

P projects both e_1 and e_2 onto the vector $\begin{bmatrix} 1/2 \\ 1/2 \\ 0 \end{bmatrix}$. The vector e_3 is mapped to itself. So

$$P = \begin{bmatrix} \frac{1}{2} & \frac{1}{2} & 0 \\ \frac{1}{2} & \frac{1}{2} & 0 \\ 0 & 0 & 1 \end{bmatrix}$$

> `P := Matrix([[1/2,1/2,0],[1/2,1/2,0],[0,0,1]]):`

Here is a picture of Hotel3d projected onto this plane:
> `Transform(P,Hotel3d);`

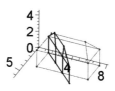

The command `Projectmat3d(n)` constructs the 3 by 3 projection matrix that projects vectors onto the plane through the origin with normal vector n. For example, if we want to project onto the yz plane

(as in Example 1E), the matrix is:
```
> Projectmat3d(<1,0,0>);
```
$$\begin{bmatrix} 0 & 0 & 0 \\ 0 & 1 & 0 \\ 0 & 0 & 1 \end{bmatrix}$$

Section 2: Kernel and Range

In this section we introduce two concepts associated with matrix transformations that are analogous to concepts you have seen in your study of functions in precalculus and calculus. Consider, for example, $f(x) = x^2 - 4$, which defines a function f from R to R. The set of "zeros" of such a function is the set of values of x such that $f(x) = 0$; in this example, the set of zeros is $\{2, -2\}$. The "range" of such a function is the set of values of $f(x)$, where x ranges over the domain R; in this example, the range is $\{y : -4 \leq y\}$. We now define the analogous concepts for the matrix transformation $T(x) = A\,x$, where A is any m by n matrix. T is thus a transformation that takes vectors in R^n to vectors in R^m.

Definition: The ***kernel*** of T is the set of vectors x in R^n such that $T(x) = 0$ (where 0 denotes the zero vector in R^m); in other words, the kernel is the set of "zeros" of T.

Definition: The ***range*** of T is the set of vectors $T(x)$, where x ranges over all vectors in R^n; in other words, the range is the set of all output (i.e., image) vectors of T in R^m.

Exercise 2.1: (By hand) Find the kernel and range for the matrix transformation T that projects vectors in R^3 onto the *xz* plane. Determine your answers by geometric reasoning.

 Student Workspace

 Answer 2.1

All the vectors on the *y* axis are projected onto the zero vector, and no other vectors are. So the kernel of T is the *y* axis. Since every vector in R^3 is projected by T onto the plane of projection, the range of T must be contained in the *xz* plane. In fact the entire *xz* plane is the range, since every vector in the *xz* plane is the projection of some vector (e.g., itself).

Both the kernel and range can be described using familiar ideas from our study of matrix equations:

- Since $T(x) = A\,x$, the kernel of T is the set of solutions of the homogeneous matrix equation $A\,x = 0$.
- Since $T(x) = A\,x$, the range of T is the set of all vectors b such that the matrix equation $A\,x = b$ has a solution.

Furthermore, the Invertibility Theorem (Theorem 7 of Chapter 3) reveals important connections between kernel, range, and invertibility. Recall that if A is an n by n matrix, then A is invertible if and only if the only solution to the homogeneous equation $A\,x = 0$ is the trivial solution $x = 0$. Also, A is invertible if and only if the matrix equation $A\,x = b$ has a solution x for every vector b in R^n. Putting these facts together with the above two facts about kernel and range, we see that

- If A is an n by n matrix and if T is the matrix transformation from R^n to R^n defined by $T(x) = A\,x$, then the following statements are equivalent to one another:
  - A is invertible;
  - the kernel of T is $\{0\}$;
  - the range of T is R^n.

======================

Example 2A: Let T be the 3 by 3 matrix transformation that rotates vectors through an angle θ about the z axis. Find the kernel and range of T.

Solution: When you rotate a vector, you cannot get the zero vector unless you start with the zero vector. So the kernel of T consists of just one vector, the zero vector itself; that is, the kernel of T is $\{0\}$. Therefore, by the above remarks on invertibility, the range of T is all of R^3. We can also see this as follow: For any vector v in R^3, choose u to be the vector that we get by rotating v back by the same angle θ; then we have $T(u) = v$, which shows that v is in the range of T.

======================

Exercise 2.2: In Section 1 we considered projections onto a plane. Now let's project onto a line. Let $T(x) = A\,x$ be the matrix transformation that projects vectors in R^3 onto the z axis.

(a) Find the kernel and range of T.
(b) Find the matrix A and apply it to the Hotel3d figure.
(c) Does the matrix that you found in (b) have an inverse? Use your answer to (a) to justify your conclusion.

 Student Workspace

 Answer 2.2

(a) The kernel of T is the entire xy plane and the range is the z axis.
(b) Both e_1 and e_2 are mapped to the zero vector, and e_3 is mapped to itself. Therefore:
```
> A := Matrix([[0,0,0],[0,0,0],[0,0,1]]);
```
$$A := \begin{bmatrix} 0 & 0 & 0 \\ 0 & 0 & 0 \\ 0 & 0 & 1 \end{bmatrix}$$
```
> Transform(A,Hotel3d,axes=none);
```

(c) The matrix A is not invertible, since its kernel is not equal to $\{0\}$.

Exercise 2.3: Let T be the projection transformation in Exercise 2.2 and U be the rotation transformation in Exercise 1.2. (U rotates vectors 90 degrees counter-clockwise about the x axis.) Find the kernel and range of the composite transformation $T\,U$. **Caution:** Keep in mind that $T\,U(x) = T(U(x))$, which means that U is applied first to x, then T.

Since the range of U is all of R^3, the range of $T\,U$ is the same as the range of T: the z axis. To find the kernel of the composite, we must find all vectors x satifying $T(U(x)) = 0$; i.e., all vectors x such that $U(x)$ is in the kernel of T. Since the kernel of T is the xy plane and since the rotation U maps the xz plane onto the xy plane, we conclude that the xz plane is the kernel of the composite.

Problems

Problem 1: Reflections in R^3

Let Rxy, Rxz, Ryz be the 3 by 3 matrices that reflect vectors across the xy, xz, yz planes, respectively.
(a) Find each of these matrices by hand.
(b) Create a single picture that displays the Hotel3d figure and its three reflections across the coordinate planes.
(c) Find the 3 by 3 matrix that reflects vectors across the vertical plane $y = -x$. (Use either the technique in Example 1B or the method of decomposition.)
(d) Write the matrix you found in (c) as a product of three matrices: a rotation, a reflection, and another rotation, where the reflection is one of the reflections you found in (a). Hint: Use a sketch of vectors in the xy plane to see how to achieve this reflection as a composite of three transformations.

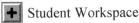

Problem 2: Projections in R^3

Let *Pxy*, *Pxz*, *Pyz* be the 3 by 3 matrices that project vectors onto the the *xy, xz, yz* planes, respectively.
(a) Find each of these matrices by hand.
(b) Create a single picture that displays the Hotel3d figure and its projections onto the three coordinate planes.
(c) Do *Pxy*, *Pxz*, and *Pyz* commute? That is, are the equations

$$Pxy\, Pxz = Pxz\, Pxy, \quad Pxy\, Pyz = Pyz\, Pxy, \quad Pxz\, Pyz = Pyz\, Pxz$$

true? Give a geometric justification for your answer by describing what each side of the equation does to a typical vector.

Student Workspace

Problem 3: Projection onto a line in R^3

Let $T(x) = A\,x$ be the transformation that projects vectors onto the line containing the vector $\begin{bmatrix} 1 \\ 0 \\ 1 \end{bmatrix}$.

(a) Find all fixed points of T. (Recall that a "fixed point" of a transformation T is a vector *x* that is mapped to itself: $T(x) = x$.)
(b) Find the kernel and range of T.
(c) Find the matrix A and apply it to the Hotel3d figure.

Student Workspace

Problem 4: Product of two reflections

(a) Compute the product of the reflection across the *yz* plane and the reflection across the *xz* plane. Apply the product to the Hotel3d figure. What familiar transformation is this product? Check your answer by using one of the special matrix construction commands in this module to construct your answer without computing a product.
(b) Compute the product of the two reflection matrices below, and apply the product to the Hotel3d figure. What familiar transformation is this product? (Hint: What are the fixed points of the composite?) Check your answer as in (a).

```
> R := Reflectmat3d(<1,1,1>);
  S := Reflectmat3d(<1,1,-2>);
```

Student Workspace

Problem 5: Kernel and range of a composite transformation

Let *P* be the 3 by 3 projection matrix that projects vectors onto the *xy* plane and *R* be the 3 by 3 rotation matrix that rotates vectors 90 degrees counter-clockwise about the *y* axis. Determine your answers to the

following questions by geometric reasoning, and explain your reasoning.
(a) Find the kernel and range of the composite transformation $P R$.
(b) Find the kernel and range of the composite transformation $R P$.

 Student Workspace

 Problem 6: Rotation about the diagonal of a cube (challenge)

Consider the three special types of rotation matrices that rotate vectors about one of the coordinate axes:
```
> Rx := theta -> Rotatemat3d(<1,0,0>,theta);
> Ry := theta -> Rotatemat3d(<0,1,0>,theta);
> Rz := theta -> Rotatemat3d(<0,0,1>,theta);
```
The "unit cube" has one vertex at $(0,0,0)$ and the opposite vertex at $(1,1,1)$. Using only the three rotation matrices Rx, Ry, Rz above, construct the matrix that rotates the unit cube through the angle θ about the diagonal connecting $(0,0,0)$ and $(1,1,1)$ in the counter-clockwise direction when looking along the diagonal toward the origin. Test your transformation on the Cube3d figure.

Hint: First rotate to line up the diagonal with one of the axes; then perform the desired rotation; and finally return the diagonal back to its original position.

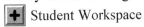 Student Workspace

Linear Algebra Modules Project
Chapter 4, Module 3

Computer Graphics

 Purpose of this module

The purpose of the module is to explore the uses of geometric matrix transformations in computer graphics. In particular, we introduce homogeneous coordinates as a means of combining translations with other geometric transformations on R^2 and R^3.

 Prerequisites

Module 1, "Geometry of Matrix Transformations of the Plane." Also, Module 2, "Geometry of Matrix Transformations of 3-Space," is required for Section 2.

 Commands used in this module

```
>  restart: with(plots): with(LinearAlgebra): with(Lamp):
   UseHardwareFloats := false: Digits:=6:
```

Tutorial

 Section 1: Homogeneous 2D Coordinates

The picture below shows a red triangle with vertices (4, 3), (5, 9), and (6, 3) that has been translated up and to the right by the vector $\begin{bmatrix} 3 \\ 2 \end{bmatrix}$. This is accomplished by adding the translation vector $\begin{bmatrix} 3 \\ 2 \end{bmatrix}$ to each of the points of the original figure.

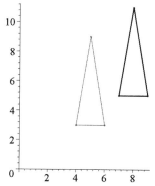

Given our experience with geometric matrix transformations in Module 1, we naturally ask if we can accomplish such a translation by multiplying by a suitable 2 by 2 matrix. The answer turns out to be: No such matrix exists. Here is why. Imagine that you do have a matrix M that translates every point by the vector $\begin{bmatrix} 3 \\ 2 \end{bmatrix}$. Then you would get, for every point (a, b) in R^2,

$$M \begin{bmatrix} a \\ b \end{bmatrix} = \begin{bmatrix} a+3 \\ b+2 \end{bmatrix} \text{ and hence, in particular, } M \begin{bmatrix} 0 \\ 0 \end{bmatrix} = \begin{bmatrix} 3 \\ 2 \end{bmatrix}.$$

This is clearly not possible, for if you multiply any matrix M by the zero vector, you always get the zero vector. So it is not possible for a 2 by 2 matrix M to accomplish this translation or indeed any translation.

However, in computer graphics animations, the most basic transformation is a translation. So we persist in asking: Is there some way in which we can use matrix multiplication both to translate figures and perform the other basic geometric transformations, such as dilations, rotations, etc.?

Yes, there <u>is</u> a way -- but we need 3 by 3 matrices to do it! The trick is to express points in R^2 in what are called "homogeneous coordinates." A typical point (x, y) becomes, in homogeneous coordinates, the point $(x, y, 1)$. While this looks like a point in R^3, we will treat it as a special coded form of the point (x, y). So when we represent a figure by a matrix whose columns are points of the figure, we replace the former 2 by n matrix by a 3 by n matrix whose last row is all 1's.

For example, our red triangle above can be represented using homogeneous coordinates by the matrix "triangle":

```
> triangle := Matrix([[4,5,6,4],[3,9,3,3],[1,1,1,1]]);
```

$$triangle := \begin{bmatrix} 4 & 5 & 6 & 4 \\ 3 & 9 & 3 & 3 \\ 1 & 1 & 1 & 1 \end{bmatrix}$$

Geometric Transformations Using Homogeneous Coordinates

Under this new setup, the 2 by 2 geometric transformations from Module 1 now become 3 by 3 matrices. For example, here is the original version of the rotation matrix R followed by the new "homogeneous" version, which we have labeled Rh.

```
> R := Rotatemat(theta);
```

$$R := \begin{bmatrix} \cos(\theta) & -\sin(\theta) \\ \sin(\theta) & \cos(\theta) \end{bmatrix}$$

```
> Rh := Rotatemat(theta,homogeneous=on);
```

$$Rh := \begin{bmatrix} \cos(\theta) & -\sin(\theta) & 0 \\ \sin(\theta) & \cos(\theta) & 0 \\ 0 & 0 & 1 \end{bmatrix}$$

Note: To get the homogeneous version of the rotation matrix, we added the option "homogeneous=on" to the **Rotatemat** command. Also observe that the original rotation matrix makes up the 2 by 2 block in the upper-left corner of the homogeneous version. The rest of the matrix is filled out with zeros, except for a 1 in the lower-right corner.

Now compare how the two rotation matrices act on the point $\begin{bmatrix} x \\ y \end{bmatrix}$ and its homogeneous form $\begin{bmatrix} x \\ y \\ 1 \end{bmatrix}$.

```
> p := Vector([x,y]);
  R.p;
```

$$p := \begin{bmatrix} x \\ y \end{bmatrix}, \quad Rp = \begin{bmatrix} \cos(\theta)x - \sin(\theta)y \\ \sin(\theta)x + \cos(\theta)y \end{bmatrix}$$

```
> ph := Vector([x,y,1]);
  Rh.ph;
```

$$ph := \begin{bmatrix} x \\ y \\ 1 \end{bmatrix}, \quad Rh\,ph = \begin{bmatrix} \cos(\theta)x - \sin(\theta)y \\ \sin(\theta)x + \cos(\theta)y \\ 1 \end{bmatrix}$$

If we drop the 1 in the homogeneous form, we get the same image point by both methods.

=====================

Example 1A: Use homogeneous coordinates to rotate the "triangle" figure by an angle of π/4 radians.
Solution: First we find the homogeneous version of the appropriate rotation matrix:
```
> M := Rotatemat(Pi/4,homogeneous=on):
```

Next we apply the **Transform** command. Notice that this command also contains the option homogeneous=on.
```
> Transform(M,triangle,homogeneous=on);
```

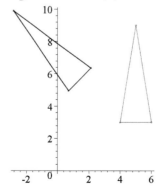

=====================

Note: Adding the option homogeneous=on to each command can become quite tedious. Alternatively, you can turn the homogeneous mode on for all future commands by executing the next line. **Caution:** For an option within a command, use equal (=) without the colon (:); but for a command that sets the

value of a variable, use colon-equal (:=).
```
> homogeneous := on;
```
$$homogeneous := on$$

You can also override this global setting in any single command by including the option homogeneous=off. To turn the homogeneous mode off, use the command **homogeneous := off**.

All the other geometric transformations from Module 1 have homogeneous versions. Here are the reflection and projection matrices in homogeneous coordinates:
```
> Reflectmat(theta);
```
$$\begin{bmatrix} \cos(2\theta) & \sin(2\theta) & 0 \\ \sin(2\theta) & -\cos(2\theta) & 0 \\ 0 & 0 & 1 \end{bmatrix}$$

```
> Projectmat(theta);
```
$$\begin{bmatrix} \frac{1}{2}\cos(2\theta)+\frac{1}{2} & \frac{1}{2}\sin(2\theta) & 0 \\ \frac{1}{2}\sin(2\theta) & \frac{1}{2}-\frac{1}{2}\cos(2\theta) & 0 \\ 0 & 0 & 1 \end{bmatrix}$$

======================

Example 1B: We use the homogeneous version of the reflection matrix to reflect the triangle figure about the y axis:
```
> M := Reflectmat(Pi/2):
> Transform(M,triangle);
```

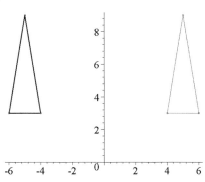

======================

Translation Using Homogeneous Coordinates

Of course the reason for switching to homogeneous coordinates is to enable us to treat translation as a matrix product. So here is the matrix that translates points by the vector $\begin{bmatrix} h \\ k \end{bmatrix}$:

Chapter 4 Matrix Transformations

```
> T := Translatemat(<h,k>);
```

$$T := \begin{bmatrix} 1 & 0 & h \\ 0 & 1 & k \\ 0 & 0 & 1 \end{bmatrix}$$

And here is proof that the translation matrix does indeed translate each point (x, y) (in homogeneous coordinates) to $(x + h, y + k)$ (in homogeneous coordinates):

```
> T.Vector([x,y,1]);
```

$$\begin{bmatrix} x + h \\ y + k \\ 1 \end{bmatrix}$$

Now let's translate the red triangle in the introduction by the vector $\begin{bmatrix} 3 \\ 2 \end{bmatrix}$:

```
> A := Translatemat(<3,2>):
> Transform(A,triangle);
```

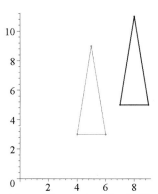

The translation matrix also provides us with much greater flexibility in applying the other geometric transformations. For example, all the rotations in Module 1 have the origin as the center of the rotation, but now we can rotate an object about any other center point.

======================

Example 1C: Rotate the triangle figure counterclockwise 45 degrees about its top vertex, (5, 9).

Solution: We break the problem into the following three steps.

- Step 1. Translate the triangle so the vertex (5, 9) is at the origin.

- Step 2. Rotate the translated figure 45 degrees about the origin.

- Step 3. Translate the figure back.

Each of these steps is accomplished (see below) by applying a matrix in homogeneous coordinates. The product of the three matrices gives us a single matrix that accomplishes the desired rotation in one step.

Step 1. Translate the triangle so the vertex (5, 9) is moved to the origin. We therefore translate the triangle by the vector $\begin{bmatrix} -5 \\ -9 \end{bmatrix}$.

```
> T := Translatemat(<-5,-9>):
> Transform(T,triangle,view=[-6..10,-6..10]);
```

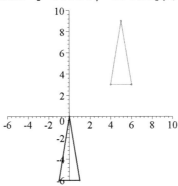

Step 2. Rotate the translated figure 45 degrees. Apply this rotation matrix R to the translated triangle, which we call "triangle1".

```
> triangle1 := T.triangle;
```

$$triangle1 := \begin{bmatrix} -1 & 0 & 1 & -1 \\ -6 & 0 & -6 & -6 \\ 1 & 1 & 1 & 1 \end{bmatrix}$$

```
> R := Rotatemat(Pi/4):
> Transform(R,triangle1,view=[-6..10,-6..10]);
```

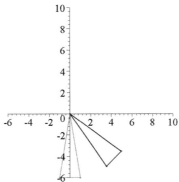

Step 3. Translate the figure so the top vertex is moved back from (0, 0) to (5, 9). Apply this translation matrix U to the rotated triangle, which we call "triangle2". (Note that the translation matrices T and U are inverses of one another.)

```
> triangle2 := R.triangle1:
> U := Translatemat(<5,9>):
> Transform(U,triangle2,view=[-6..10,-6..10]);
```

244 Chapter 4 Matrix Transformations

Now we multiply all three matrices together and thereby construct a single matrix that performs the rotation about the vertex (5, 9) with a single multiplication.

```
> M := U.R.T:
> Transform(M,triangle,view=[-6..10,-6..10]);
```

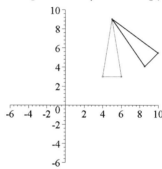

==========================

Making Movies

We can also apply the product matrix M repeatedly to create a simple animation:

```
> Movie(M,triangle,frames=20);
```

Execute the next line to see the individual frames used in the animation. Also try the option snapshot=on, which produces a nonanimated plot for printing.

```
> Movie(M,triangle,frames=20,trace=on);
```

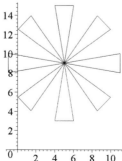

Exercise 1.1: Modify the commands in Example 1C to do each of the following:
(a) Rotate the triangle figure about its top vertex by a smaller angle, say π/8. Run **Movie** to check your matrix. (The commands that created the matrix *M* in Example 1C are repeated in the Student Workspace. Start by modifying those commands.)
(b) Rotate about one of the other vertices by the same angle as in (a) but rotate clockwise. Remember to adjust both translation matrices.
(c) The Lamph figure is shown below. Find a single matrix *N* that rotates this figure by 90 degrees counter-clockwise about the center point (7, 7.5). Test your matrix by using the **Transform** command.

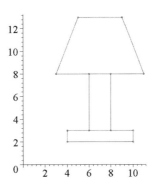

+ Student Workspace
– Answer 1.1

(a) Rotate the triangle figure about its top vertex by a smaller angle, say π/8.
```
> homogeneous := on:
> T := Translatemat(<-5,-9>):
> R := Rotatemat(Pi/8):
> U := Translatemat(<5,9>):
> M := U.R.T:
> Movie(M,triangle,frames=16);
```
(b) Rotate about one of the other vertices by the same angle as in (a), but rotating clockwise.
```
> T := Translatemat(<-6,-3>):
> R := Rotatemat(-Pi/8):
> U := Translatemat(<6,3>):
> M := U.R.T:
> Movie(M,triangle,frames=16);
```
(c) Find a matrix *N* that rotates the Lamph figure by 90 degrees counter-clockwise about its center point (7, 7.5).
```
> T := Translatemat(<-7,-7.5>):
> R := Rotatemat(Pi/2):
> U := Translatemat(<7,7.5>):
```

```
> M := U.R.T:
> Transform(M,Lamph);
```

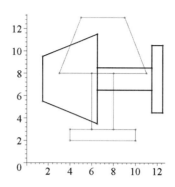

 Section 2: Homogeneous 3D Coordinates

We define the homogeneous coordinates for a point (x, y, z) in R^3 to be $(x, y, z, 1)$. The geometric transformations of R^3 are then represented by 4 by 4 matrices. For example, here are scaling and reflection matrices from Module 2, side by side with their homogeneous versions. Notice that in each case the original transformation makes up the 3 by 3 block in the upper-left corner of the matrix. The rest of the matrix is filled out with zeros, except for a 1 in the lower-right corner.

$$\begin{bmatrix} 2 & 0 & 0 \\ 0 & 3 & 0 \\ 0 & 0 & 5 \end{bmatrix} \quad \begin{bmatrix} 2 & 0 & 0 & 0 \\ 0 & 3 & 0 & 0 \\ 0 & 0 & 5 & 0 \\ 0 & 0 & 0 & 1 \end{bmatrix} \quad \begin{bmatrix} -1 & 0 & 0 \\ 0 & 1 & 0 \\ 0 & 0 & 1 \end{bmatrix} \quad \begin{bmatrix} -1 & 0 & 0 & 0 \\ 0 & 1 & 0 & 0 \\ 0 & 0 & 1 & 0 \\ 0 & 0 & 0 & 1 \end{bmatrix}$$

Let's be certain we are still in homogeneous mode:
```
> homogeneous := on;
```
Throughout this section we will apply matrices to the figure Hotel3dh, which is the Hotel3d figure in Module 2 expressed in homogeneous coordinates. Move it around to become familiar with its shape and location:
```
> Drawmatrix(Hotel3dh);
```

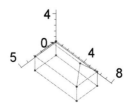

Note: The names of all figures that we use in the remainder of this module will end in an "h" to signify that they are expressed in homogeneous coordinates. Some of the 2D figures that you used in Module 1 have been redefined as figures lying in the *xy* plane of R^3. For example, here is the Bug sitting in the *xy* plane of R^3:

```
> Drawmatrix(Bug3dh);
```

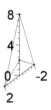

Other figures are "House3dh", "Lamp3dh", and a new figure, "Chevron3dh", that lies in the *xz* plane:

```
> Drawmatrix(Chevron3dh);
```

========================

Example 2A: The 3D translation matrix is similar to the one in Section 1. We construct the matrix corresponding to translation by the vector $\begin{bmatrix} a \\ b \\ c \end{bmatrix}$ by using **Translatemat3d(<a,b,c>)**. For example:

```
> T := Translatemat3d(<10,-6,8>);
```

$$T := \begin{bmatrix} 1 & 0 & 0 & 10 \\ 0 & 1 & 0 & -6 \\ 0 & 0 & 1 & 8 \\ 0 & 0 & 0 & 1 \end{bmatrix}$$

Here is the result of applying this translation to a point with homogeneous coordinates (*x*, *y*, *z*, 1):

```
> T.Vector([x,y,z,1]);
```
$$\begin{bmatrix} x+10 \\ -6+y \\ 8+z \\ 1 \end{bmatrix}$$

Now we apply T to Hotel3dh by using the **Transform** command. Move the plot around until you can see the translation from a good point of view. Try looking at it straight down each of the coordinate axes.

```
> Transform(T,Hotel3dh);
```

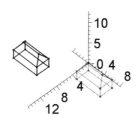

Rotation Transformations in Homogeneous Coordinates

As in Module 2, we use the command **Rotatemat3d** to produce rotation matrices for R^3. For example, to produce the nonhomogeneous and homogeneous versions of the matrix that rotates vectors about the z axis, we use:

```
> Rotatemat3d(<0,0,1>,theta,homogeneous=off);
```
$$\begin{bmatrix} \cos(\theta) & -\sin(\theta) & 0 \\ \sin(\theta) & \cos(\theta) & 0 \\ 0 & 0 & 1 \end{bmatrix}$$

```
> Rotatemat3d(<0,0,1>,theta,homogeneous=on);
```
$$\begin{bmatrix} \cos(\theta) & -\sin(\theta) & 0 & 0 \\ \sin(\theta) & \cos(\theta) & 0 & 0 \\ 0 & 0 & 1 & 0 \\ 0 & 0 & 0 & 1 \end{bmatrix}$$

Example 2B: Let's rotate the Chevron3dh figure by 30 degrees about the z axis. (Since the homogeneous mode is still "on" globally, we do not need to include it in each command.)

```
> Rz := Rotatemat3d(<0,0,1>,Pi/6):
> Transform(Rz,Chevron3dh);
```

Next, let's use the **Movie** command to see what happens when this rotation is applied 12 times:
```
> Movie(Rz,Chevron3dh,frames=12);
```

Explore ~ Here is an animation based on a composition of two matrices. Can you predict what the animation will look like before running it? Try modifying the commands to reverse the direction of the rotation or the translation (or both).
```
> T := Translatemat3d(<0,0,1>):
> Rz := Rotatemat3d(<0,0,1>,Pi/6):
> M := Rz.T:
> Movie(M,Chevron3dh,frames=24);
```

Example 2C: We rotate the Hotel3dh figure about the *y* axis, first by a single rotation, then in an animation with the same rotation applied repeatedly, and then part of the same animation with all the frames visible:
```
> Ry := Rotatemat3d(<0,1,0>,Pi/12):
> Transform(Ry,Hotel3dh);
```

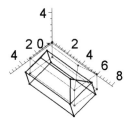

```
> Movie(Ry,Hotel3dh,frames=24);
> Movie(Ry,Hotel3dh,frames=4,trace=on,snapshot=on);
```

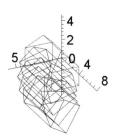

Exercise 2.1: Find a 4 by 4 matrix that reflects vectors across the *yz* plane using homogeneous coordinates. Test your matrix on the Hotel3dh figure.

+ Student Workspace

− Answer 2.1

```
> R := Diagmat([-1,1,1,1]):
```
Or use :
```
> R := Reflectmat3d(<1,0,0>):
> Transform(R,Hotel3dh);
```

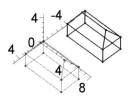

Turn the figure around to observe its symmetry.

Exercise 2.2: Find a 4 by 4 matrix that projects vectors onto the *xz* plane. Test your matrix on the Hotel3d figure.

+ Student Workspace

− Answer 2.2

```
> P := Diagmat([1,0,1,1]):
```
Or use:
```
> P := Projectmat3d(<0,1,0>):
> Transform(P,Hotel3dh);
```

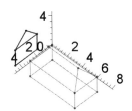

 Section 3: Moving an Object Along a Curve (optional)

If we want to create a realistic animation of a car moving along a curved road, we face two problems:
- How can we keep the car on the road as the road curves?
- How can we keep the car pointing in the direction of motion?

We address these problems one at a time in the following two subsections. First, let's again be certain that we are in homogeneous mode.

```
> homogeneous := on;
```
$$homogeneous := on$$

Staying on the Curve

Here is the parametrized curve we will use:
```
> x := t -> cos(t);
  y := t -> sin(2*t);
```
$$x := \cos$$
$$y := t \to \sin(2\,t)$$

And here is the definition of the car in homogeneous coordinates:
```
> car := Matrix([[.15,-.15,-.15,.15],[0,.075,-.075,0],[1,1,1,1]]);
```
$$car := \begin{bmatrix} .15 & -.15 & -.15 & .15 \\ 0 & .075 & -.075 & 0 \\ 1 & 1 & 1 & 1 \end{bmatrix}$$

We plot both the curve and the car in a single picture. Note that the car is pointing in the direction of the positive x axis.
```
> pict := Drawmatrix(car):
  pathpic := plot([x(t),y(t),t=0..2*Pi],color=black):
  display([pict,pathpic]);
```

To create the illusion of motion along the curve, we need to place the car at numerous points along the curve. We do this by translating the car from the origin to points on the curve. For example, let's translate the car to the point corresponding to time $t = 2$:

```
> T := Translatemat(<x(2),y(2)>):
> carpict := Drawmatrix(T.car):
  display([carpict,pathpic]);
```

Next, we write a loop to create and save a similar picture for 33 values of t, ranging from $t = 0$ to $t = 2\pi$ in increments of $\pi/16$. We call the pictures pic_1, pic_2, etc., so we can display them afterwards.

Maple note: When writing a "do loop", you begin with a "for" statement that tells how many times to execute the loop. The rest of the loop consists of a sequence of Maple commands preceded by "do" and followed by "end do:" All the lines of code should follow a single Maple prompt (>).

```
> for k from 0 to 32
  do
  tvalue := k*Pi/16:
  T := Translatemat(<x(tvalue),y(tvalue)>);
  carpict := Drawmatrix(T.car):
  pic[k+1] := display([carpict,pathpic]):
  end do:
```

Exercise 3.1: Explain in words what each of the five lines in the loop does.

 Student Workspace

 Answer 3.1

```
tvalue := k*Pi/16:
```
sets the current value for *t*.
```
T := Translatemat(<x(tvalue),y(tvalue)>);
```
calculates the translation matrix for the position corresponding to this value of *t*.
```
carpict := Drawmatrix(T.car):
```
creates picture of the car after it has been translated to the current position.
```
pic[k+1] := display([carpict,pathpic]):
```
puts pictures of car and curve together and labels this as the (*k* + 1)th frame of the animation.

Now we can see the pictures all at once by using the **display** command:
```
> display([seq(pic[k],k=1..33)]);
```

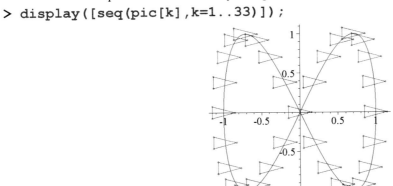

Or, by adding the instruction "insequence=true" to the previous **display** command, we can create an animation:
```
> display([seq(pic[k],k=1..33)],insequence=true);
```

Pointing in the Direction of Motion

We are halfway there! Now to keep the car pointing in the direction of motion, we should rotate the car at each step of the animation. If we can find the appropriate rotation matrix *R* for each value of *t*, we will have solved our last remaining problem.

Recall that the velocity (or tangent) vector to the parametrized curve always points in the direction of motion. We calculate the velocity vector by taking the derivative of each of the component functions of the curve with respect to *t*:
```
> vel := Vector([diff(x(t),t),diff(y(t),t)]);
```
$$vel := \begin{bmatrix} -\sin(t) \\ 2\cos(2t) \end{bmatrix}$$

Next we define the unit vector T in the direction of *vel*, namely $\frac{vel}{|vel|}$:

```
> T := Expand(1/Mag(vel)*vel);
```

$$T := \begin{bmatrix} -\dfrac{\sin(t)}{\sqrt{\sin(t)^2 + 4\cos(2t)^2}} \\ 2\dfrac{\cos(2t)}{\sqrt{\sin(t)^2 + 4\cos(2t)^2}} \end{bmatrix}$$

Recall that the columns of a 2 by 2 matrix transformation are the images of the standard unit vectors e_1 and e_2. Since the car points in the direction of e_1 and we want it to point in the direction of the unit vector T, the image of e_1 is T. Furthermore, if we were to rotate e_1 by 90 degrees counter-clockwise, we would get e_2; so rotating T by 90 degrees counter-clockwise, we get the image of e_2. Let's call this image N:

```
> N := Rotatemat(Pi/2,homogeneous=off).T;
```

$$N := \begin{bmatrix} -2\dfrac{\cos(2t)}{\sqrt{\sin(t)^2 + 4\cos(2t)^2}} \\ -\dfrac{\sin(t)}{\sqrt{\sin(t)^2 + 4\cos(2t)^2}} \end{bmatrix}$$

The rotation matrix we seek has columns T and N:

```
> R := Matrix([T,N]):
```

We must also construct the corresponding homogeneous version of this rotation matrix:

```
> Rh := Copyinto(R,Idmat(3),1,1):
```

Since the rotation matrix is a function of t, let's express it as such:

```
> Rot := tval -> map(evalf,subs(t=tval,Rh));
```

$$Rot := tval \rightarrow map(evalf, subs(t = tval, Rh))$$

For example, here is the rotation matrix when $t = 2$:

```
> Rot(2);
```

$$\begin{bmatrix} -.571012 & .820938 & 0. \\ -.820938 & -.571012 & 0. \\ 0. & 0. & 1. \end{bmatrix}$$

We are ready to test our results. We repeat the definition of the curve, and then we repeat the loop that we used earlier but with one modification. Can you see where we have made the change?

```
> pathpic := plot([x(t),y(t),t=0..2*Pi],color=black):
> for k from 0 to 32
    do
    tvalue := k*Pi/16:
    T1 := Translatemat(<x(tvalue),y(tvalue)>);
    carpict := Drawmatrix(T1.Rot(tvalue).car):
```

```
    pic[k+1] := display([carpict,pathpic]):
end do:
```
Now try out the animation:
```
> display([seq(pic[k],k=1..33)],insequence=true,scaling=constrained);
```
You can see the individual elements of the picture by executing the next line.
```
> display([seq(pic[k],k=1..33)],scaling=constrained);
```

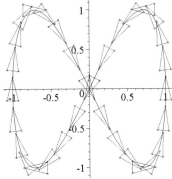

Problem 5 asks you to apply the ideas in this section to motion along a curve in R^3. However, for that problem, you will also need to know how to use the unit normal vector to find the rotation matrix. That approach is addressed in the following appendix.

Appendix: Using the Unit Normal Vector

We introduce in this section an alternative method for finding the rotation matrix *Rot* we found above. Recall from calculus that the unit normal vector *N* for a curve is perpendicular to the unit tangent vector *T*. Furthermore, *N* is found by computing the derivative of *T* and then converting the derivative to a unit vector. This is quite tedious to do by hand, but routine for Maple. In the region below, we calculate the derivative of *T* and call it *dT*. Then we define the unit normal vector $N = \frac{dT}{|dT|}$. We have suppressed the output of these expressions since they are quite messy. Change the colons to semicolons if you would like to see the output.
```
> dT := map(simplify,map(diff,T,t)):
> N := Expand(1/Mag(dT)*dT):
```
Finally, we build the rotation matrix from *T* and *N* exactly as we did earlier:
```
> R := Matrix([T,N]):
> Rh := Copyinto(R,Idmat(3),1,1):
> Rot := tval -> map(evalf,subs(t=tval,Rh));
```

Problems

Problem 1: What matrix produces the movie?

The 2D animation below was created using the command

Movie(M,triangle,frames=18)

for some matrix *M*. The triangle dimensions are decreasing by 10% in each frame. Find the matrix *M*. Hint: *M* is the product of four matrices. In your answer, list the four matrices that are used to construct *M*, and indicate the order in which they are multiplied. Check your answer by constructing *M* and then entering and executing the above **Movie** command.
Reminder: If you want to print your animation, use the option snapshot=on.

```
> triangle := Matrix([[4,5,6,4],[3,9,3,3],[1,1,1,1]]):
```

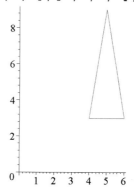

 Student Workspace

Problem 2: Translation and rotation

(a) Enter the 3 by 4 matrix *tri* whose columns are the homogeneous coordinates for the triangle with vertices (-1, -1), (1, -1), (0, 1). Use the command **Drawmatrix(tri,homogeneous=on)** to draw the triangle.
(b) Use **Translatemat** and **Transform** to show the triangle *tri* translated to the right 10 units.
(c) (By hand) Find the vertices of the translated triangle in (b) by multiplying the homogeneous coordinates in (a) by the translation matrix in (b).
(d) Use the **Movie** command to make a 36-frame animation of the triangle *tri* in which each frame is the result of applying a rotation about the origin of $\dfrac{\pi}{18}$ radians followed by a translation by the vector $\begin{bmatrix} .4 \\ .1 \end{bmatrix}$.
Where does the triangle *tri* end up?

(e) (Challenge) Explain why, in the part (d) animation, the triangle *tri* ends up where it does; that is, explain why the repeated translations do not translate the triangle.

 Student Workspace

Problem 3: Twister -- the movie

The animation below was created using the command **Movie(M,Hotel3dh,frames=16)** for some matrix *M*. Find *M*. Hint: You may want to view the animation from several points of view. It may also help to advance it frame by frame using the button marked ->| .

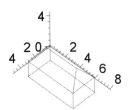

 Student Workspace

Problem 4: Programming a 3D animation (optional)

Note: Study the discussion of programming in Section 3 before attempting this problem.

Execute the all the commands below to generate an animation. After studying the animation, answer the questions that follow.

```
> homogeneous := on:
> p := 'p':
> L := Diagmat([1,0,1,1]):
> M := Diagmat([0,1,1,1]):
> N := Translatemat3d(<-3.5,0,-2>):
> for j from 0 to 12
  do
    R := Rotatemat3d(<0,1,0>,j*Pi/6):
    W := N^(-1).R.N.Hotel3dh:
    part1 := Drawmatrix(W,figcolor=red):
    part2 := Drawmatrix(L.W,figcolor=blue):
    part3 := Drawmatrix(M.W,figcolor=blue):
    p[j+1] := display([part1,part2,part3]):
```

258 Chapter 4 Matrix Transformations

```
    end do:
> display([seq(p[k],k=1..13)],insequence=true);
```
(a) Describe in words what is happening in the animation.
(b) Explain the purpose for each of the commands above.
(c) Modify the code to create a similar animation based on a rotation about a line that is parallel to the *x* axis (instead of the *y* axis) and that cuts the Hotel3d figure through a central point.

 Student Workspace

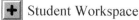 Problem 5: Flying along a curve in 3-Space (optional)

In this problem you will extend the methods in Section 3 to the corresponding problem of moving a figure along a curve in 3-space. You are given a space curve and a figure called Jet3dh. The figure is oriented so that its wings are in the *xy* plane and its nose is in the direction of e_1. Your task is to create an animation that flies Jet3dh along the curve in such a way that:

- Jet3dh always points in the direction of the tangent vector; and

- the wings of Jet3dh always lie in the "osculating plane" (the plane containing the unit tangent vector *T* and unit normal vector *N*).

Hints: Start with a copy of the commands in Section 3 and make whatever modifications are required. To draw a space curve, use the **spacecurve** command. To show Jet3dh without its corner points, use the option points=off in the **Drawmatrix** command. Calculate *N* as in the appendix to Section 3. It will help to calculate a third unit vector $B = T \times N$ (the cross product of *T* and *N*).
Here is the curve that Jet3dh will fly along:
```
> x := t -> cos(t)/2;
  y := t -> sin(t);
  z := t -> sin(2*t)/2;
```
We plot both the curve and Jet3dh in a single picture:
```
> part1 := Drawmatrix(Jet3dh,points=off,linecolor=blue,axes=framed):
  part2 := spacecurve([x(t),y(t),z(t),t=0..2*Pi],color=red):
  display3d([part1,part2]);
```
 Student Workspace

 Problem 6: The Art of linear algebra (optional)

Was there a picture or animation in this module that you found particularly appealing? Did it make you want to try to create something of your own ? This problem is an invitation for you to be creative with the mathematical tools that have been presented in this module. The only requirement is that you annotate your work so that someone else (such as another student) can understand what you did.

 Student Workspace

Chapter 5: Vector Spaces

Module 1. Subspaces

Module 2. Basis and Dimension

Module 3. Subspaces Associated with a Matrix

Module 4. Loops and Spanning Trees -- An Application

Commands used in this chapter

Column(A,i); selects the ith column of the matrix A.
Matrix([[a,b],[c,d]]); defines the matrix with rows $[a, b]$ and $[c, d]$.
Matrix([u,v,w]); produces the matrix whose columns are the vectors u, v, w.
subs({a=3,b=13},expr); substitutes the values for a and b in the expression *expr*.
Vector([a,b,c]); defines the vector $\langle a, b, c \rangle$.

LAMP commands:
Basisgrid(u,v,[a,b]); draws xy and uv coordinate grids and the point $a\,u + b\,v$.
Matsolve(A,b,free=t); solves the matrix-vector equation $A\,x = b$ for x. The solution is expressed in terms of the free variable t.
Nullbasis(A); produces a basis for the null space of the matrix A.
Reduce(M); produces a row echelon form of the matrix M.
Reduce(M,form=rref); produces the <u>reduced</u> row echelon form of the matrix M.

Linear Algebra Modules Project
Chapter 5, Module 1

Subspaces

 Special instructions

The Tutorial for this module is designed to be done by hand without the use of Maple. Likewise, most of the Problems can and should be done without the use of Maple. The answers to the Exercises are all at the end of the Tutorial.

 Purpose of this module

The purpose of this module is to introduce the concept of a subspace of R^n and to study the principal examples of subspaces.

 Prerequisites

Linear combinations, span, lines and planes, linear systems, matrix algebra.

```
> restart: with(LinearAlgebra): with(plots): with(Lamp):
  UseHardwareFloats := false: Digits :=6:
```

Tutorial

 Section 1: Definition of Subspace

A simple example of a subspace is a plane in R^3 passing through the origin:

================

Example 1A: Consider the plane through the origin defined by the homogeneous equation $x - 3y + 2z = 0$. For convenience we will use the letter S to denote the set consisting of all vectors in this plane. In other words, S is the solution set for this homogeneous equation.

We decompose the solution set S, and thus find two direction vectors that span the plane:

$$\begin{bmatrix} x \\ y \\ z \end{bmatrix} = \begin{bmatrix} 3y - 2z \\ y \\ z \end{bmatrix} = y \begin{bmatrix} 3 \\ 1 \\ 0 \end{bmatrix} + z \begin{bmatrix} -2 \\ 0 \\ 1 \end{bmatrix} = y v_1 + z v_2$$

$$\text{where } v_1 = \begin{bmatrix} 3 \\ 1 \\ 0 \end{bmatrix} \text{ and } v_2 = \begin{bmatrix} -2 \\ 0 \\ 1 \end{bmatrix}$$

The vectors v_1, v_2 give us an alternative way of describing S, namely S = Span$\{v_1, v_2\}$. This description of the plane has the advantage that we can use v_1 and v_2 to construct all of the vectors in the plane. In contrast, the equation $x - 3y + 2z = 0$ describes the plane S only implicitly; that is, this equation allows us to test whether a vector is in the plane, but it doesn't give us a direct way to construct these vectors.

==================

We now note two simple, but important, algebraic properties that hold for vectors in the set S:

- Property (1) If you choose any vector u in S and multiply it by any real number k, the resulting vector ku will also be in S.

- Property (2) If you choose any two vectors u and v in S and add them, the resulting vector $u + v$ will also be in S.

Try visualizing these properties in the plane S, using the geometry of vectors in 3-space. Proving these properties is straightforward using vector algebra :

Proof of property (1): Let u be any vector in S and k be any scalar. Since S = Span$\{v_1, v_2\}$, we can express u as a linear combination of the vectors v_1 and v_2: $u = av_1 + bv_2$ (where a and b are real numbers). Now multiply this equation by k; we get

$$ku = k(av_1 + bv_2) = (ka)v_1 + (kb)v_2$$

So ku is also a linear combination of the vectors v_1 and v_2, and hence ku is contained in **S**.

Proof of property (2): Let u and v be any vectors in S. Each can be expressed as a linear combination of v_1 and v_2: $u = av_1 + bv_2$ and $v = cv_1 + dv_2$. Now add the two equations; we get

$$u + v = (av_1 + bv_2) + (cv_1 + dv_2) = (a+c)v_1 + (b+d)v_2$$

So $u + v$ is also a linear combination of v_1 and v_2, and hence $u + v$ is contained in S.

The two properties above are the defining characteristics of a "subspace" of R^n:

> ***Definition:*** Let S be a set of vectors in R^n. S is a ***subspace*** of R^n if:
> - (1) ku is in S whenever u is a vector in S and k is a scalar; and
> - (2) $u + v$ is in S whenever u and v are vectors in S; and
> - (3) The zero vector of R^n is contained in S.

Properties (1) and (2) are often referred to as the "closure" properties for a subspace. We say that a subspace is "closed" under the operations of scalar multiplication and vector addition in the sense that these operations always produce vectors that are still within the subspace. Property (3) is included simply to guarantee that a subspace contains at least one vector; an empty subspace would be of no use. In fact, the subset $\{0\}$, which contains only the zero vector, is an example of a subspace. (Check that all three properties hold for this subspace.) This trivial example is called the "zero subspace" and is the only subspace that contains finitely many vectors.

Our first example of a subspace, the plane through the origin in Example 1A, turns out to be a very good model for many of the subspaces that you will encounter later in this chapter. The next two theorems describe the two ways in which subspaces typically arise.

Theorem 1: If $v_1, v_2, ..., v_p$ are vectors in R^n, then Span$\{v_1, v_2, ..., v_p\}$ is a subspace of R^n.

Exercise 1.1: The proof of Theorem 1 is a natural generalization of the approach we used in proving the closure properties (1) and (2) for Span$\{v_1, v_2\}$ in Example 1A. Using the proof in Example 1A as a model, write down the proof for the case where $p = 3$; i.e., show that if v_1, v_2, v_3 are any three vectors in R^n, then Span$\{v_1, v_2, v_3\}$ is a subspace of R^n.

Student Workspace

Now let's generalize the first way we described the subspace S in Example 1A, namely by a homogeneous linear equation. The generalization is the solution set of any homogenous linear system:

Theorem 2: Let A be any m by n matrix. The set of solutions of the homogeneous equation $A x = 0$ is a subspace of R^n.

At first glance, Theorems 1 and 2 may seem quite unrelated. However, there is a close connection between them, as Exercise 1.2 illustrates.

Exercise 1.2: Below is a 3 by 4 matrix A and the solution set S of the homogeneous equation $A x = 0$:

$$A = \begin{bmatrix} 1 & 3 & 0 & 4 \\ 1 & 2 & 2 & -4 \\ 2 & 6 & 0 & 8 \end{bmatrix} \quad x = \begin{bmatrix} -6x_3 + 20x_4 \\ 2x_3 - 8x_4 \\ x_3 \\ x_4 \end{bmatrix}$$

(a) Decompose the above solution set S into the span of a set of vectors.
(b) Use the result from (a) to show that S is a subspace of R^4.

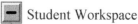

It is easy to see how the procedure used in answering Exercise 1.2 could be applied to any homogeneous equation $A x = 0$. But how could we prove Theorem 2 directly from the definition of subspace? In Problem 1 you will use the rules of matrix algebra to write a short proof of Theorem 2 without solving the homogeneous system or using Theorem 1.

Summary: The principal ways in which subspaces of R^n arise are as the span of a set of vectors and as the solution set of a homogeneous linear system.

In fact, any subspace of R^n can be expressed in either of these two ways.

================

Example 1B: Suppose S is a subspace of R^n, and suppose we choose four vectors from this subspace at random. Call these vectors w_1, w_2, w_3, w_4. Now we form a new vector p by taking a linear combination of these four vectors. For example, let $p = 2 w_1 + 5 w_2 + 4 w_3 + 7 w_4$. Explain why p must also be a vector in the subspace S.

Solution: Each of the four terms in the linear combination is a vector in S by property (1), since each term is a scalar multiple of a vector in S. So let's rewrite p as the sum of these four vectors, $p = y_1 + y_2 + y_3 + y_4$, where $y_1 = 2 w_1$, $y_2 = 5 w_2$, etc. We need to show that the sum of these four vectors in S is a vector in S. The sum of the first two vectors, $y_1 + y_2$, and the sum of the last two vectors, $y_3 + y_4$, are each vectors in S by property (2). So we can rewrite p as $p = z_1 + z_2$, where $z_1 = y_1 + y_2$ and $z_2 = y_3 + y_4$ are each vectors in S. Therefore p is a vector in S, since it is a sum of two vectors in S.

================

Example 1B illustrates a general fact about subspaces that we summarize as follows:

- If S is a subspace of R^n and $\{v_1, v_2, ..., v_p\}$ is any set of vectors in S, then every linear combination of these vectors is in S.

Section 2: Geometry of Subspaces

What subsets of R^2 are subspaces of R^2? What subsets of R^3 are subspaces of R^3? We will see that the only subsets that are subspaces are the "linear" subsets through the origin, such as lines and planes through the origin. This leaves a great many subsets that are not subspaces. For example, subsets of R^2 that are not subspaces include curves that are not lines, lines that do not go through the origin, or any bounded set (such as a circle, polygon, or line segment) except for the zero subspace. Subsets of R^3 that are not subspaces include curves that are not lines, surfaces that are not planes, lines or planes that do not go through the origin, or any bounded set (such as a sphere, circle, or polyhedron) except for the zero subspace.

Exercise 2.1: For each of the following subsets of R^2, show that the subset is not a subspace. To show that a subset is not a subspace, give a counter-example to one of the three properties in the definition of subspace. Also draw the subset (by hand, if you like), and include in your drawing the vectors that provide your counter-example.

(a) The set of all points on the line $x + y = 1$.
(b) The set of all points on the half-line $y = x$, where $0 \leq x$.
(c) The set of all points on the graph of the equation $y = x^2$.

 Student Workspace

================

Example 2A: Find all possible subspaces of R^2.

Solution: We will build up all possible subspaces systematically from the smallest to the largest.

Case 1: The set $\{0\}$ containing just the zero vector of R^2 is a subspace of R^2: (1) Any scalar times the zero vector is the zero vector; (2) the zero vector plus the zero vector is the zero vector.

Case 2: If a subspace S contains something more than just the zero vector, we can select a nonzero vector u in S. Then, by property (1) for subspaces, S must contain all vectors $a\,u$, where a is any scalar. That is, S must contain all vectors in Span$\{u\}$. If S = Span$\{u\}$, then S is a subspace of R^2, by Theorem 1. Note that S is a line through the origin.

Case 3: If a subspace S contains something more than just a line through the origin, i.e., more than Span$\{u\}$, we can select a vector v that is in S but not in Span$\{u\}$. Since S is a subspace, S must contain Span$\{u, v\}$. Since v is not in Span$\{u\}$, the vectors u and v are not parallel; therefore Span$\{u, v\}$ is all of R^2. So S = R^2.

Conclusion: The subspaces of R^2 are: The zero subspace, $\{0\}$; the lines through the origin (that is, each such line is a subspace); and all of R^2.

================

In Problem 2, we will ask you to carry out a similar analysis for subspaces of R^3.

 Answers to Exercises

Answer 1.1:

Let v_1, v_2, v_3 be any three vectors in R^n. We show that S = Span$\{v_1, v_2, v_3\}$ is a subspace of R^n by showing that the three properties of the definition are satisfied.

Proof of property (1): Let u be any vector in S and k be any scalar. Since S = Span$\{v_1, v_2, v_3\}$, we can express u as a linear combination of the vectors v_1, v_2, v_3: $u = a v_1 + b v_2 + c v_3$ (where a, b, c are real numbers). Now multiply this equation by k; we get:

$$k u = k (a v_1 + b v_2 + c v_3) = (k a) v_1 + (k b) v_2 + (k c) v_3$$

So $k u$ is also a linear combination of the vectors v_1, v_2, v_3, and hence $k u$ is contained in S.

Proof of property (2): Let u and v be any vectors in S. Each can be expressed as a linear combination of v_1, v_2, v_3: $u = a v_1 + b v_2 + c v_3$ and $v = d v_1 + e v_2 + f v_3$. Now add these two equations; we get:

$$u + v = (a v_1 + b v_2 + c v_3) + (d v_1 + e v_2 + f v_3) = (a + d) v_1 + (b + e) v_2 + (c + f) v_3$$

So $u + v$ is also a linear combination of v_1, v_2, v_3, and therefore $u + v$ is a vector in S.

Proof of property (3): The zero vector is included in the span of any set of vectors, as we see by choosing all the weights to be zero: $0 v_1 + 0 v_2 + 0 v_3 = 0$.

Answer 1.2:
Here is the decomposition of the given solution set into the set of all linear combinations of two vectors, u and v, in R^4:

$$x = \begin{bmatrix} -6 x_3 + 20 x_4 \\ 2 x_3 - 8 x_4 \\ x_3 \\ x_4 \end{bmatrix} = x_3 \begin{bmatrix} -6 \\ 2 \\ 1 \\ 0 \end{bmatrix} + x_4 \begin{bmatrix} 20 \\ -8 \\ 0 \\ 1 \end{bmatrix} = x_3 u + x_4 v$$

(b) Since $S = \text{Span}\{u, v\}$, where u and v are the two vectors above, S is a subspace of R^4 by Theorem 1.

Answer 2.1:
(a) Since $0 + 0 \neq 1$, the vector $\langle 0, 0 \rangle$ is not in the subset, which violates property (3) of the definition. (Properties (1) and (2) are also violated.)
(b) The vector $\langle 1, 1 \rangle$ is in the subset, but $-2 \langle 1, 1 \rangle$ is not in the subset, which violates property (1) of the definition. (Neither property (2) nor (3) is violated.)
(c) The vectors $\langle 1, 1 \rangle$ and $\langle -1, 1 \rangle$ are in the subset, but their sum $\langle 0, 2 \rangle$ is not in the subset, which violates property (2) of the definition. (Property (1) is also violated but not (3).)

Problems

Problem 1: Solution set of $A x = 0$ is a subspace

Prove Theorem 2 directly from the definition of subspace. That is, let A be an m by n matrix, and let S be the set of all vectors x in R^n such that $A x = 0$. Use the rules of matrix algebra to show that S satisfies properties (1), (2), and (3).

Note: To help you see the type of proof that is wanted here, we provide the first few steps of the proof of property (2):

Let u and v be any vectors in S. Then $A\,u = 0$ and $A\,v = 0$. Now use the rules of matrix algebra to show that $u + v$ is in S. (Use a similar method for the proof of property (1).)

Student Workspace

Problem 2: Subspaces of R^3

What are all the subspaces of R^3? Model your method of analysis on the method in Example 2A.

Student Workspace

Problem 3: Subsets that are not subspaces

For each of the following subsets of R^n, show that the subset is not a subspace by giving a counter-example to one of the three properties in the definition of subspace. A drawing might help your explanation.

(a) The subset of R^3 consisting of all the solutions to the equation $2x - 3y + 4z = 5$.

(b) The subset of R^2 consisting of all the rational points on the line $y = x$; that is, x (and hence y) is a ratio of integers, m/n ($n \neq 0$).

(c) The subset of R^2 consisting of all the points on one or both of the lines $y = x$ and $y = -x$ (i.e., the "union" of these two lines).

(d) The subset of R^5 consisting of all the solutions of the nonhomogeneous equation $A\,x = b$, where A is a given 4 by 5 matrix and b is a given nonzero vector in R^4.

Student Workspace

Problem 4: Two ways of representing a subspace

Let $S_1 = \mathrm{Span}\{u, v\}$, where u and v are given below, and let S_2 be the solution set of the homogeneous equation $A\,x = 0$, where A is given below. (Suggestion: You may want to use Maple in solving this problem.)

(a) Determine whether the vectors p and q below are in S_1 or not; also determine whether they are in S_2 or not.

(b) Show that S_1 and S_2 are the same subspace; that is, show that every vector in S_1 is also in S_2, and every vector in S_2 is also in S_1.

```
> u := Vector([2,-1,0,2]);
  v := Vector([0,1,-4,2]);
```

```
> A := Matrix([[0,2,1,1],[2,2,0,-1]]);
> p := Vector([1,-1,2,0]);
  q := Vector([1,0,2,0]);
```
+ Student Workspace

Problem 5: Vectors Ax form a subspace

(a) Let A be the 4 by 3 matrix defined below. Let S denote the subset of R^4 consisting of all the vectors Ax, where x ranges over all the vectors in R^3. Explain why S is a subspace of R^4. Hint: If you can write S as the span of a set of vectors, then you can simply apply Theorem 1.

(b) Let A be a m by n matrix. Let S denote the subset of R^m consisting of all the vectors Ax, where x ranges over all the vectors in R^n. Prove that S is a subspace of R^m. See hint in (a).

$$A = \begin{bmatrix} 0 & -5 & 5 \\ -1 & -5 & 4 \\ 2 & -3 & 5 \\ 5 & 5 & 0 \end{bmatrix}$$

+ Student Workspace

Problem 6: Fixed points form a subspace

Let A be any n by n matrix and let S be the set of "fixed points" of A; that is, S is the set of all vectors x in R^n such that $Ax = x$. Prove that S is a subspace of R^n. Hint: Use rules of matrix algebra to change the problem into one for which you can simply apply Theorem 2.

+ Student Workspace

Linear Algebra Modules Project
Chapter 5, Module 2

Basis and Dimension

 Purpose of this module

The purpose of this module is to introduce the concepts of basis and dimension and to gain an understanding of them through both geometric and algebraic examples.

 Prerequisites

Subspaces, linear independence, span, lines and planes, linear systems, matrix algebra.

 Commands used in this module

```
> restart: with(LinearAlgebra): with(plots): with(Lamp):
  UseHardwareFloats := false: Digits := 6:
```

Tutorial

 Section 1: Definition of Basis and Dimension

We all have an intuitive sense of the meaning of "dimension." We say that a line has dimension one (length), a plane has dimension two (length and width), and 3-space has dimension three (length, width, and height). For dimensions greater than 3, geometric intuition is less reliable, and so our first task in this module will be to develop a general and unambiguous definition of dimension. Specifically, we will define the concept of the dimension of a subspace.

Consider, for example, a plane through the origin in R^3, which is a subspace of R^3. We expect this subspace should have dimension two. Note, not coincidentally, that this plane is the span of exactly two vectors, Span$\{u, v\}$. It can be also written as the span of three or more vectors, but two is the fewest number of vectors that can span the plane. Furthermore, $\{u, v\}$ is linearly independent, and any set of more than two vectors in the plane would be linearly dependent. This observation leads us to make the following definitions:

Definition: Let S be any subspace of R^n. A set of vectors $\{v_1, ..., v_p\}$ in S is called a ***basis*** of S if $\{v_1, ..., v_p\}$ spans S and $\{v_1, ..., v_p\}$ is linearly independent.

Definition: Let S be any subspace of R^n other than the zero subspace $\{0\}$. If S has a basis consisting of p vectors, then p is called the ***dimension*** of S. By convention, we say that the dimension of the zero subspace is 0.

Module 5.2 Basis and Dimension

Summary: A basis of a subspace S is a linearly independent set of vectors in S that spans S. The dimension of S is the number of vectors in a basis of S.

Note: As we will see, a subspace can have many different bases, but all the bases for a given subspace have the same number of vectors. (See Theorem 4 in the Appendix to this module.) This assures us that the dimension of a subspace does not depend on which basis we choose for that subspace.

===============

Example 1A: Let $S = \text{Span}\{u\}$, where $u = \langle 1, 3, -2 \rangle$. So S is a line through the origin. Since $\{u\}$ is linearly independent and spans S, the dimension of S is 1 (the number of vectors in the basis $\{u\}$).

===============

Example 1B: Let $S = \text{Span}\{u, v\}$, where $u = \langle 1, 3, -2 \rangle$ and $v = \langle 2, -1, 0 \rangle$. So S is a plane through the origin. Since $\{u, v\}$ is linearly independent and spans S, the dimension of S is 2 (the number of vectors in the basis $\{u, v\}$).

===============

Exercise 1.1: (By hand) Let $S = \text{Span}\{u, v, w\}$, where

$$u = \langle 1, 3, -1, 3 \rangle, \quad v = \langle 2, -1, 0, -2 \rangle, \quad w = \langle 4, 5, -2, 4 \rangle$$

(a) Check that $w = 2u + v$. Is the set of three vectors $\{u, v, w\}$ a basis for S? Why or why not?
(b) Explain why $\text{Span}\{u, v, w\} = \text{Span}\{u, v\}$.
(c) What is the dimension of S ?

+ Student Workspace

− Answer 1.1

(a) The set $\{u, v, w\}$ cannot be a basis since it is not linearly independent.
(b) Every linear combination of u, v, w can be written as a linear combination of just the vectors u and v:

$$a u + b v + c w = a u + b v + c (2 u + v) = (a + 2 c) u + (b + c) v$$

(c) Since $\{u, v\}$ is linearly independent and $\text{Span}\{u, v\} = S$, $\{u, v\}$ is a basis for S. Therefore the dimension of S is 2.

===============

Example 1C: It should come as no surprise that the dimension of R^n is n. Here is the proof. The following vectors span R^n and are linearly independent:

$$e_1 = \begin{bmatrix} 1 \\ 0 \\ . \\ . \\ . \\ 0 \end{bmatrix}, \quad e_2 = \begin{bmatrix} 0 \\ 1 \\ 0 \\ . \\ . \\ 0 \end{bmatrix}, \quad \ldots, \quad e_n = \begin{bmatrix} 0 \\ . \\ . \\ 0 \\ 0 \\ 1 \end{bmatrix}$$

(The vector e_i has a 1 in the *i*th component and 0's elsewhere; you might recognize these vectors as the columns of the *n* by *n* identity matrix.) So $\{e_1, e_2, ..., e_n\}$ is a basis of R^n (called the "standard basis" of R^n). Since this set of vectors contains exactly *n* vectors, R^n has dimension *n*. For example, the standard basis of R^3 consists of the vectors

$$e_1 = \begin{bmatrix} 1 \\ 0 \\ 0 \end{bmatrix}, \quad e_2 = \begin{bmatrix} 0 \\ 1 \\ 0 \end{bmatrix}, \quad e_3 = \begin{bmatrix} 0 \\ 0 \\ 1 \end{bmatrix}$$

which are sometimes denoted *i, j, k* instead of e_1, e_2, e_3.

Language note: We express the concepts of span and basis in a variety of phrases. If, for example, a subspace S equals $\mathrm{Span}\{v_1, ..., v_k\}$, we refer to $\{v_1, ..., v_k\}$ as a *spanning set* for S, or we say that S is *spanned by* $\{v_1, ..., v_k\}$ or that $\{v_1, ..., v_k\}$ *spans* S. Since a basis of S is a spanning set of S and has the property that no fewer vectors can span S, we refer to a basis of S as a *minimal spanning set* of S. We will use these phrases freely whenever the occasion warrants.

 Section 2: How to Find a Basis

As we saw in Module 1, the two typical forms for a subspace are the solution set of a homogeneous matrix equation and the span of a set of vectors. We will see how to find a basis in each of these two circumstances.

Example 2A: Solve the linear system $Ax = 0$, where A is given below, and find a basis for the solution set. What is the dimension of the solution set?

```
> A := Matrix([[-1,-1,-2,3,1],[-9,5,-4,-1,-5],[7,-5,2,3,5]]);
```

$$A := \begin{bmatrix} -1 & -1 & -2 & 3 & 1 \\ -9 & 5 & -4 & -1 & -5 \\ 7 & -5 & 2 & 3 & 5 \end{bmatrix}$$

Solution: We use **Matsolve** to solve $Ax = 0$ and then decompose the solution to find a basis:

```
> Matsolve(A,zero,free=x);
```

$$\begin{bmatrix} -x_3 + x_4 \\ -x_3 + 2x_4 + x_5 \\ x_3 \\ x_4 \\ x_5 \end{bmatrix}$$

The decomposition of this solution is

$$\begin{bmatrix} -x_3+x_4 \\ -x_3+2x_4+x_5 \\ x_3 \\ x_4 \\ x_5 \end{bmatrix} = x_3 \begin{bmatrix} -1 \\ -1 \\ 1 \\ 0 \\ 0 \end{bmatrix} + x_4 \begin{bmatrix} 1 \\ 2 \\ 0 \\ 1 \\ 0 \end{bmatrix} + x_5 \begin{bmatrix} 0 \\ 1 \\ 0 \\ 0 \\ 1 \end{bmatrix} = x_3\,u + x_4\,v + x_5\,w$$

So the solution set is Span$\{u, v, w\}$. We must also check that $\{u, v, w\}$ is a linearly independent set; that is, we must solve the equation $c_1 u + c_2 v + c_3 w = 0$ and check that the only solution is when all the coefficients are zero. We rewrite the equation as $B x = 0$, where B is the matrix whose columns are the vectors u, v, w; then we solve $B x = 0$:

```
> u := Vector([-1,-1,1,0,0]):
  v := Vector([1,2,0,1,0]):
  w := Vector([0,1,0,0,1]):
> B := Matrix([u,v,w]);
```

$$B := \begin{bmatrix} -1 & 1 & 0 \\ -1 & 2 & 1 \\ 1 & 0 & 0 \\ 0 & 1 & 0 \\ 0 & 0 & 1 \end{bmatrix}$$

```
> Matsolve(B,zero,free=x);
```

Since the only solution to $B x = 0$ is the trivial solution $x = 0$, the set of vectors $\{u, v, w\}$ is linearly independent. (In fact, we saw in Module 4 of Chapter 2 that whenever we decompose the solution set of a homogeneous linear system, the spanning vectors produced by the decomposition are <u>always</u> linearly independent. See Example 3A and Problem 12 of that module.) Therefore $\{u, v, w\}$ is a basis for the solution set of $A x = 0$, and so the solution set has dimension 3 (the number of vectors in this basis).

===============

Next we consider a subspace that is given as a span of a set of vectors.

===============

Example 2B: Find a basis for $S = \text{Span}\{v_1, v_2, v_3, v_4\}$ (see below). What is the dimension of S?

```
> v1 := Vector([1,3,2,-5]):
  v2 := Vector([0,1,5,-3]):
  v3 := Vector([4,1,1,-1]):
  v4 := Vector([-2,5,3,-9]):
```

Solution:
(1) We begin by finding out whether the set $\{v_1, v_2, v_3, v_4\}$ is linearly independent. If it is, then it is a basis for S and we are done. We solve the vector equation $c_1 v_1 + c_2 v_2 + c_3 v_3 + c_4 v_4 = 0$ by writing it in

the form $Cx = 0$, where C is the matrix whose columns are the vectors v_1, v_2, v_3, v_4:

```
> C := Matrix([v1,v2,v3,v4]);
```

$$C := \begin{bmatrix} 1 & 0 & 4 & -2 \\ 3 & 1 & 1 & 5 \\ 2 & 5 & 1 & 3 \\ -5 & -3 & -1 & -9 \end{bmatrix}$$

```
> Matsolve(C,zero,free=x);
```

$$\begin{bmatrix} -2x_4 \\ 0 \\ x_4 \\ x_4 \end{bmatrix}$$

Since $Cx = 0$ has nontrivial solutions, $\{v_1, v_2, v_3, v_4\}$ is linearly dependent. Setting $x_4 = 1$ yields the solution $c_1 = -2$, $c_2 = 0$, $c_3 = 1$, $c_4 = 1$ and hence the linear dependence relation $-2v_1 + v_3 + v_4 = 0$.

(2) We use the linear dependence relation to solve for one of the v_i in terms of the others; we choose to solve for v_3: $v_3 = 2v_1 - v_4$. Since v_3 is a linear combination of v_1 and v_4, we can eliminate it. In other words, Span$\{v_1, v_2, v_4\}$ = Span$\{v_1, v_2, v_3, v_4\}$.

(3) Is the subset consisting of the three vectors $\{v_1, v_2, v_4\}$ a basis for S? We once again check for linear independence.

```
> A := Matrix([v1,v2,v4]):
  Matsolve(A,zero,free=x);
```

$$\begin{bmatrix} 0 \\ 0 \\ 0 \end{bmatrix}$$

Since the set $\{v_1, v_2, v_4\}$ is linearly independent and spans S, it is a basis for S, which therefore has dimension 3.

================

Exercise 2.1: Use the linear dependence relation that we found in Example 2B to find two other bases for the subspace S in Example 2B.

[+] Student Workspace

[−] Answer 2.1

We can use the linear dependence relation $-2v_1 + v_3 + v_4 = 0$ to solve for v_1 or v_4 in terms of the other two vectors. Therefore other possible bases are: $\{v_2, v_3, v_4\}$ and $\{v_1, v_2, v_3\}$. Why is v_2 in all three of the bases we have found?

Section 3: Coordinates Relative to a Basis

One way in which we will find a basis useful is that it gives us an alternative coordinate system. The usual perpendicular coordinate axes in the plane, for example, are related to the standard basis vectors, $i = \langle 1, 0\rangle$ and $j = \langle 0, 1\rangle$, which are unit vectors that lie on the axes, point in the positive direction, and provide the unit of measurement. If we now have another basis $\{u, v\}$ of R^2, we can think of this basis as determining an alternative coordinate system with axes that are not necessarily perpendicular and units of measurement on the two axes that may be different.

=====================

Example 3A: The vector $w = \langle 5, 6\rangle$ can be written as a linear combination of the standard basis vectors i and j, $w = 5\,i + 6\,j$, where the coefficients are just the components of w. Let's consider another basis of R^2, $\{u, v\}$, where $u = \langle 3, 1\rangle$ and $v = \langle -1, 4\rangle$. The vector w can also be written as a linear combination of u and v: $w = 2\,u + v$. The coefficients in this linear combination, 2 and 1, are no longer the components of w; however, we can think of them as "the coordinates of w relative to the basis $\{u, v\}$," or, more succinctly, "the uv coordinates of w."

Execute the **Basisgrid** command below to see an illustration of these coordinates. The picture shows the standard xy coordinate grid with the uv coordinate grid superimposed. The point (5, 6) with uv coordinates (2, 1) is shown as a small blue circle.

```
> u := Vector([3,1]);
  v := Vector([-1,4]);
```

$$u := \begin{bmatrix} 3 \\ 1 \end{bmatrix}, \quad v := \begin{bmatrix} -1 \\ 4 \end{bmatrix}$$

```
> Basisgrid(u,v,[2,1]);
```

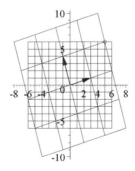

=====================

Explore ~~ Enter other coordinates in the **Basisgrid** command and check the resulting picture. Can you see, for example, what the xy coordinates of the point with uv coordinates (-1, –2) are, and what the uv coordinates of the point with xy coordinates (–7, 2) are? Use **Basisgrid** to check your answers. Try similar explorations with the basis $\{\langle -2, -1\rangle, \langle 1, 3\rangle\}$.

Chapter 5 Vector Spaces

Student Workspace

Definition: Suppose $B = \{v_1, ..., v_p\}$ is a basis of a subspace S and w is a vector in S. Then the weights $c_1, ..., c_p$ in the linear combination $w = c_1 v_1 + ... + c_p v_p$ are called the ***coordinates of w relative to the basis B***, or more succinctly, the ***B-coordinates*** of w.

========================

Example 3B: Find the coordinates of the vector w relative to the basis $\{v_1, v_2, v_3\}$ (see below).

```
> w := Vector([5,8,-10,2]);
> v1 := Vector([2,-3,0,4]);
  v2 := Vector([5,1,-2,0]);
  v3 := Vector([1,-1,2,-2]);
```

$$w := \begin{bmatrix} 5 \\ 8 \\ -10 \\ 2 \end{bmatrix}, \quad v1 := \begin{bmatrix} 2 \\ -3 \\ 0 \\ 4 \end{bmatrix}, \quad v2 := \begin{bmatrix} 5 \\ 1 \\ -2 \\ 0 \end{bmatrix}, \quad v3 := \begin{bmatrix} 1 \\ -1 \\ 2 \\ -2 \end{bmatrix}$$

Solution: We must solve the equation $c_1 v_1 + c_2 v_2 + c_3 v_3 = w$ for the weights c_1, c_2, c_3. This equation can be written $A x = w$, where A is the matrix whose columns are v_1, v_2, v_3, and x is the column vector of unknown weights. Thus:

```
> A := Matrix([v1,v2,v3]);
```

$$A := \begin{bmatrix} 2 & 5 & 1 \\ -3 & 1 & -1 \\ 0 & -2 & 2 \\ 4 & 0 & -2 \end{bmatrix}$$

```
> Matsolve(A,w,free=x);
```

$$\begin{bmatrix} -1 \\ 2 \\ -3 \end{bmatrix}$$

Therefore the coordinates of w relative to the given basis are (-1, 2, –3).

========================

Exercise 3.1: (a) Find the coordinates of the vector u below relative to the basis in Example 3B.
(b) What are the coordinates of $3u$? What are the coordinates of $3u - 2w$? Try to answer these questions without a calculation.

```
> u := Vector([-10,-3,0,8]);
```

$$u := \begin{bmatrix} -10 \\ -3 \\ 0 \\ 8 \end{bmatrix}$$

Student Workspace

Answer 3.1

> `Matsolve(A,u,free=x);`

$$\begin{bmatrix} 1 \\ -2 \\ -2 \end{bmatrix}$$

(a) So the coordinates of u are $(1, -2, -2)$.

(b) The coordinates of $3\,u$ are found by multiplying the coordinates of u by 3: $(3, -6, -6)$. The coordinates of $3\,u - 2\,w$ are found by applying the indicated operations to the coordinates:

$$3\,(1, -2, -2) - 2\,(-1, 2, -3) = (5, -10, 0)$$

Appendix: Theorems on Basis and Dimension

The following theorems provide the fundamental properties of bases and dimension. Although we will not prove these theorems in any of the modules, your instructor may have you study the proofs in class or in a textbook. We will use these theorems freely.

Theorem 3: Suppose a subspace S is spanned by $\{v_1, v_2, ..., v_p\}$ and that $\{w_1, w_2, ..., w_q\}$ is another set of vectors in S. If $p < q$, then $\{w_1, w_2, ..., w_q\}$ is linearly dependent.

Theorem 4: If $\{v_1, v_2, ..., v_p\}$ and $\{w_1, w_2, ..., w_q\}$ are bases of a subspace S, then $p = q$.

Theorem 5: If a nonzero subspace S is spanned by $\{v_1, v_2, ..., v_p\}$, then some subset of $\{v_1, v_2, ..., v_p\}$ is a basis of S.

Theorem 6: Suppose S is a subspace of dimension p. Then:
(a) Any set of more than p vectors in S is linearly dependent.
(b) Any set of fewer than p vectors in S will not span S.
(c) Any set of exactly p vectors in S is linearly independent if and only if it spans S.

Theorem 7: If S is a nonzero subspace of R^n, then S has a basis.

Problems

Problem 1: Find bases by hand

(a) (By hand) Find a basis for the solution set of $2x - y + 4z = 0$ (a plane). What is the dimension of the solution set?

(b) (By hand) Find a basis for the solution set of the system of equations $\{2x - y + 4z = 0,\ x + 3z = 0\}$ (the line of intersection of two planes). What is the dimension of the solution set?

Student Workspace

Problem 2: Two bases for one subspace

(a) Find a basis for the solution set of the homogeneous linear system below two different ways: (i) Use **Matsolve** to solve $Ax = 0$, where A is the coefficient matrix of the linear system, and (ii) use the **solve** command. Then decompose the solution set resulting from each method. (Some of the commands you will need are supplied below.)

(b) Express each vector in the first basis as a linear combination of the vectors in the second basis, and express each vector in the second basis as a linear combination of the vectors in the first basis.

(c) In fact, whenever you have two different bases for one subspace, it is always possible to express each vector in the first basis as a linear combination of the vectors in the second basis, and each vector in the second basis as a linear combination of the vectors in the first basis. Why?

```
> eqn1 := -2*x[1]+6*x[2]+4*x[3]-x[4]+x[5] = 0;
  eqn2 := 2*x[2]+2*x[3]-x[4]+x[5] = 0;
  eqn3 := 2*x[1]-x[2]-x[4]+x[5] = 0;
  eqn4 := x[1]+5*x[2]+7*x[3]-4*x[4]+4*x[5] = 0;
> A := Genmatrix([eqn1,eqn2,eqn3,eqn4],
    [x[1],x[2],x[3],x[4],x[5]],augmented=false);
> solve({eqn1,eqn2,eqn3,eqn4},{x[1],x[2],x[4]});
```

Student Workspace

Problem 3: Find a basis for the span of a set of vectors

(a) Find a basis for $S = \text{Span}\{u, v, w, z\}$ (see below). What is the dimension of this subspace S?
```
> u := Vector([-2, 3, -2, -1]):
  v := Vector([1, -1, 2, 2]):
  w := Vector([3, -3, -2, -4]):
  z := Vector([-3, 5, -2, 0]):
```
(b) Use the linear dependence relation you found in (a) to find all the subsets of $\{u, v, w, z\}$ that are bases for S. For example, is $\{u, v, w, z\}$ a basis of S? Is $\{u, v, w\}$ a basis of S? $\{u, v, z\}$? $\{u, w, z\}$? $\{v, w, z\}$? $\{u, v\}$? (and so on). Explain.

Student Workspace

Problem 4: Maximum dimension of a subspace

Explain your reasoning in answering the following questions:

(a) Suppose S is the span of 4 vectors in R^5. What is the largest dimension that S could have? Give an example of such a subspace S.

(b) Suppose S is the span of 4 vectors in R^3. What is the largest dimension that S could have? Give an example of such a subspace S.

+ Student Workspace

Problem 5: Constructing new bases from old

(a) Let S = Span{u, v, w}, where u, v, w are given below. Check that {u, v, w} is linearly independent (and hence is a basis of S).
(b) Find a basis of S that includes {$u + v, u + v + w$}.
(c) If you are given a set of vectors {p, q, r} in S (where S is as given above), describe an efficient method for determining whether {p, q, r} is a basis of S. Explain why your method works.

```
> u := Vector([2,4,4,-3]):
  v := Vector([-4,2,-4,1]):
  w := Vector([4,1,1,-3]):
```

+ Student Workspace

Problem 6: Interchangeable bases?

Each of the sets of vectors {u_1, u_2}, {v_1, v_2}, {w_1, w_2}, {z_1, z_2, z_3} (see below) is a basis for a subspace of R^4. Two of these bases are bases for the same subspace of R^4. (a) Which two? (There is only one correct answer.) (b) Explain why your conclusion in (a) is correct by describing how the four subspaces are related to one another geometrically.

```
> u1 := Vector([2,0,4,-3]):
  u2 := Vector([-1,2,-4,1]):
> v1 := Vector([1,2,0,-2]):
  v2 := Vector([0,-4,4,1]):
> w1 := Vector([-1,2,-4,1]):
  w2 := Vector([-2,3,1,0]):
> z1 := Vector([1,2,0,-2]):
  z2 := Vector([3,2,4,-5]):
  z3 := Vector([-3,0,1,3]):
```

+ Student Workspace

Problem 7: Coordinates with respect to two bases

The set of vectors {u_1, u_2, u_3} is a basis for a subspace S of R^4, and {v_1, v_2, v_3} is a basis for the same subspace. A vector w in S has the coordinates (2, 4, −3) relative to the first basis. Find its coordinates relative to the second basis.

```
> u1 := Vector([1,-2,0,3]):
  u2 := Vector([0,-1,3,2]):
  u3 := Vector([2,-3,1,0]):
> v1 := Vector([1,-1,-3,1]):
  v2 := Vector([-1,1,1,1]):
  v3 := Vector([1,-2,2,1]):
```
[+] Student Workspace

Linear Algebra Modules Project
Chapter 5, Module 3

Subspaces Associated with a Matrix

Purpose of this module

The purpose of this module is to introduce the column space, row space, and null space of a matrix and to study the relationships among their dimensions. Also, the methods for finding bases that were introduced in Module 2, "Basis and Dimension," are explored in greater depth, and additional methods are introduced.

Prerequisites

Basis and dimension of a subspace; matrix transformations; solution of linear systems.

Commands used in this module

```
> restart: with(LinearAlgebra): with(plots): with(Lamp):
  UseHardwareFloats := false: Digits := 6:
```

Tutorial

Section 1: Definitions and Introduction

Definition: Suppose A is an m by n matrix. Three important subspaces are associated with A:
- The *column space* of A is the span of the column vectors of A, which is a subspace of R^m.
- The *row space* of A is the span of the row vectors of A, which is a subspace of R^n.
- The *null space* of A is the solution set of $A\,x = 0$, which is a subspace of R^n.

==================

Example 1A: Let A be the 2 by 3 matrix $\begin{bmatrix} 1 & 1 & 3 \\ 1 & 2 & 5 \end{bmatrix}$. Describe the three subspaces associated with this matrix, and find their dimensions.

Solution: The column space of A is the span of the column vectors $\begin{bmatrix} 1 \\ 1 \end{bmatrix}, \begin{bmatrix} 1 \\ 2 \end{bmatrix}, \begin{bmatrix} 3 \\ 5 \end{bmatrix}$. Note that these vectors are not a basis for the column space, since they are linearly dependent ($v_1 + 2\,v_2 = v_3$). However, any two of these three vectors can serve as a basis for the column space, which therefore has dimension 2. The column space is thus all of R^2.

The row space of A is the span of the row vectors $\langle 1, 1, 3 \rangle$ and $\langle 1, 2, 5 \rangle$. Since these two vectors are

Chapter 5 Vector Spaces

linearly independent, they constitute a basis for the row space. Their span is therefore a subspace of dimension 2 (i.e., a plane through the origin) in R^3.

The null space of A is the solution set of the equation $A\,x = 0$. We find the solution below.

```
> A := Matrix([[1,1,3],[1,2,5]]);
```

$$A := \begin{bmatrix} 1 & 1 & 3 \\ 1 & 2 & 5 \end{bmatrix}$$

```
> Matsolve(A,zero,free=x);
```

$$\begin{bmatrix} -x_3 \\ -2\,x_3 \\ x_3 \end{bmatrix}$$

So the null space consists of all multiples of the vector $\langle -1, -2, 1 \rangle$. This vector is a basis for the null space, which therefore has dimension 1. The null space is thus a line through the origin in R^3.

=================

Dimensions of the Column, Row, and Null Spaces

The dimensions of the column space, row space, and null space are closely related to one another and to the rank of the matrix. Recall (from Module 2 of Chapter 1) that the rank of a matrix A is the number of nonzero rows in any echelon form of A and is also equal to the number of pivot columns of A. In this module, we will see why the relationships described in the next two theorems are true.

Theorem 8: The column space and row space of a matrix A have the same dimension. This dimension is equal to r, the rank of A.

Theorem 9: The dimension of the null space of a matrix A is $n - r$, where n is the number of columns of A and r is the rank of A.

For the 2 by 3 matrix A in Example 1A, we saw that both the column space and row space have dimension 2 (although one is a subspace of R^2 and the other is a subspace of R^3!). The null space, which is a subspace of R^3, has dimension 1, and we observe that this dimension is equal to $n - r = 3 - 2$.

Exercise 1.1: Let $C = \begin{bmatrix} 1 & 2 & -1 \\ 2 & 4 & -2 \end{bmatrix}$. Find a basis for the column space of C, the row space of C, and the null space of C. Check that their dimensions agree with the theorems above. Describe each of the subspaces geometrically.

[+] Student Workspace

[−] Answer 1.1

A basis for the column space of C is $\left\{ \begin{bmatrix} 1 \\ 2 \end{bmatrix} \right\}$, and therefore the column space of C has dimension 1.

The column space is thus a line through the origin in R^2.

A basis for the row space of C is $\{\langle 1, 2, -1 \rangle\}$, and therefore the row space of C has dimension 1. The row space is thus a line through the origin in R^3. Note that while the column and row spaces are quite different from one another, they do have the same dimension. Also, an echelon form of C is $\begin{bmatrix} 1 & 2 & -1 \\ 0 & 0 & 0 \end{bmatrix}$, and hence C has one pivot column; so the rank r of C is also 1.

The null space of C is the set of solutions of $x_1 + 2x_2 - x_3 = 0$. The solutions are:

$$\begin{bmatrix} -2x_2 + x_3 \\ x_2 \\ x_3 \end{bmatrix} = x_2 \begin{bmatrix} -2 \\ 1 \\ 0 \end{bmatrix} + x_3 \begin{bmatrix} 1 \\ 0 \\ 1 \end{bmatrix}$$

Therefore the null space of C is $\text{Span}\{\langle -2, 1, 0 \rangle, \langle 1, 0, 1 \rangle\}$. Since these two vectors are linearly independent, they constitute a basis for the null space of C, which therefore has dimension 2. Note that $2 = 3 - 1$; that is, the dimension of the null space of C does indeed equal $n - r$. The null space is a plane through the origin in R^3.

Note: When referring to the rows of a matrix in Example 1A and Exercise 1.1, we naturally wrote the rows horizontally to match the way they appear in the matrix. However, we make no distinction between row vectors and column vectors as mathematical objects; for example, $\langle 2, 3, -1 \rangle$ and $\begin{bmatrix} 2 \\ 3 \\ -1 \end{bmatrix}$ are considered to be the same vector in R^3. Of course, when we compute a matrix-vector product $A\,v$, we always write the vector v vertically.

Section 2: Geometric Examples

In this section, we study the column space and null space of matrix transformations. We will see that these concepts are related to range and kernel of a transformation, which were introduced in Chapter 4, Module 2, Section 2. For those who may have omitted that module, we repeat part of that discussion.

If A is a matrix, we define the corresponding matrix transformation T by $T(x) = A\,x$. If A is a 2 by 2 matrix, then T is a function from R^2 to R^2. If A is an m by n matrix, then T is a function from R^n to R^m.

Some concepts associated with matrix transformations are analogous to concepts you have seen in your study of functions in precalculus and calculus. Consider, for example, $f(x) = x^2 - 4$, which defines a function f from R to R. The set of "zeros" of such a function is the set of values of x such that $f(x) = 0$; in this example, the set of zeros is $\{2, -2\}$. The "range" of such a function is the set of values of $f(x)$, where x ranges over the domain R; in this example, the range is $\{y : -4 \leq y\}$.

We introduce now the analogous concepts for a matrix transformation T from R^n to R^m:

Definition: The **kernel** of T is the set of vectors x in R^n such that $T(x) = 0$ (where 0 denotes the zero vector in R^m); in other words, the kernel is the set of "zeros" of T.

Definition: The **range** of T is the set of vectors $T(x)$, where x ranges over all vectors in R^n; in other words, the range is the set of all output (or image) vectors of T in R^m.

Since $T(x) = A x$, the kernel of the transformation T is equal to the null space of the matrix A. Also, the range of T is the set of all vectors $A x$, where x ranges over all the vectors in R^n. Now recall that $A x$ can be written as a linear combination of the columns of A where the weights are the components of x. Therefore the range of T is the set of all linear combinations of the columns of A, which is the column space of A. Thus:

Theorem 10: Suppose A is an m by n matrix and T is the corresponding transformation from R^n to R^m defined by $T(x) = A x$. Then (a) the kernel of T is the null space of A, and (b) the range of T is the column space of A.

Note: In the examples below we will use the same letter for the matrix and for the corresponding matrix transformation, rather than use different letters such as A and T above.

================

Example 2A: Use geometry to find the null space and column space of the 2 by 2 projection matrix P that projects vectors onto the x axis. (See the illustration below, which shows two vectors, u and v, and their images under P.)

Solution: Since P projects vectors that are on the y axis to 0, the kernel of the transformation P is the y axis and therefore the null space of the matrix P is also the y axis. The range of the transformation P is the x axis, and therefore the column space of the matrix P is also the x axis.

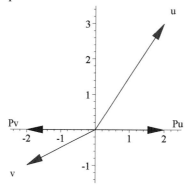

================

Note: We were able to solve Example 2A without referring to the numerical entries of the matrix P, and you will be asked to do the same in the exercises below. However, if you have worked through Module 1 of Chapter 4, "Geometry of Matrix Transformations of the Plane," you can use the matrix commands from that module to check your answers. Here are some examples:

```
> Projectmat(0);
```
$$\begin{bmatrix} 1 & 0 \\ 0 & 0 \end{bmatrix}$$

```
> Reflectmat(Pi/3);
```
$$\begin{bmatrix} -\dfrac{1}{2} & \dfrac{1}{2}\sqrt{3} \\ \dfrac{1}{2}\sqrt{3} & \dfrac{1}{2} \end{bmatrix}$$

```
> Rotatemat(Pi/4);
```
$$\begin{bmatrix} \dfrac{1}{2}\sqrt{2} & -\dfrac{1}{2}\sqrt{2} \\ \dfrac{1}{2}\sqrt{2} & \dfrac{1}{2}\sqrt{2} \end{bmatrix}$$

Exercise 2.1: Let Q be the 2 by 2 projection matrix that projects vectors onto the line $y = x$. Reasoning geometrically as in Example 2A, find the null space and column space of Q. Again, you should be able to answer this question without referring to the numerical entries of the matrix Q.

 Student Workspace

 Answer 2.1

> A projection sends vectors that are perpendicular to the line of projection to the zero vector. So the kernel of Q is the set of vectors on the perpendicular line $y = -x$. In other words, the null space is Span$\{\langle -1, 1 \rangle\}$. The range of Q is the set of points on the line of projection; in other words, the column space of Q is Span$\{\langle 1, 1 \rangle\}$.

Exercise 2.2:
(a) Let S be any 2 by 2 reflection matrix. Reasoning geometrically, find the null space and column space of S.
(b) Let T be any 2 by 2 rotation matrix. Reasoning geometrically, find the null space and column space of T.
(c) What are all the 2 by 2 matrices that have null space equal to $\{0\}$ and column space equal to R^2?

 Student Workspace

 Answer 2.2

> (a) When you reflect a vector, you cannot get the zero vector unless you start with the zero vector. So the kernel (hence null space) of S is $\{0\}$. Every vector is the reflection of its reflection. So the range (hence column space) of S is all of R^2.

(b) Similarly, the null space of T is $\{0\}$ and the column space of T is R^2. Specifically, if T is the rotation through an angle θ, then any vector v can be written as $T u$, where u is the vector obtained by rotating v back by the angle θ; this shows that the range of a rotation is R^2.

(c) If A is any 2 by 2 invertible matrix, its null space is $\{0\}$ and its column space is R^2.

What are the null space and column space of an n by n invertible matrix?

Exercise 2.3: Let P be the 3 by 3 projection matrix that projects vectors in R^3 onto the yz plane. So P maps the point (x, y, z) to the point $(0, y, z)$. Reasoning geometrically, find the null space and column space of P.

 Student Workspace

 Answer 2.3

The projection P sends vectors that are perpendicular to the plane of projection to the zero vector. So the kernel of P is the set of vectors on the x axis. In other words, the null space is $\text{Span}\{\langle 1, 0, 0\rangle\}$. The range of P is the set of points on the plane of projection; in other words, the column space of P is $\text{Span}\{\langle 0, 1, 0\rangle, \langle 0, 0, 1\rangle\}$.

 Section 3: Method for Finding a Null Space Basis

In Example 2A of Module 2, "Basis and Dimension," you learned a method for finding a basis for the solution set of a homogeneous linear system $A x = 0$. Note that this solution set is also the null space of the matrix A. So let's review the method in Example 2A of Module 2:

==================

Example 3A: Find a basis for the null space of the matrix A below:

```
> A := Matrix([[-1,-1,-2,3,1],[-9,5,-4,-1,-5],[7,-5,2,3,5]]);
```

$$A := \begin{bmatrix} -1 & -1 & -2 & 3 & 1 \\ -9 & 5 & -4 & -1 & -5 \\ 7 & -5 & 2 & 3 & 5 \end{bmatrix}$$

Solution: We use **Matsolve** to solve $A x = 0$ and then decompose the solution to find a basis:

```
> Matsolve(A,zero,free=x);
```

$$\begin{bmatrix} -x_3 + x_4 \\ -x_3 + 2 x_4 + x_5 \\ x_3 \\ x_4 \\ x_5 \end{bmatrix}$$

Therefore the decomposition of this solution is:

$$x = \begin{bmatrix} -x_3 + x_4 \\ -x_3 + 2x_4 + x_5 \\ x_3 \\ x_4 \\ x_5 \end{bmatrix} = x_3 \begin{bmatrix} -1 \\ -1 \\ 1 \\ 0 \\ 0 \end{bmatrix} + x_4 \begin{bmatrix} 1 \\ 2 \\ 0 \\ 1 \\ 0 \end{bmatrix} + x_5 \begin{bmatrix} 0 \\ 1 \\ 0 \\ 0 \\ 1 \end{bmatrix} = x_3 u + x_4 v + x_5 w \quad [1]$$

So the null space of A is Span$\{u, v, w\}$, and $\{u, v, w\}$ is a basis of the null space. Alternatively, we can use the **Nullbasis** command:

> Nullbasis(A);

$$\left[\begin{bmatrix} -1 \\ -1 \\ 1 \\ 0 \\ 0 \end{bmatrix}, \begin{bmatrix} 1 \\ 2 \\ 0 \\ 1 \\ 0 \end{bmatrix}, \begin{bmatrix} 0 \\ 1 \\ 0 \\ 0 \\ 1 \end{bmatrix} \right]$$

==================

It is not entirely obvious that the spanning set in Example 3A, $\{u, v, w\}$, is linearly independent. However, in Module 4 of Chapter 2, we saw that the set of vectors produced by the method of decomposition is <u>always</u> linearly independent. Here is the proof for the set of vectors $\{u, v, w\}$ in Example 3A:

To show that $\{u, v, w\}$ is linearly independent, we must show that the vector equation $c_1 u + c_2 v + c_3 w = 0$ has only the trivial solution $c_1 = c_2 = c_3 = 0$. Let's rewrite the left side of this vector equation as a single vector:

$$c_1 \begin{bmatrix} -1 \\ -1 \\ 1 \\ 0 \\ 0 \end{bmatrix} + c_2 \begin{bmatrix} 1 \\ 2 \\ 0 \\ 1 \\ 0 \end{bmatrix} + c_3 \begin{bmatrix} 0 \\ 1 \\ 0 \\ 0 \\ 1 \end{bmatrix} = \begin{bmatrix} -c_1 + c_2 \\ -c_1 + 2c_2 + c_3 \\ c_1 \\ c_2 \\ c_3 \end{bmatrix} = \begin{bmatrix} 0 \\ 0 \\ 0 \\ 0 \\ 0 \end{bmatrix} \quad [2]$$

From the last three components of this equation, we see that $c_1 = c_2 = c_3 = 0$; therefore $\{u, v, w\}$ is linearly independent. Note especially that the weights c_1, c_2, c_3 correspond precisely to the three free variables, x_3, x_4, x_5, in the solution set of the linear system $A\,x = 0$; in fact, equation [2] is just equation [1] written in the opposite order. This explains why a similar proof works in the general case: the weights in the linear combination correspond to the free variables in the solution set of $A\,x = 0$.

We can now see why the following theorem, which was first stated in Section 1, is true:

Theorem 9: The dimension of the null space of a matrix A is $n - r$, where n is the number of columns of A and r is the rank of A.

Here's why. As we saw in the discussion above, the dimension of the null space of A equals the number of vectors produced by decomposing the solution set of $A\,x = 0$, and this number also equals the number of free variables in the solution set of $A\,x = 0$. Furthermore, Theorem 3 of Chapter 1 says that the number of free variables in the solution set of $A\,x = 0$ is $n - r$, where n is the number of columns of A and r is the rank of A. Therefore the dimension of the null space of A is the number of free variables, $n - r$.

Exercise 3.1: (a) For the matrix B below, use the command **Reduce(B)** and Theorem 9 to determine the dimension of the null space of B.
(b) Use **Matsolve** as in Example 3A to find a basis for the null space of B.

```
> B := Matrix([[2,2,-3,-8,5,-8],[-7,-7,-5,-3,-2,-3],[1,1,4,7,-3,7]]);
```

$$B := \begin{bmatrix} 2 & 2 & -3 & -8 & 5 & -8 \\ -7 & -7 & -5 & -3 & -2 & -3 \\ 1 & 1 & 4 & 7 & -3 & 7 \end{bmatrix}$$

 Student Workspace

Answer 3.1

```
> Reduce(B);
```

$$\begin{bmatrix} 2 & 2 & -3 & -8 & 5 & -8 \\ 0 & 0 & \dfrac{-31}{2} & -31 & \dfrac{31}{2} & -31 \\ 0 & 0 & 0 & 0 & 0 & 0 \end{bmatrix}$$

(a) Since $r = 2$ and $n = 6$, the dimension of the null space of B is $6 - 2 = 4$.
(b) Solve $B\,x = 0$:

```
> Matsolve(B,zero,free=x);
```

$$\begin{bmatrix} -x_2 + x_4 - x_5 + x_6 \\ x_2 \\ -2x_4 + x_5 - 2x_6 \\ x_4 \\ x_5 \\ x_6 \end{bmatrix}$$

This solution decomposes as follows:

$$\begin{bmatrix} -x_2+x_4-x_5+x_6 \\ x_2 \\ -2x_4+x_5-2x_6 \\ x_4 \\ x_5 \\ x_6 \end{bmatrix} = x_2 \begin{bmatrix} -1 \\ 1 \\ 0 \\ 0 \\ 0 \\ 0 \end{bmatrix} + x_4 \begin{bmatrix} 1 \\ 0 \\ -2 \\ 1 \\ 0 \\ 0 \end{bmatrix} + x_5 \begin{bmatrix} -1 \\ 0 \\ 1 \\ 0 \\ 1 \\ 0 \end{bmatrix} + x_6 \begin{bmatrix} 1 \\ 0 \\ -2 \\ 0 \\ 0 \\ 1 \end{bmatrix}$$

$$= x_2 v_1 + x_4 v_2 + x_5 v_3 + x_6 v_4$$

Alternatively, we can use **Nullbasis(B)**:

> Nullbasis(B);

$$\left[\begin{bmatrix} -1 \\ 1 \\ 0 \\ 0 \\ 0 \\ 0 \end{bmatrix}, \begin{bmatrix} 1 \\ 0 \\ -2 \\ 1 \\ 0 \\ 0 \end{bmatrix}, \begin{bmatrix} -1 \\ 0 \\ 1 \\ 0 \\ 1 \\ 0 \end{bmatrix}, \begin{bmatrix} 1 \\ 0 \\ -2 \\ 0 \\ 0 \\ 1 \end{bmatrix} \right]$$

So $\{v_1, v_2, v_3, v_4\}$ is a basis for the null space of B, which confirms that the dimension of the null space is 4.

Section 4: Method for Finding a Column Space Basis

================

Example 4A: Find a basis for the column space of the matrix C below:

$$C = \begin{bmatrix} 1 & 2 & 3 & 1 \\ 0 & 3 & 3 & 3 \\ -1 & 0 & -1 & 1 \end{bmatrix}$$

Solution method 1: It just so happens that we can spot some linear dependence relations among the columns, C_1, C_2, C_3, C_4, of C:

$$C_3 = C_1 + C_2 \quad \text{and} \quad C_4 = C_2 - C_1$$

Since $\{C_1, C_2\}$ is linearly independent, this set of two vectors is a basis for the column space of C.

Solution method 2: This is a systematic method that does not depend on being able to spot linear dependence relations among the columns of a matrix; instead, we use a basis for the null space of the matrix to find them:

> C := Matrix([[1,2,3,1],[0,3,3,3],[-1,0,-1,1]]);

$$C := \begin{bmatrix} 1 & 2 & 3 & 1 \\ 0 & 3 & 3 & 3 \\ -1 & 0 & -1 & 1 \end{bmatrix}$$

> Nullbasis(C);

$$\left[\begin{bmatrix}-1\\-1\\1\\0\end{bmatrix},\begin{bmatrix}1\\-1\\0\\1\end{bmatrix}\right]$$

Each of these two null space vectors satisfies the matrix-vector equation $Cx = 0$. Therefore, since the product Cx can be written as a linear combination of the columns of C, we get two equations of the form $x_1 C_1 + x_2 C_2 + x_3 C_3 + x_4 C_4 = 0$:

$$-C_1 - C_2 + C_3 = 0 \quad \text{and} \quad C_1 - C_2 + C_4 = 0$$

These are precisely the linear dependence relations we spotted in Method 1. Again, $\{C_1, C_2\}$ is a basis for the column space of C.

Summary of solution method 2: We find a basis for the null space of the matrix and use each basis vector to write down a linear dependence relation for the columns of the matrix. As we will see further on, if we <u>solve for the rightmost column</u> in each linear dependence relation, the remaining columns will form a basis for the column space.

===================

Exercise 4.1: Use solution method 2 of Example 4A to find a basis for the column space of the matrix A below:

```
> A := Matrix([[-1,-1,-2,3,1],[-9,5,-4,-1,-5],[7,-5,2,3,5]]);
```

$$A := \begin{bmatrix} -1 & -1 & -2 & 3 & 1 \\ -9 & 5 & -4 & -1 & -5 \\ 7 & -5 & 2 & 3 & 5 \end{bmatrix}$$

 Student Workspace

Answer 4.1

We first find a basis for the null space of A:
```
> Nullbasis(A);
```

$$\left[\begin{bmatrix}-1\\-1\\1\\0\\0\end{bmatrix},\begin{bmatrix}1\\2\\0\\1\\0\end{bmatrix},\begin{bmatrix}0\\1\\0\\0\\1\end{bmatrix}\right]$$

Each of these three null space vectors satisfies the matrix-vector equation $Ax = 0$. Therefore, since the product Ax can be written as a linear combination of the columns of A, we get three linear dependence relations of the form $x_1 A_1 + ... + x_5 A_5 = 0$:

$$-A_1 - A_2 + A_3 = 0, \quad A_1 + 2 A_2 + A_4 = 0, \quad A_2 + A_5 = 0$$

We solve for the <u>rightmost</u> vector in each of the above equations:

$$A_3 = A_1 + A_2, \quad A_4 = -A_1 - 2A_2, \quad A_5 = -A_2$$

In each case, we expressed a column in terms of the columns A_1 and A_2. Therefore, $\{A_1, A_2\}$ spans the column space of A. Since $\{A_1, A_2\}$ is linearly independent, $\{A_1, A_2\}$ is a basis for the column space of A. So the basis we have found is:

$$\left\{ \begin{bmatrix} -1 \\ -9 \\ 7 \end{bmatrix}, \begin{bmatrix} -1 \\ 5 \\ -5 \end{bmatrix} \right\}$$

Quick Method for Finding a Column Space Basis

For the matrix B below, we can find a column space basis very quickly.

$$B = \begin{bmatrix} 1 & 0 & 0 & -2 \\ 0 & 1 & 0 & 1 \\ 0 & 0 & 1 & -1 \\ 0 & 0 & 0 & 0 \end{bmatrix}$$

Here's how: Let's denote the columns of B by B_1, B_2, B_3, B_4. Then $\{B_1, B_2, B_3\}$ (the set of pivot columns) is clearly a basis for the column space of B, and we can express the fourth column as a linear combination of the basis columns: $B_4 = -2B_1 + B_2 - B_3$.

This was quick because B is in reduced row echelon form. Furthermore, this observation suggests a quick method for finding a basis for the column space for any matrix A. Recall that every matrix A is row equivalent to a matrix R in reduced row echelon form, and that the equations $Ax = 0$ and $Rx = 0$ have the same solution set, since row operations do not change the solutions of a linear system. Therefore the columns of A and the columns of R satisfy the same linear dependence relations. Since the pivot columns of a matrix in reduced row echelon form are clearly a basis for the column space of that matrix, the corresponding columns of any row equivalent matrix are a basis for the column space of that matrix. We have thus shown:

- The pivot columns of a matrix A constitute a basis for the column space of A. Therefore the dimension of the column space of A equals the number of pivot columns of A, i.e., the rank of A.

Not only have we found a quick method for finding a column space basis, but we also proved part of Theorem 8! (See Section 1.)

====================

Example 4B: Use the quick method above to find a basis for the column space of the matrix A (repeated below) in Example 4A. Also, express each column of A as a linear combination of these basis columns.

```
> A := Matrix([[-1,-1,-2,3,1],[-9,5,-4,-1,-5],[7,-5,2,3,5]]);
```

$$A := \begin{bmatrix} -1 & -1 & -2 & 3 & 1 \\ -9 & 5 & -4 & -1 & -5 \\ 7 & -5 & 2 & 3 & 5 \end{bmatrix}$$

Chapter 5 Vector Spaces

Solution:

> R := Reduce(A,form=rref);

$$R := \begin{bmatrix} 1 & 0 & 1 & -1 & 0 \\ 0 & 1 & 1 & -2 & -1 \\ 0 & 0 & 0 & 0 & 0 \end{bmatrix}$$

Maple note: The option "form=rref" asks for the <u>reduced</u> row echelon form (rref).

We see that the first and second columns of the reduced row echelon form R are the pivot columns of R; therefore the first and second columns of A constitute a basis for the column space of A. Furthermore, we see that the third column of R is the sum of the first two columns of R; therefore the same is true for the columns of A: $A_3 = A_1 + A_2$. Similarly, $A_4 = -A_1 - 2A_2$ and $A_5 = -A_2$.

Warning: Row operations <u>change</u> the column space of a matrix. For example, the matrices A and R above do not have the same column space, even though their columns satisfy the same linear dependence relations.

===================

Exercise 4.2: Use the quick method above to find a basis for the column space of the matrix N below. Also, express each column of N as a linear combination of these basis columns.

> N := Matrix([[-4,-2,-3,3,-3],[-4,6,1,-1,1],
   [3,1,2,-2,3],[1,-1,0,0,1]]);

$$N := \begin{bmatrix} -4 & -2 & -3 & 3 & -3 \\ -4 & 6 & 1 & -1 & 1 \\ 3 & 1 & 2 & -2 & 3 \\ 1 & -1 & 0 & 0 & 1 \end{bmatrix}$$

+ Student Workspace

− Answer 4.2

> Reduce(N,form=rref);

$$\begin{bmatrix} 1 & 0 & \frac{1}{2} & \frac{-1}{2} & 0 \\ 0 & 1 & \frac{1}{2} & \frac{-1}{2} & 0 \\ 0 & 0 & 0 & 0 & 1 \\ 0 & 0 & 0 & 0 & 0 \end{bmatrix}$$

Since the first, second, and fifth columns of N are the pivot columns of N, these columns form a basis for the column space of N:

$$\left\{ \begin{bmatrix} -4 \\ -4 \\ 3 \\ 1 \end{bmatrix}, \begin{bmatrix} -2 \\ 6 \\ 1 \\ -1 \end{bmatrix}, \begin{bmatrix} -3 \\ 1 \\ 3 \\ 1 \end{bmatrix} \right\}$$

From the echelon form of N, we see how to express the remaining columns of N in terms of these columns:
$$N_3 = \frac{N_1}{2} + \frac{N_2}{2}, \quad N_4 = -\frac{N_1}{2} - \frac{N_2}{2}$$

Section 5: Method for Finding a Row Space Basis

Since the row space of A is also the column space of A^T, we could use the method of Section 4 to find a row space basis. However, **Reduce(A,form=rref)** gives us a row space basis immediately:

================

Example 5A: Find a basis for the row space of the matrix B below:
> B := Matrix([[-1,-2,-3,2],[4,-5,-1,5],[-2,2,0,-2]]);

$$B := \begin{bmatrix} -1 & -2 & -3 & 2 \\ 4 & -5 & -1 & 5 \\ -2 & 2 & 0 & -2 \end{bmatrix}$$

> R := Reduce(B,form=rref);

$$R := \begin{bmatrix} 1 & 0 & 1 & 0 \\ 0 & 1 & 1 & -1 \\ 0 & 0 & 0 & 0 \end{bmatrix}$$

The nonzero rows of R form a basis for the row space of B:
$$\{\langle 1, 0, 1, 0 \rangle, \langle 0, 1, 1, -1 \rangle\}$$

================

To see why the method of Example 5A works, we will explain two facts:
- Row operations do not change the row space of a matrix;
- The nonzero rows of a reduced row echelon form matrix are linearly independent.

Row operations merely replace rows of a matrix by linear combinations of the original rows. For example, let's denote the rows of the matrix B above by b_1, b_2, b_3. The first step in reducing matrix B to matrix R above might be to replace the second row of B by $4 b_1 + b_2$. Thus, after one row operation, we have a new matrix C whose rows are b_1, c, b_3, where $c = 4 b_1 + b_2$. These vectors are certainly in the row space of B. Also the rows of B are in the row space of C, since $b_2 = c - 4 b_1$. A similar argument holds for any row operation, which shows that row operations do not change the row space of a matrix.

One way to see that the nonzero rows of the reduced matrix R above are linearly independent is to look at just their entries in the pivot columns, which are columns 1 and 2. These are the standard basis vectors in R^2, $\langle 1, 0 \rangle$ and $\langle 0, 1 \rangle$, which we know are linearly independent. This argument will work for any reduced row echelon matrix: If you look at just the entries of the nonzero rows that are in the pivot columns, you will have the standard basis vectors, which are linearly independent.

Chapter 5 Vector Spaces

We therefore conclude:

- The nonzero rows of the reduced row echelon form of a matrix A constitute a basis for the row space of A. Therefore the dimension of the row space of A is the number of nonzero rows of the reduced row echelon form of A.

Since the number of nonzero rows of a row echelon form of A equals the rank of A, we see from this conclusion and the similar conclusion in Section 4 that we have proved the other of our two main theorems of this module:

Theorem 8: The column space and row space of a matrix A have the same dimension. This dimension is equal to r, the rank of A.

Exercise 5.1: Find a row space basis for the matrix T below by the method of Example 5A.

```
> T := Matrix([[-4,-4,3,1],[-2,6,1,-1],[-3,1,2,0],
  [3,-1,-2,0],[-3,1,3,1]]);
```

$$T := \begin{bmatrix} -4 & -4 & 3 & 1 \\ -2 & 6 & 1 & -1 \\ -3 & 1 & 2 & 0 \\ 3 & -1 & -2 & 0 \\ -3 & 1 & 3 & 1 \end{bmatrix}$$

+ Student Workspace

− Answer 5.1

```
> R := Reduce(T,form=rref);
```

$$R := \begin{bmatrix} 1 & 0 & 0 & \frac{5}{8} \\ 0 & 1 & 0 & \frac{-1}{8} \\ 0 & 0 & 1 & 1 \\ 0 & 0 & 0 & 0 \\ 0 & 0 & 0 & 0 \end{bmatrix}$$

So a basis for the row space of T is

$$\{\langle 1, 0, 0, \tfrac{5}{8}\rangle, \langle 0, 1, 0, -\tfrac{1}{8}\rangle, \langle 0, 0, 1, 1\rangle\}$$

Notice that the matrix T is the transpose of the matrix N in Exercise 4.2. The method of this section produced quite a different basis than the one found in Exercise 4.2.

Problems

Problem 1: Find a null space basis

(a) Using the method of decomposition as in Example 3A, find a basis for the null space of the matrix A below. Check by using the **Nullbasis** command.

(b) What is the dimension of the null space of A? Use the definition of dimension to explain why your answer is correct.

(c) Apply the command **Reduce(A)** and use the result to confirm the dimension of the null space of A. Explain how it confirms the dimension. Hint: We can see the rank of the matrix from its echelon form.

```
> A := Matrix([[-1,-2,-3,-2,-3],[-6,4,-2,-2,-5],
    [4,-4,0,2,3],[-1,-1,-2,3,0]]);
```

Student Workspace

Problem 2: Find a column space basis

(a) Using solution method 2 of Example 4A, find a basis for the column space of the matrix A in Problem 1 (repeated below).

(b) Write out the linear dependence relations that the null space vectors give you, and use them to write every column of A as a linear combination of the basis vectors in (a).

(c) Check your answer in (a) by using **Reduce(A,form=rref)**. Explain what the reduced row echelon form tells you. Hint: Look for the pivot columns.

```
> A := Matrix([[-1,-2,-3,-2,-3],[-6,4,-2,-2,-5],
    [4,-4,0,2,3],[-1,-1,-2,3,0]]):
```

Student Workspace

Problem 3: Projection example in R^3

Suppose P is the 3 by 3 matrix transformation that projects every vector in R^3 to the xy plane.

(a) (By hand) Reasoning geometrically, as in Section 2, find the null space and column space of P and their dimensions. Explain your reasoning.

(b) Find the matrix P. Hint: Use the fact that

$$P\begin{bmatrix} x \\ y \\ z \end{bmatrix} = \begin{bmatrix} x \\ y \\ 0 \end{bmatrix} \text{ for all points } \begin{bmatrix} x \\ y \\ z \end{bmatrix}$$

(c) Confirm your answers to (a) by applying appropriate Maple commands to the matrix P.

Student Workspace

Problem 4: Another projection example in R^3

Suppose Q is the 3 by 3 matrix that projects every vector in R^3 to the plane $x - 2y + 3z = 0$.
(a) (By hand) What is the null space of Q and what is the column space of Q? Find a basis for each of these two subspaces, and find their dimensions.
(b) Find the matrix Q as follows. Suppose the vectors you found in (a) are u, v, w. (You should have found three vectors, if you take the two bases together.) Then Qu, Qv, Qw should be easy to figure out geometrically. Therefore you can write down the matrix product $Q[u, v, w]$ and then can solve an equation of the form $QA = B$ to find Q.
(c) Confirm your answers to (a) by applying appropriate Maple commands to the matrix Q.

+ Student Workspace

Problem 5: Name that dimension (by hand)

Let A by an m by n matrix of rank r. Express the following numbers in terms of m, n and r.
(a) The dimension of the column space of A.
(b) The dimension of the row space of A.
(c) The dimension of the null space of A.
(d) The dimension of the null space of A^T.
(e) The number of free variables in the solution set of $Ax = 0$.
(f) The number of free variables in the solution set of $A^T x = 0$.

+ Student Workspace

Problem 6: Spaces associated with invertible matrices

If A is a n by n invertible matrix, what are the null space, column space, and row space of A? Explain.

+ Student Workspace

Problem 7: Two bases for a column space

(a) Find a basis for the column space of the matrix C below by the method of Section 4.
(b) Find a basis for the column space of C by the method of Example 5A applied to C^T.
(c) Just knowing that the bases you found in (a) and (b) are bases for the same subspace, how must the two sets of basis vectors be related to one another algebraically?

```
> C := Matrix([[-2,0,2,1],[3,-1,-5,-3],[4,4,2,2],
    [4,-2,-6,-3],[2,0,-2,-1]]);
```

+ Student Workspace

Problem 8: True/False questions

For each of the following statements, say whether the statement is true or false, and justify your conclusion. That is, if the statement is true, refer to a theorem that shows why it is true; if it is false, give a counter-example. If you are unsure whether a statement is true or false, you might use **Randmat** to construct some examples that you can test. For example, the command below constructs a random 5 by 3 matrix of rank 2.

```
> A := Randmat(5,3,rank=2);
```

(a) If A is a 4 by 3 matrix, A must have a nonzero null space.
(b) If A is a 3 by 4 matrix, A must have a nonzero null space.
(c) If A is a 4 by 4 matrix whose null space is $\{0\}$, its column space must be all of R^4.
(d) If A is a 4 by 3 matrix whose null space is $\{0\}$, its column space must be all of R^4.

Student Workspace

Problem 9: Construct that matrix

For each of the following descriptions, either construct a matrix that satisfies the description or explain why no such matrix can exist.

(a) A is a 3 by 4 matrix of rank 4. (b) A is a 3 by 4 matrix of rank 2.

(c) A is a 3 by 4 matrix of rank 3 and $A \begin{bmatrix} 1 \\ 2 \\ -1 \\ 0 \end{bmatrix} = 0$.

(d) A is a 4 by 3 matrix of rank 3 and $A \begin{bmatrix} 1 \\ 2 \\ -1 \end{bmatrix} = 0$.

Student Workspace

Problem 10: Rank patterns for n by n matrices

For each of the following collections of n by n matrices, find the rank of all such matrices and justify your conclusion. For example, you might state a basis for the row space, column space, or null space of the matrices. You may use the special commands **LetterN(n)**, **LetterL(n)**, and **Bandmat(n)** provided in the Student Workspace to construct and experiment with examples of these matrices. However, you may be able to see what the ranks must be without experimenting at all.

(a) The n by n matrices of 0's and 1's, where the 1's form the shape of the letter N.
(b) The n by n matrices of 0's and 1's, where the 1's form the shape of the letter L.
(c) The n by n matrices of 0's and 1's, where the 1's form two diagonal bands just above and below the main diagonal.

Student Workspace

Linear Algebra Modules Project
Chapter 5, Module 4

Loops and Spanning Trees -- An Application

Purpose of This Module

The purpose of this module is to provide a few examples of the ways in which bases arise and are used in practice.

Prerequisites

Null space and column space of a matrix; basis and dimension; solution of linear systems.

Commands used in this module

```
> restart; with(LinearAlgebra): with(plots): with(Lamp):
  UseHardwareFloats := false: Digits := 6:
```

Tutorial

Section 1: Traffic Flow Problem

The figure below shows traffic circulation and inflow/outflow from a two-block area with six intersections (called "nodes") and seven one-way streets (called "arcs") between them.
[This figure is taken from *Introduction to Linear Algebra, 3rd ed.*, by Johnson, Riess, and Arnold, Addison-Wesley, 1993, page 48.]

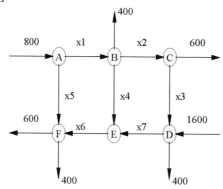

The numbers and variables in the diagram represent the number of vehicles per hour that pass along each arc. For example, every hour, 800 vehicles flow into Node A from outside the network (i.e., outside the two-block area) and must head toward Node B or Node F. The unknown traffic flow x_1 is the number of vehicles per hour that travel from Node A to Node B. (If x_1 is negative, the vehicles actually travel from B to A.)

We can find a system of linear equations satisfied by the seven unknown flows $x_1, ..., x_7$ by applying the "conservation of flow" principle:

- The number of vehicles that enter a node each hour must equal the number of vehicles that leave this node.

For example, the number of vehicles that enter Node A in an hour is 800, and the number of vehicles leaving A is $x_1 + x_5$; therefore:

Node A: $x_1 + x_5 = 800$

However, for reasons that will become clear later, we multiply the Node A equation by -1:

Node A: $-x_1 - x_5 = -800$

The situation at Node B is more complicated: x_1 vehicles enter B, while 400, x_2, and x_4 vehicles exit B. We could therefore say that $x_1 = x_2 + x_4 + 400$; but, as we want all unknowns to be on the left side of the equation, we write:

Node B: $x_1 - x_2 - x_4 = 400$

Similarly, for the other four nodes we have:

Node C: $x_2 - x_3 = 600$
Node D: $x_3 - x_7 = -1200$
Node E: $x_4 - x_6 + x_7 = 0$
Node F: $x_5 + x_6 = 1000$

Clearly, with six equations and seven unknown flows, there must be at least one free variable. So if our system of equations has any solution at all, it has infinitely many; that is, there are infinitely many flow patterns that accommodate the given inflows and outflows. We solve the system by writing it in the form $Mx = b$, and discover that there are indeed solutions and that the system has two free variables:

```
> M := Matrix([[-1,0,0,0,-1,0,0],[1,-1,0,-1,0,0,0],
   [0,1,-1,0,0,0,0],[0,0,1,0,0,0,-1],[0,0,0,1,0,-1,1],
   [0,0,0,0,1,1,0]]);
```

$$M := \begin{bmatrix} -1 & 0 & 0 & 0 & -1 & 0 & 0 \\ 1 & -1 & 0 & -1 & 0 & 0 & 0 \\ 0 & 1 & -1 & 0 & 0 & 0 & 0 \\ 0 & 0 & 1 & 0 & 0 & 0 & -1 \\ 0 & 0 & 0 & 1 & 0 & -1 & 1 \\ 0 & 0 & 0 & 0 & 1 & 1 & 0 \end{bmatrix}$$

Chapter 5 Vector Spaces

```
> b := Vector([-800,400,600,-1200,0,1000]);
```

$$b := \begin{bmatrix} -800 \\ 400 \\ 600 \\ -1200 \\ 0 \\ 1000 \end{bmatrix}$$

```
> soln := Matsolve(M,b,free=x);
```

$$soln := \begin{bmatrix} -200 + x_6 \\ -600 + x_7 \\ -1200 + x_7 \\ x_6 - x_7 \\ 1000 - x_6 \\ x_6 \\ x_7 \end{bmatrix}$$

We decompose the solution:

$$soln = \begin{bmatrix} -200 \\ -600 \\ -1200 \\ 0 \\ 1000 \\ 0 \\ 0 \end{bmatrix} + x_6 \begin{bmatrix} 1 \\ 0 \\ 0 \\ 1 \\ -1 \\ 1 \\ 0 \end{bmatrix} + x_7 \begin{bmatrix} 0 \\ 1 \\ 1 \\ -1 \\ 0 \\ 0 \\ 1 \end{bmatrix} = p + x_6 u + x_7 v$$

where p is a particular solution of the linear system, and $\{u, v\}$ is a basis for the solution set of the corresponding homogeneous system $Mx = 0$.

```
> p := Vector([-200,-600,-1200,0,1000,0,0]):
  u := Vector([1,0,0,1,-1,1,0]):
  v := Vector([0,1,1,-1,0,0,1]):
```

Interpreting the particular solution p is instructive: Since the values of x_4, x_6, and x_7 are zero, no traffic passes through Node E in any direction. At Node D, on the other hand, there is a net inflow of $1600 - 400 = 1200$ vehicles, which flows backward ($x_3 = -1200$) to Node C. Then at Node C, 600 vehicles exit and the remaining 600 flow backward ($x_2 = -600$) to Node B. At Node B, 400 exit and the remaining 200 vehicles flow backward ($x_1 = -200$) to Node A. At Node A, 800 vehicles enter, joining the remaining 200, and the total flow forward ($x_5 = 1000$) to Node F. Finally, at Node F, the remaining $1000 = 600 + 400$ vehicles exit. So the flow represented by p is counterclockwise around the upper part of the loop DCBAF.

Since the solution set of the homogeneous system $Mx = 0$ is also the null space of M, we can find a basis for this system another way:

```
> N := Nullbasis(M);
```

$$N := \begin{bmatrix} 1 & 0 \\ 0 & 1 \\ 0 & 1 \\ 1 & -1 \\ -1 & 0 \\ 1 & 0 \\ 0 & 1 \end{bmatrix}$$

These null space basis vectors u and v are also interesting to interpret. Keep in mind that these vectors represent flows where there is no traffic moving into or out of the network, since the right sides of the system of equations $Mx = 0$ are all zero.

We now describe the traffic flow represented by the null space vector u. First note that, since $x_2 = 0$, $x_3 = 0$, and $x_7 = 0$, no traffic flows through Nodes C or D. The first component, $x_1 = 1$, tells us that one vehicle goes from Node A to Node B; since $x_4 = 1$, this vehicle goes next to Node E; since $x_6 = 1$, this vehicle goes next to Node F; and since $x_5 = -1$, the vehicle returns to Node A. Summarizing, we can say that the vector u represents one vehicle traveling clockwise around the left-hand loop, ABEF.

Exercise 1.1: Describe similarly the traffic flow represented by the null space vector v.

+ Student Workspace

 Answer 1.1

The vector v represents one vehicle traveling clockwise around the right-hand loop, BCDE.

The remaining loop, ABCDEF, is discussed in Problem 1.

The two loop basis vectors u, v can give us useful insights about the solutions of our traffic flow example, since all solutions can be written in the form $p + a\,u + b\,v$, where p is the particular solution:

```
> p;
```

$$\begin{bmatrix} -200 \\ -600 \\ -1200 \\ 0 \\ 1000 \\ 0 \\ 0 \end{bmatrix}$$

Suppose, therefore, we decide to modify the flow represented by p as follows. Let's send another 1200 vehicles clockwise around the loop BCDE but no vehicles around the loop ABEF. That is, let's form the linear combination $p + 0\,u + 1200\,v$. Since p sends 1200 vehicles from Node D to Node C and 1200 v sends 1200 vehicles from Node C to Node D, the new value of x_3 will be zero. (So we can repair the potholes in that street.) Check this reasoning by executing the following command:

```
> p+1200*v;
```
$$\begin{bmatrix} -200 \\ 600 \\ 0 \\ -1200 \\ 1000 \\ 0 \\ 1200 \end{bmatrix}$$

Exercise 1.2: Find a solution that diverts all traffic from the arcs AB and BC, i.e., a solution in which x_1 and x_2 are zero.

 Student Workspace

Answer 1.2

One method is purely algebraic: We look at the solutions x and make the obvious choices $x_6 = 200$ and $x_7 = 600$ to zero out the first two components:

```
> soln;
```
$$\begin{bmatrix} -200 + x_6 \\ -600 + x_7 \\ -1200 + x_7 \\ x_6 - x_7 \\ 1000 - x_6 \\ x_6 \\ x_7 \end{bmatrix}$$

```
> subs({x[6]=200,x[7]=600},soln);
```
$$\begin{bmatrix} 0 \\ 0 \\ -600 \\ -400 \\ 800 \\ 200 \\ 600 \end{bmatrix}$$

Another method is more geometric: We look at the particular solution p and observe that we need to send 200 vehicles clockwise around the loop ABEF and 600 vehicles clockwise around the loop BCDE to change the values of x_1 and x_2 to zero. Thus, we want the linear combination $p + 200\,u + 600\,v$:

Module 5.4 Loops and Spanning Trees -- An Application   301

> p;

$$\begin{bmatrix} -200 \\ -600 \\ -1200 \\ 0 \\ 1000 \\ 0 \\ 0 \end{bmatrix}$$

> p+200*u+600*v;

$$\begin{bmatrix} 0 \\ 0 \\ -600 \\ -400 \\ 800 \\ 200 \\ 600 \end{bmatrix}$$

Finally, let's examine the form of the coefficient matrix of our linear system:

> M;

$$\begin{bmatrix} -1 & 0 & 0 & 0 & -1 & 0 & 0 \\ 1 & -1 & 0 & -1 & 0 & 0 & 0 \\ 0 & 1 & -1 & 0 & 0 & 0 & 0 \\ 0 & 0 & 1 & 0 & 0 & 0 & -1 \\ 0 & 0 & 0 & 1 & 0 & -1 & 1 \\ 0 & 0 & 0 & 0 & 1 & 1 & 0 \end{bmatrix}$$

Notice that each column consists of a single entry of -1 and a single entry of 1, with all remaining entries being 0. This is not accidental. The columns of M correspond to the unknowns $x_1, ..., x_7$, respectively, and the rows correspond to the nodes A, ..., F, respectively. So the entry -1 in the first column represents the fact that x_1 vehicles are leaving node A, and the entry 1 in the first column represents x_1 vehicles entering node B. We will see more matrices having this structure as we go on. (Aside: Earlier, when we multiplied the Node A equation by -1, we were helping to ensure this pattern of 1's and -1's. We can always obtain this pattern by consistently putting a minus sign before an unknown when it's an outflow and a plus sign before an unknown when it's an inflow.)

Section 2: Oil Pipeline Construction Problem

The figure below shows a network consisting of an oil refinery at node 1 and oil fields at nodes 2 through 6. The arcs connecting them represent pipelines that could be constructed between them. Our goal in this problem is to find a smallest subset of arcs that are sufficient to connect all the oil fields to the refinery. The figure shows a flow of x_1 barrels of oil flowing from node 2 to node 1, x_2 barrels of oil flowing from node 4 to node 1, etc. In addition, imagine another arc coming out of node 1, which would represent the sum of the flows x_1, x_2, x_3 flowing into the refinery; this sum represents the total oil flow from all the oil fields. Also, at each of the nodes 2, 3, 4, 5, and 6, imagine oil flowing into that node from the corresponding oil field.

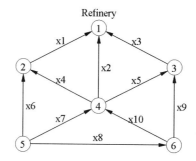

If we had provided numerical values for the flows from each of the five oil fields, we could write, as we did in Section 1, six equations (one for each node) in the ten unknown flows, $x_1, ..., x_{10}$. However, we do have enough information to write down the coefficient matrix P of this system:

```
> P := Matrix([[1,1,1,0,0,0,0,0,0,0],[-1,0,0,1,0,1,0,0,0,0],
  [0,0,-1,0,1,0,0,0,1,0],[0,-1,0,-1,-1,0,1,0,0,1],
  [0,0,0,0,0,-1,-1,-1,0,0],[0,0,0,0,0,0,0,1,-1,-1]]);
```

$$P := \begin{bmatrix} 1 & 1 & 1 & 0 & 0 & 0 & 0 & 0 & 0 & 0 \\ -1 & 0 & 0 & 1 & 0 & 1 & 0 & 0 & 0 & 0 \\ 0 & 0 & -1 & 0 & 1 & 0 & 0 & 0 & 1 & 0 \\ 0 & -1 & 0 & -1 & -1 & 0 & 1 & 0 & 0 & 1 \\ 0 & 0 & 0 & 0 & 0 & -1 & -1 & -1 & 0 & 0 \\ 0 & 0 & 0 & 0 & 0 & 0 & 0 & 1 & -1 & -1 \end{bmatrix}$$

Note that the coefficient matrix P has the same structure as the coefficient matrix M in Section 1. For example, the first column of P has the entry -1 in the second row and the entry 1 in the first row; as in Section 1, this represents the flow x_1 going from node 2 to node 1.

Exercise 2.1: Describe the meaning of column 4 of this matrix.

 Student Workspace

 Answer 2.1

Column 4 says that the flow x_4 goes from node 4 (-1 in the 4th row) to node 2 (1 in the second row).

Now let's see how we can use the matrix P to find a smallest collection of arcs that will connect all the oil fields to the refinery. Notice that loops are undesirable, since they represent more than one route connecting two nodes. So we want to remove arcs that contribute to loops, but we do not want to remove arcs whose removal would disconnect the network. Removing an arc corresponds to removing a column of P. Therefore, since loops come from null space vectors (as we saw in Section 1), we want to remove

Module 5.4 Loops and Spanning Trees -- An Application

columns of P until the null space is $\{0\}$. When the null space is $\{0\}$, the remaining columns will be linearly independent. However, if we remove a column of P that is not a linear combination of the remaining columns of P, removal of the corresponding arc would disconnect the network. So we want to remove only columns of P that are linear combinations of remaining columns of P. In other words, we want to remove columns of P until we are left with a basis of the column space of P!

```
> Reduce(P,form=rref);
```

$$\begin{bmatrix} 1 & 0 & 0 & -1 & 0 & 0 & 1 & 0 & 1 & 1 \\ 0 & 1 & 0 & 1 & 1 & 0 & -1 & 0 & 0 & -1 \\ 0 & 0 & 1 & 0 & -1 & 0 & 0 & 0 & -1 & 0 \\ 0 & 0 & 0 & 0 & 0 & 1 & 1 & 0 & 1 & 1 \\ 0 & 0 & 0 & 0 & 0 & 0 & 0 & 1 & -1 & -1 \\ 0 & 0 & 0 & 0 & 0 & 0 & 0 & 0 & 0 & 0 \end{bmatrix}$$

Since the pivot columns of P are 1, 2, 3, 6, 8, these columns of P form a basis for the column space of P. We construct the submatrix of P that has just these columns:

```
> Q := Matrix([Column(P,1),Column(P,2),Column(P,3),
    Column(P,6),Column(P,8)]);
```

$$Q := \begin{bmatrix} 1 & 1 & 1 & 0 & 0 \\ -1 & 0 & 0 & 1 & 0 \\ 0 & 0 & -1 & 0 & 0 \\ 0 & -1 & 0 & 0 & 0 \\ 0 & 0 & 0 & -1 & -1 \\ 0 & 0 & 0 & 0 & 1 \end{bmatrix}$$

Since the columns of Q are linearly independent, the null space of Q is $\{0\}$. Let's check:

```
> Matsolve(Q,zero,free=x);
```

$$\begin{bmatrix} 0 \\ 0 \\ 0 \\ 0 \\ 0 \end{bmatrix}$$

On your paper copy of the pipeline network, darken the five arcs corresponding to the five columns we chose above. Note the following facts about the smaller pipeline network you have drawn: Each node is connected to each of the other nodes, either directly by a single arc or by a sequence of adjoining arcs (so oil can flow from every oil field to the refinery); and no arc can be removed without cutting off at least one oil field from the refinery. Such a network is called a "spanning tree." So we have now designed a more efficient pipeline route. (Aside: A more realistic version of this problem would take into account the costs of constructing each of the pipelines, and we would seek a spanning tree that also minimizes the total construction costs.)

Problems

Problem 1: Another traffic loop

(a) Find a null space vector that represents the loop "one vehicle traveling counterclockwise around the two-block area FEDCBA."

(b) Express this vector as a linear combination of u and v, where $\{u, v\}$ is the null space basis we found in Section 1.

Student Workspace

Problem 2: Another traffic diversion

Find a solution to the traffic flow example in Section 1 that diverts all traffic from arcs AF and CD.

Student Workspace

Problem 3: Another spanning tree

By hand, draw a different spanning tree for our pipeline problem than the one we found in Section 2. Find the corresponding submatrix of P and check that its null space is $\{0\}$. Also check that its column space has the same dimension as the column space of P.

Student Workspace

Problem 4: Pipeline computation

(a) Solve the system of linear equations $P\,x = b$, where P is the pipeline matrix (repeated below) and b is the vector defined below. We can think of the first component of b as the total number of barrels of oil (in millions of barrels) that reach the refinery from all of the oil fields per year, and the remaining components as the number of barrels of oil that leave each of the oil fields.

```
> P := Matrix([[1,1,1,0,0,0,0,0,0],[-1,0,0,1,0,1,0,0,0],
   [0,0,-1,0,1,0,0,0,1,0],[0,-1,0,-1,-1,0,1,0,0,1],
   [0,0,0,0,0,-1,-1,-1,0,0],[0,0,0,0,0,0,1,-1,-1]]);
> b := Vector([20,-2,-3,-1,-6,-8]);
```

(b) Write down the conservation of flow equation for node 2, and check that it corresponds to row 2 of the augmented matrix of the linear system $P\,x = b$.

(c) Decompose your answer in (a), as we did in our solution of the traffic flow problem in Section 1. On a drawing of the pipeline arcs, mark the loops corresponding to each of the null space basis vectors in your decomposition.

(d) Below is the matrix Q from Section 2 that corresponds to the spanning tree we found. Remove one of the columns of Q to form a new matrix M. By hand, draw the pipelines corresponding to the columns of

M. Why is this not an acceptable solution to the pipeline problem? Also, solve $Mx = b$ and explain why you get the result that you do.

```
> Q := Matrix([Column(P,1),Column(P,2),Column(P,3),
    Column(P,6),Column(P,8)]);
```

+ Student Workspace

Problem 5: Dimensions, dimensions everywhere

Suppose E is an m by n matrix arising from a network of m nodes and n arcs in the same way as matrix M of Section 1 and matrix P of Section 2. So each column of E corresponds to one of the arcs and each row to one of the nodes; also, each column of E has one -1, one 1, and the remaining entries 0. (E is called the *incidence matrix* of the network.) Suppose also that E arises from a connected network. (A network is *connected* if each node can be joined to each of the other nodes by a sequence of adjoining arcs of the network.) It is not hard to see that any spanning tree for a connected network has exactly $m - 1$ arcs. (Two nodes are connected by one arc, three nodes by two arcs, etc.) After you read the following questions, you might want to perform some calculations on the additional examples below to see what's going on.

(a) What is the rank of E? Explain.
(b) What is the dimension of the null space of E? Explain.
(c) What is the dimension of the null space of E^T? Find a basis for the null space of E^T.

```
> E1 := Matrix([[1,1,0],[-1,0,-1],[0,-1,1]]);
> E2 := Matrix([[1,1,1,0,0,0],[-1,0,0,1,1,0],
    [0,-1,0,-1,0,-1],[0,0,-1,0,-1,1]]);
```

+ Student Workspace

Chapter 6: Eigenvalues and Eigenvectors

Module 1. Introduction to Eigenvalues and Eigenvectors

Module 2. The Characteristic Polynomial

Module 3. Eigenvector Bases and Discrete Dynamical Systems

Module 4. Diagonalization and Similarity

Module 5. Complex Eigenvalues and Eigenvectors

 Commands used in this chapter

abs(x); produces the absolute value of x.
argument(z); produces the polar angle (i.e., argument) of the complex number z.
Column(A,i); selects the ith column of the matrix A.
conjugate(z); produces the complex conjugate of the complex number or vector z.
display([pict1,pict2]); displays together a group of previously defined pictures.
evalf(expr); evaluates the expression $expr$ as a decimal approximation.
factor(expr); factors the algebraic expression $expr$.
Im(z); produces the imaginary part of the complex number or vector z.
map(simplify,M); applies the **simplify** command to every entry in the matrix or vector M.
Matrix([[a,b],[c,d]]); defines the matrix with rows $[a, b]$ and $[c, d]$.
Re(z); produces the real part of the complex number or vector z.
subs({a=3,b=13},expr); substitutes the values for a and b in the expression $expr$.
Vector([a,b,c]); defines the vector $\langle a, b, c \rangle$.

LAMP commands:
Basisgrid(u,v,[a,b]); draws xy and uv coordinate grids and the point $a u + b v$.
Charpoly(A,t); produces **Det(A-t*I);**, the characteristic polynomial of the matrix A.
Clock(A); an animation which simultaneously displays a rotating unit vector v and its image $A v$, where A is a 2 by 2 matrix.
Componentplot(A,x0,points=n); plots (in an animation) the components of the vectors $A^k x_0$ for k from 0 to n.
Det(M); calculates the determinant of the matrix M.
Diagmat([a,b,c]); produces a diagonal matrix with diagonal entries a, b, c.

Drawvec(u, [v,w]); draws the vector u with tail at the origin and the vector with tail at v and head at w. (Vectors are in 2-space.)

Evalues(A); produces the eigenvalues of the matrix A.

Evectors(A); produces the eigenvalues and eigenvectors of the matrix A.

Expand(expr); expands and evaluates the vector or matrix expression *expr*.

Headtail(A,vector=n); displays n equally-spaced unit vectors v and their images $A\,v$, where A is a 2 by 2 matrix.

Idmat(n); produces the n by n identity matrix.

Matsolve(A,b,free=t); solves the matrix-vector equation $A\,x = b$ for x. The solution is expressed in terms of the free variable t.

Nullbasis(A); produces a basis for the null space of the matrix A.

Projectmat(theta); produces the 2 by 2 matrix that projects vectors onto the line making the angle θ with the x axis.

Reflectmat(theta); produces the 2 by 2 matrix that reflects vectors across the line making the angle θ with the x axis.

Rotatemat(theta); produces the 2 by 2 matrix that rotates vectors through the angle θ.

Trajectory(A,x0,n); plots the points $x_k = A^k x_0$ (in an animation) for k from 0 to n.

Trajectory3d(A,x0,n); plots the points $x_k = A^k x_0$ (in an animation) for k from 0 to n.

Transform(M,F); draws the 2d or 3d figure whose vertices are the columns of the matrix F and also draws the transformed figure $M\,F$.

Linear Algebra Modules Project
Chapter 6, Module 1

Introduction to Eigenvalues and Eigenvectors

■ Purpose of this module

The purpose of this module is to introduce the concepts of eigenvector and eigenvalue from both algebraic and geometric points of view.

■ Prerequisites

Linear systems; geometric transformations (Module 1 of Chapter 4); linear independence, basis and dimension; null space of a matrix.

+ Commands used in this module

```
> restart; with(LinearAlgebra): with(plots): with(Lamp):
  UseHardwareFloats := false: Digits := 6:
```

Tutorial

■ Section 1: Definition and Examples of Eigenvectors and Eigenvalues

Definition: If A is a square matrix, we say that a nonzero vector x is an ***eigenvector*** of A and a scalar λ is the associated ***eigenvalue*** if the following equation holds:

$$A x = \lambda x$$

In other words, when x is multiplied by A, the result is just a scalar multiple of x.

==================

Example 1A: The vector $u = \begin{bmatrix} 3 \\ 2 \end{bmatrix}$ is an eigenvector of the matrix $A = \begin{bmatrix} 5 & -3 \\ -4 & 9 \end{bmatrix}$, since multiplying u by A results in a scalar multiple of u:

$$\begin{bmatrix} 5 & -3 \\ -4 & 9 \end{bmatrix} \begin{bmatrix} 3 \\ 2 \end{bmatrix} = \begin{bmatrix} 9 \\ 6 \end{bmatrix} = 3 \begin{bmatrix} 3 \\ 2 \end{bmatrix}$$

So the eigenvalue associated with $u = \begin{bmatrix} 3 \\ 2 \end{bmatrix}$ is $\lambda = 3$. Here is the same computation in Maple:

```
> A := Matrix([[5,-3],[-4,9]]);
```

```
> u := Vector([3,2]);
```

$$A := \begin{bmatrix} 5 & -3 \\ -4 & 9 \end{bmatrix}$$

$$u := \begin{bmatrix} 3 \\ 2 \end{bmatrix}$$

```
> A.u;
```

$$\begin{bmatrix} 9 \\ 6 \end{bmatrix}$$

Exercise 1.1: For the matrix B below, which of the vectors u, v, w are eigenvectors of B? For each one that is an eigenvector, what is its associated eigenvalue?

```
> B := Matrix([[0,-1,1],[-2,-1,-1],[-2,1,-3]]);
> u := Vector([1, 2, 0]);
  v := Vector([-1, 1, 1]);
  w := Vector([1, 1, 1]);
```

 Student Workspace

Answer 1.1

```
> u, B.u;
```

$$\begin{bmatrix} 1 \\ 2 \\ 0 \end{bmatrix}, \begin{bmatrix} -2 \\ -4 \\ 0 \end{bmatrix}$$

u is an eigenvector of B with associated eigenvalue $\lambda = -2$, since $B\,u = -2\,u$.

```
> v, B.v;
```

$$\begin{bmatrix} -1 \\ 1 \\ 1 \end{bmatrix}, \begin{bmatrix} 0 \\ 0 \\ 0 \end{bmatrix}$$

v is an eigenvector of B with associated eigenvalue $\lambda = 0$, since $B\,u = 0\,u$. (Note that while eigenvectors, by definition, cannot be zero, the scalar 0 is a legitimate eigenvalue.)

```
> w, B.w;
```

$$\begin{bmatrix} 1 \\ 1 \\ 1 \end{bmatrix}, \begin{bmatrix} 0 \\ -4 \\ -4 \end{bmatrix}$$

w is not an eigenvector of B, since $B\,w$ is not a scalar multiple of w.

If u is an eigenvector of a matrix A with associated eigenvalue λ, then it turns out that every nonzero multiple of u is also an eigenvector with the same eigenvalue. That is, if $A\,u = \lambda\,u$ then we also have $A\,v = \lambda\,v$ for every vector $v = c\,u$, where c is any nonzero scalar. Test this fact in the next example.

Example 1B: In Example 1A, we saw that the vector $u = \begin{bmatrix} 3 \\ 2 \end{bmatrix}$ is an eigenvector of the matrix A with associated eigenvalue $\lambda = 3$. Change the value of c (below) and observe that each nonzero multiple of u is also an eigenvector of A and that the associated eigenvalue is still 3.

```
> A := Matrix([[5,-3],[-4,9]]);
> u := Vector([3,2]);
```

$$A := \begin{bmatrix} 5 & -3 \\ -4 & 9 \end{bmatrix}$$

$$u := \begin{bmatrix} 3 \\ 2 \end{bmatrix}$$

```
> c := 2;
  v := c*u;
```

$$c := 2$$

$$v := \begin{bmatrix} 6 \\ 4 \end{bmatrix}$$

```
> A.v;
```

$$\begin{bmatrix} 18 \\ 12 \end{bmatrix}$$

So $A\,v = 3\,v$, which says that v is also an eigenvector with eigenvalue 3.

Exercise 1.2: (By hand) Use the rules of matrix algebra to prove the following statement:

- If A is a square matrix with eigenvector u and associated eigenvalue λ, then $A\,v = \lambda\,v$ for every vector $v = c\,u$, where c is any nonzero scalar; i.e., every nonzero multiple of an eigenvector is also an eigenvector with the same eigenvalue.

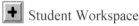 Student Workspace

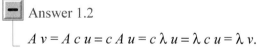 Answer 1.2

$A\,v = A\,c\,u = c\,A\,u = c\,\lambda\,u = \lambda\,c\,u = \lambda\,v.$

As we have seen, if we are given an eigenvector x of a matrix A, then finding its associated eigenvalue is simply a matter of comparing $A\,x$ with x. We now consider the reverse problem. If we are given a eigenvalue λ for a matrix A, how can we find a corresponding eigenvector x so that $A\,x = \lambda\,x$? To help us solve the equation $A\,x = \lambda\,x$, we rewrite it in the form of a homogeneous matrix-vector equation:

$$A\,x = \lambda\,x$$

Write $\lambda\,x$ as $\lambda\,I\,x$ and move it to the left side:

$$A\,x - \lambda\,I\,x = 0$$

Factor out x (distributive law):
$$(A - \lambda I)x = 0$$
Let's describe what we have just deduced:

> **Theorem 1:** A number λ is an eigenvalue of the square matrix A if and only if the equation $(A - \lambda I)x = 0$ has a nontrivial solution x. The nontrivial solutions are the eigenvectors associated with λ.

==================

Example 1C: The matrix A defined below has an eigenvalue $\lambda = 3$. Use Theorem 1 to find the associated eigenvectors.

```
> A := Matrix([[1,2],[2,1]]);
```

$$A := \begin{bmatrix} 1 & 2 \\ 2 & 1 \end{bmatrix}$$

Solution: We begin by calculating the matrix $A - 3I$. Note that this is simply the matrix A with 3 subtracted from each of its diagonal entries:

$$A - 3I = \begin{bmatrix} 1 & 2 \\ 2 & 1 \end{bmatrix} - \begin{bmatrix} 3 & 0 \\ 0 & 3 \end{bmatrix} = \begin{bmatrix} -2 & 2 \\ 2 & -2 \end{bmatrix}$$

We now solve the homogeneous linear system $(A - 3I)x = 0$, which we do using **Matsolve**:

```
> I := Idmat(2);
> Matsolve(A-3*I,zero,free=x);
```

$$I := \begin{bmatrix} 1 & 0 \\ 0 & 1 \end{bmatrix}$$

$$\begin{bmatrix} x_2 \\ x_2 \end{bmatrix}$$

As expected, the equation $(A - 3I)x = 0$ has nontrivial solutions; each of them is an eigenvector with associated eigenvalue $\lambda = 3$. Below, we enter the eigenvector we get by choosing $x_2 = 1$ and then check that it works.

```
> u := Vector([1,1]);
> A.u;
```

$$u := \begin{bmatrix} 1 \\ 1 \end{bmatrix}$$

$$\begin{bmatrix} 3 \\ 3 \end{bmatrix}$$

We can therefore describe the set of all eigenvectors associated with the eigenvalue 3 as the set of all nonzero multiples of the vector $\begin{bmatrix} 1 \\ 1 \end{bmatrix}$.

==================

Exercise 1.3: (a) The matrix A in Example 1C has a second eigenvalue $\lambda = -1$. Find the set of eigenvectors associated with this eigenvalue.
(b) As Problem 6 shows, we cannot have more than two eigenvalues for a 2 by 2 matrix. Confirm, for example, that $\lambda = 2$ is not an eigenvalue of A.

```
> A := Matrix([[1,2],[2,1]]);
```

$$A := \begin{bmatrix} 1 & 2 \\ 2 & 1 \end{bmatrix}$$

+ Student Workspace

− Answer 1.3

(a)
```
> I := Idmat(2);
> Matsolve(A+I,zero,free=x);
```

$$I := \begin{bmatrix} 1 & 0 \\ 0 & 1 \end{bmatrix}, \begin{bmatrix} -x_2 \\ x_2 \end{bmatrix}$$

So the eigenvectors associated with $\lambda = -1$ are all nonzero multiples of the vector $\begin{bmatrix} -1 \\ 1 \end{bmatrix}$.

(b)
```
> Matsolve(A-2*I,zero,free=x);
```

$$\begin{bmatrix} 0 \\ 0 \end{bmatrix}$$

Since the only vector that satisfies the equation $(A - 2I)x = 0$ is $x = 0$, the scalar $\lambda = 2$ is not an eigenvalue of A.

Exercise 1.4: Eigenvalues of diagonal matrices are particularly easy to spot; they are the diagonal entries of the matrix. For the diagonal matrix K below, find an eigenvector for each of the eigenvalues (i.e., for each of the diagonal entries). You may be able to do this in your head by observing that certain familiar vectors are eigenvectors.

```
> K := Diagmat([3,2,-5]);
```

$$K := \begin{bmatrix} 3 & 0 & 0 \\ 0 & 2 & 0 \\ 0 & 0 & -5 \end{bmatrix}$$

+ Student Workspace

− Answer 1.4

The standard basis vectors, e_1, e_2, e_3, are eigenvectors of K. The first standard basis vector is an eigenvector whose associated eigenvalue is the first diagonal entry of K; and similarly for the other two standard basis vectors:

```
[ > K.Vector([1,0,0]);
[ > K.Vector([0,1,0]);
[ > K.Vector([0,0,1]);
```

$$\begin{bmatrix} 3 \\ 0 \\ 0 \end{bmatrix}, \begin{bmatrix} 0 \\ 2 \\ 0 \end{bmatrix}, \begin{bmatrix} 0 \\ 0 \\ -5 \end{bmatrix}$$

Note: The standard basis vectors, $e_1, \ldots, e_n$, in R^n are eigenvectors for any n by n diagonal matrix K. For example, $K e_1 = d_1 e_1$, where d_1 is the first diagonal entry of K; so d_1 is the eigenvalue associated with e_1. Similarly for the remaining standard basis vectors.

Since the diagonal entries of a diagonal matrix are its eigenvalues, we see that an n by n diagonal matrix has at most n distinct eigenvalues (since some of the diagonal entries may be repeated). As we will see in the next module, this generalizes to arbitrary matrices, diagonal or not:

Theorem 2: An n by n matrix has at most n distinct eigenvalues.

 Section 2: The Geometry of Eigenvectors

The definition of eigenvector has a simple geometric interpretation. If $A x = \lambda x$ then the vector $A x$ will point in the same or opposite direction as x (depending on whether $0 < \lambda$ or $\lambda < 0$), and the magnitude of the eigenvalue will correspond to the "stretch/shrink" factor. For example, recall that the matrix $A = \begin{bmatrix} 5 & -3 \\ -4 & 9 \end{bmatrix}$ has the eigenvector $u = \begin{bmatrix} 3 \\ 2 \end{bmatrix}$ with associated eigenvalue $\lambda = 3$. Here is a picture of the vectors u and $A u$ drawn with red head and blue head respectively. Note that $A u$ is in the same direction as u but is 3 times as long.

```
> A := Matrix([[5,-3],[-4,9]]):
  u := Vector([3,2]):
> Drawvec(u,[A.u,headcolor=blue],view=[-2..10,-2..10]);
```

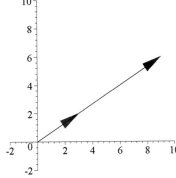

Example 2A: Here is an example where we can find the eigenvalues and eigenvectors of a matrix simply by picturing how the matrix behaves as a geometric transformation. Let R be the matrix that reflects vectors across the line $y = x$:

$$R = \begin{bmatrix} 0 & 1 \\ 1 & 0 \end{bmatrix}$$

The figure below shows what R does to several vectors. Use this picture to determine all eigenvectors of R and their associated eigenvalues.

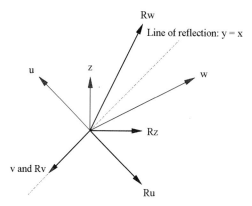

Solution: R reflects vectors that lie along the line $y = -x$ to their negatives, and it keeps vectors that lie along the line $y = x$ fixed in place. Thus every nonzero vector along the line $y = -x$ (such as u above) is an eigenvector with eigenvalue $\lambda = -1$, and every nonzero vector along the line $y = x$ (such as v above) is an eigenvector with eigenvalue $\lambda = 1$.

Even if we cannot figure out the eigenvalues of a matrix from our knowledge of its behavior as a geometric transformation, we will find pictures such as the one above quite helpful.

Let's consider the matrix A below as a matrix transformation. The vectors u and v below are eigenvectors of A with different eigenvalues. We will use **Drawvec** to draw the input vectors u and v with red heads and their corresponding output vectors with blue heads and thick shafts. See if you can guess the eigenvalue for each eigenvector just by looking at the picture.

```
> A := Matrix([[1,1],[1/2,3/2]]);
> u := Vector([-2,1]);
> v := Vector([1,1]);
> Drawvec(u,v,[A.u,headcolor=blue,thickness=4],
    [A.v,headcolor=blue,thickness=4]);
```

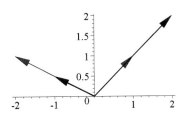

Sometimes it is easier to visualize eigenvectors by putting the tail of the output vector Ax at the head of the input vector x (again, the input vectors have thin shafts and the output vectors thick shafts):

```
> Drawvec([u,shaftcolor=red],[v,shaftcolor=red],
   [u,u+A.u,shaftcolor=blue,headcolor=blue,thickness=4],
   [v,v+A.v,shaftcolor=blue,headcolor=blue,thickness=4]);
```

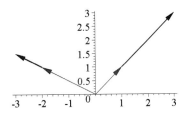

Notice that the thick blue output vector in the second quadrant is half the length of the corresponding thin red input vector, and the other output vector is twice the length of its input. Hence the eigenvalues associated with u and v are $1/2$ and 2, respectively.

In the next visualization, we show a large number of equally-spaced input vectors with their corresponding output vectors. The input vectors are the red vectors of length 1 with tail at the origin, and each output vector has its tail at the head of the corresponding input vector and has a thick blue shaft. Can you spot roughly which input vectors are eigenvectors, and can you estimate their eigenvalues? (Recall that if two vectors are in the same or opposite direction, they are multiples of one another.)

```
> Headtail(A);
```

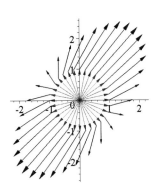

Chapter 6 Eigenvalues and Eigenvectors

The next visualization is an animation that displays a rotating unit input vector (red) with its corresponding output vector (blue). Whenever the hands of this "clock" cross, we have an eigenvector with a positive eigenvalue. When they point in opposite directions, we have an eigenvector with a negative eigenvalue.

```
> Clock(A);
```

Both the **Headtail** and **Clock** commands accept an optional second argument that allows you to specify the number of input vectors. (The defaults are 32 and 64, respectively.) For example, try:

```
> Headtail(A,vectors=42);
```

In the next two exercises, you will use **Headtail** or **Clock** to investigate the eigenvectors and eigenvalues of some of the geometric transformations you studied in Module 1 of Chapter 4, "Geometry of Matrix Transformations of the Plane." Use the commands **Rotatemat(θ)**, **Reflectmat(θ)**, and **Projectmat(θ)** to construct these transformations as you need them.

Exercise 2.1: Using **Headtail** and/or **Clock**, experiment with the reflection matrix below, and then answer the following questions. What are the eigenvectors and eigenvalues of this reflection matrix? What are the eigenvectors and eigenvalues of every reflection matrix? (Try reasoning geometrically from your knowledge of what a reflection matrix does to vectors.)

```
> R := Reflectmat(Pi/4);
```

$$R := \begin{bmatrix} 0 & 1 \\ 1 & 0 \end{bmatrix}$$

 Student Workspace

 Answer 2.1

Let R be a reflection matrix that reflects vectors across a line through the origin. Then R reflects vectors on this line to themselves; thus every nonzero vector on the line of reflection is an eigenvector with associated eigenvalue 1. The transformation R reflects vectors perpendicular to the line of reflection to their negatives; thus every nonzero vector on the line through the origin perpendicular to the line of reflection is an eigenvector with associated eigenvalue -1.

Exercise 2.2: Using **Headtail** and/or **Clock**, experiment with the projection matrix below, and then answer the following questions. What are the eigenvectors and eigenvalues of this projection matrix? What are the eigenvectors and eigenvalues of every projection matrix? (Try reasoning geometrically from your knowledge of what a projection matrix does to vectors.)

```
> P := Projectmat(Pi/4);
```

$$P := \begin{bmatrix} \frac{1}{2} & \frac{1}{2} \\ \frac{1}{2} & \frac{1}{2} \end{bmatrix}$$

 Student Workspace

 Answer 2.2

Let P be a projection matrix that projects vectors onto a line through the origin. Then P projects vectors on this line to themselves; thus every nonzero vector on this line of projection is an eigenvector with associated eigenvalue 1. The transformation P projects vectors perpendicular to the line of projection to the zero vector; thus every nonzero vector on the line through the origin perpendicular to the line of projection is an eigenvector with associated eigenvalue 0.

Note: Recall that, by definition, the zero vector is never an eigenvector. However, the zero scalar can be an eigenvalue, as we saw in Exercises 1.1 and 2.2.

 Section 3: Eigenspaces

Recall Theorem 1 from Section 1:

- A number λ is an eigenvalue of the square matrix A if and only if the equation $(A - \lambda I)x = 0$ has a nontrivial solution x. The nontrivial solutions are the eigenvectors associated with λ.

Since a square matrix M is singular if and only if $Mx = 0$ has a nontrivial solution (Theorem 6 in Chapter 3), we can restate Theorem 1 as follows:

Theorem 3: A number λ is an eigenvalue of a square matrix A if and only if the matrix $A - \lambda I$ is singular.

Equivalently, a number λ is an eigenvalue of a square matrix A if and only if the nullspace of the matrix $A - \lambda I$ is not $\{0\}$. This nonzero nullspace is the subject of the next definition.

Definition: If A is a square matrix with eigenvalue λ, then the null space of the matrix $A - \lambda I$ is called the *eigenspace* associated with the eigenvalue λ.

Note that the eigenspace associated with λ includes the zero vector, although the zero vector is not an eigenvector. All the other vectors in the eigenspace, however, are eigenvectors associated with λ.

Definition: The dimension of the eigenspace associated with an eigenvalue λ is called the *geometric multiplicity* of the eigenvalue λ. (That is, the dimension of the null space of $A - \lambda I$ is the geometric multiplicity of λ.)

So the geometric multiplicity of an eigenvalue λ is the number of linearly independent eigenvectors associated with λ.

=====================

Example 3A: For the matrix A below, find a basis for the eigenspace associated with the eigenvalue $\lambda = -6$. What is the geometric multiplicity of this eigenvalue?

```
> A := Matrix([[4,10,-10],[-5,-11,5],[-5,-5,-1]]);
```

__Chapter 6 Eigenvalues and Eigenvectors__

$$A := \begin{bmatrix} 4 & 10 & -10 \\ -5 & -11 & 5 \\ -5 & -5 & -1 \end{bmatrix}$$

Solution: We solve the equation $A\,x = -6\,x$ by writing it in the form $(A + 6\,I)\,x = 0$:

```
> I := Idmat(3);
  Matsolve(A+6*I,zero,free=x);
```

$$I := \begin{bmatrix} 1 & 0 & 0 \\ 0 & 1 & 0 \\ 0 & 0 & 1 \end{bmatrix}, \begin{bmatrix} -x_2 + x_3 \\ x_2 \\ x_3 \end{bmatrix}$$

To find a basis of this eigenspace, we decompose the solution in the usual way:

$$\begin{bmatrix} -x_2 + x_3 \\ x_2 \\ x_3 \end{bmatrix} = x_2 \begin{bmatrix} -1 \\ 1 \\ 0 \end{bmatrix} + x_3 \begin{bmatrix} 1 \\ 0 \\ 1 \end{bmatrix}$$

So $\{\langle -1, 1, 0\rangle, \langle 1, 0, 1\rangle\}$ is a basis for the eigenspace associated with $\lambda = -6$. Alternatively, we can use the **Nullbasis** command, which produces a basis for the null space in one step:

```
> Nullbasis(A+6*I);
```

$$\left\{ \begin{bmatrix} -1 \\ 1 \\ 0 \end{bmatrix}, \begin{bmatrix} 1 \\ 0 \\ 1 \end{bmatrix} \right\}$$

Since the eigenspace basis has two vectors, the geometric multiplicity of the eigenvalue -6 is 2.

======================

Exercise 3.1: For the matrix B below, find a basis for the eigenspace associated with the eigenvalue $\lambda = 5$. What is the geometric multiplicity of this eigenvalue?

```
> B := Matrix([[5,1,0,0],[0,5,0,0],[0,0,5,0],[0,0,0,3]]);
```

$$B := \begin{bmatrix} 5 & 1 & 0 & 0 \\ 0 & 5 & 0 & 0 \\ 0 & 0 & 5 & 0 \\ 0 & 0 & 0 & 3 \end{bmatrix}$$

➕ Student Workspace

➖ Answer 3.1

```
> Nullbasis(B-5*Idmat(4));
```

$$\left\{ \begin{bmatrix} 1 \\ 0 \\ 0 \\ 0 \end{bmatrix}, \begin{bmatrix} 0 \\ 0 \\ 1 \\ 0 \end{bmatrix} \right\}$$

The output from the **Nullbasis** command is a basis for the eigenspace associated with $\lambda = 5$. Since this basis has two vectors, the geometric multiplicity of this eigenvalue is 2.

Finding Eigenvalues and Eigenspaces by Geometry

Example 3B: Here is an example where we can figure out the eigenvalues and eigenspaces by purely geometric reasoning. Let P be the 3 by 3 projection matrix that projects vectors in R^3 onto the xy plane. Find the eigenvalues and corresponding eigenspaces for P.

Solution: If v is any vector in the xy plane, then P projects v to itself: $P v = 1 v$. So 1 is an eigenvalue whose associated eigenspace is the xy plane. Since a plane has dimension 2, the geometric multiplicity of $\lambda = 1$ is 2. If v is any vector on the z axis, then P projects v to the zero vector: $P v = 0 v$. So 0 is an eigenvalue whose associated eigenspace is the z axis. Since a line has dimension 1, the geometric multiplicity of $\lambda = 0$ is 1.

Exercise 3.2: Let R be the 3 by 3 reflection matrix that reflects vectors in R^3 across the xy plane. Find the eigenvalues and corresponding eigenspaces for R.

 Student Workspace

 Answer 3.2

If v is any vector in the xy plane, then R reflects v to itself: $R v = 1 v$. So 1 is an eigenvalue whose associated eigenspace is the xy plane. Since a plane has dimension 2, the geometric multiplicity of $\lambda = 1$ is 2. If v is any vector on the z axis, then R reflects v to $-v$: $R v = -1 v$. So -1 is an eigenvalue whose associated eigenspace is the z axis. Since a line has dimension 1, the geometric multiplicity of $\lambda = -1$ is 1.

Evalues Command

Determining eigenvalues by geometric reasoning is often not feasible. In the next module we will discover a general formula for the eigenvalues of a matrix (although it is not always easy to use). For now we can simply use the **Evalues** command:

Example 3C: We apply the **Evalues** command to the matrix A below and then interpret the output.

```
> A := Matrix([[4,10,-10],[-5,-11,5],[-5,-5,-1]]);
```

$$A := \begin{bmatrix} 4 & 10 & -10 \\ -5 & -11 & 5 \\ -5 & -5 & -1 \end{bmatrix}$$

```
> Evalues(A);
```

$$\begin{bmatrix} -6 \\ -6 \\ 4 \end{bmatrix}$$

So –6 and 4 are eigenvalues of A. (In Module 2, we will learn why –6 is shown twice.)

================

Exercise 3.3: Use the **Evalues** command to find the eigenvalues of the matrix B below. Then use the **Nullbasis** command to confirm that these are eigenvalues and to determine the geometric multiplicity of each eigenvalue.

> `B := Matrix([[2,0,0,1],[1,2,0,1],[1,0,2,1],[1,0,0,2]]):`

+ Student Workspace

 Answer 3.3

> `Evalues(B);`

$$\begin{bmatrix} 3 \\ 2 \\ 2 \\ 1 \end{bmatrix}$$

> `I := Idmat(4):`
   `Nullbasis(B-3*I);`

$$\begin{bmatrix} \begin{bmatrix} 1 \\ 2 \\ 2 \\ 1 \end{bmatrix} \end{bmatrix}$$

Thus $\lambda = 3$ has geometric multiplicity 1, since the solution set of $(B - 3I)x = 0$ is 1-dimensional.

> `Nullbasis(B-2*I);`

$$\begin{bmatrix} \begin{bmatrix} 0 \\ 1 \\ 0 \\ 0 \end{bmatrix}, \begin{bmatrix} 0 \\ 0 \\ 1 \\ 0 \end{bmatrix} \end{bmatrix}$$

Thus $\lambda = 2$ has geometric multiplicity 2.

> `Nullbasis(B-I);`

$$\begin{bmatrix} \begin{bmatrix} -1 \\ 0 \\ 0 \\ 1 \end{bmatrix} \end{bmatrix}$$

Thus $\lambda = 1$ has geometric multiplicity 1.

Problems

Problem 1: Find eigenvectors from eigenvalues

(a) The matrix A below has the eigenvalue $\lambda = -4$. Find all the associated eigenvectors.
(b) The columns of A are linearly dependent. What eigenvalue must A therefore have? Find all the eigenvectors associated with this eigenvalue.

```
> A := Matrix([[-3,-1,2],[1,-5,2],[2,-2,0]]);
```

Student Workspace

Problem 2: Work backwards from eigenvalues and eigenvectors

(a) Find a 2 by 2 matrix A that has eigenvalues 2 and 3 with associated eigenvectors $\begin{bmatrix} 1 \\ 4 \end{bmatrix}$ and $\begin{bmatrix} -2 \\ 3 \end{bmatrix}$, respectively. (Hint: Write down the two equations of the form $A\,x = \lambda\,x$ that we want A to satisfy; then rewrite these two equations in the form of a single matrix equation $A\,B = C$, where B and C are known 2 by 2 matrices. Then solve this equation for A.)

(b) Now consider the general problem: Find a 2 by 2 matrix A with eigenvalues λ_1 and λ_2 and associated eigenvectors v_1 and v_2, respectively, where $\{v_1, v_2\}$ is linearly independent. Will this problem always have a solution? Explain.

Student Workspace

Problem 3: Matrix transformations of R^3

By geometric reasoning alone, determine the eigenvalues and eigenvectors of the 3 by 3 matrix transformations below. (Such transformations are discussed in Section 1 of Module 2 of Chapter 4.) Each of the problems refers to a given line L, which is always assumed to be a line through the origin; we can therefore express this line as the span of a single nonzero vector: $L = \text{Span}\{v\}$.

(a) Let R be the 3 by 3 matrix that reflects vectors in R^3 across a plane through the origin perpendicular to the line L. What are the eigenvalues of R and their associated eigenspaces?

(b) Let M be the 3 by 3 matrix that rotates vectors in R^3 about the line L through the angle θ (where θ is not a multiple of π). This matrix has one real eigenvalue. What is this eigenvalue and its associated eigenspace?

(c) Let P be the 3 by 3 matrix that projects vectors in R^3 onto the line L. This matrix has two eigenvalues. What are the eigenvalues of P and their associated eigenspaces?

Student Workspace

Problem 4: Rotation matrices in R^2

(a) Let M be the 2 by 2 matrix that rotates vectors through an angle θ in the counterclockwise direction. Use geometry to explain why most rotation matrices have no eigenvalues or eigenvectors. To stimulate your geometric insight, use **Clock** on one or two specific rotation matrices.

(b) A few rotation matrices, however, do have eigenvalues and eigenvectors. Which ones are these? What are the eigenvalues and their corresponding eigenvectors?

(c) Apply the **Evalues** command to the general rotation matrix: $M = \begin{bmatrix} \cos(\theta) & -\sin(\theta) \\ \sin(\theta) & \cos(\theta) \end{bmatrix}$.

```
> M := Rotatemat(theta);
```

Notice that there are real eigenvalues for certain values of θ only. What are these values of θ and what eigenvalues do they produce? Compare this answer to your answer in (b).

Student Workspace

```
> R := Rotatemat(Pi/4);
> Clock(R);
```

Problem 5: Upper triangular matrices

(a) Find the eigenvalues of the upper-triangular matrix U below; also find the eigenspace and geometric multiplicity of each eigenvalue. (If you do not see what the eigenvalues must be, you may use the **Evalues** command.)

```
> U := Matrix([[3,2,-1],[0,3,0],[0,0,-2]]);
```

(b) What are the eigenvalues of any upper-triangular matrix? Explain why your answer is correct.

Student Workspace

Problem 6: Linear independence of eigenvectors

(a) Suppose v_1 and v_2 are eigenvectors of a square matrix A and that λ_1 and λ_2 are the associated eigenvalues. If $\lambda_1 \neq \lambda_2$, prove that $\{v_1, v_2\}$ is linearly independent. (Hint: If $\{v_1, v_2\}$ were linearly dependent, one of the vectors would be a multiple of the other. Show that this implies that both vectors have the same eigenvalue.)

(b) Show that every 2 by 2 matrix has at most two distinct eigenvalues. (Hint: Assume that a 2 by 2 matrix A has three distinct eigenvalues with associated eigenvectors v_1, v_2, v_3. These eigenvectors are linearly dependent (why?), and we can write $v_1 = a\, v_2 + b\, v_3$ (why?). Apply A to both sides of this equation to reach a contradiction.)

Student Workspace

Problem 7: Eigenvalues of $A, A^2, A^3, \ldots$

Once we find the eigenvectors and eigenvalues of a matrix A, we can deduce the eigenvectors and eigenvalues of a great many other matrices that are derived from A. For example, as you will see, every eigenvector of A is also an eigenvector for the matrices A^2, A^3, etc.

(a) If x is an eigenvector of A with associated eigenvalue λ, what is the associated eigenvalue for A^2? You can check your guess by finding the eigenvalues of A and A^2 for some of the matrices below.

(b) Prove your answer in (a). Hint: What happens when you multiply an eigenvector x of A by A^2:
$$A^2 x = A\,A\,x = ?$$

(c) State a corresponding result (i.e., theorem) for A and A^n.

```
> A := Matrix([[1,2],[2,1]]);
> B := Matrix([[1,1],[1/2,3/2]]);
> C := Projectmat(Pi/4);
```
Student Workspace

Problem 8: Nilpotent matrices

A matrix A is said to be *nilpotent* if $A^n = O$ for some positive integer n. Check that the matrices A and B below are nilpotent.

(a) What are the eigenvalues for every nilpotent matrix? You can check your guess by finding the eigenvalues of the two matrices below.

(b) Show that your answer in (a) is correct. Hint: Problem 7 is helpful.

```
> A := Matrix([[1,-1,2],[3,-3,6],[1,-1,2]]);
> B := Matrix([[1,0,0,-1],[1,1,0,-4],[0,1,1,-6], [0,0,1,-3]]);
```
Student Workspace

Problem 9: Use $(A - tI)^{(-1)}$ to find eigenvalues

According to Theorem 3, a scalar t is an eigenvalue of a square matrix A if and only if $A - tI$ is singular (i.e., noninvertible). So we can determine the eigenvalues of A by computing the inverse of the symbolic matrix $A - tI$ and determining the values of t for which the symbolic inverse is undefined. Apply this method to the matrices A and B below. (You may use the **Evalues** command only to check your answer.)

```
> A := Matrix([[4,-1,1],[-1,4,-1],[1,-1,4]]);
> B := Matrix([[-1,-1,2,-3],[4,3,0,6],[0,0,5,9],[0,0,-1,-1]]);
```
Student Workspace

Problem 10: Visualizing Eigenvectors in R^3

In the Tutorial, you used the **Headtail** command to detect eigenvectors of 2 by 2 matrices. In this problem, you will use the **Mapcolor** command to detect eigenvectors of 3 by 3 matrices. However, unlike **Headtail**, the **Mapcolor** command does not display a sample of input vectors and their corresponding outputs, as that picture would be far too complicated to interpret. Instead, **Mapcolor** uses color to indicate the presence or absence of eigenvectors.

While our goal is to visualize eigenvectors in 3-space, it is easier to explain how **Mapcolor** works by first applying it to 2 by 2 matrices:
```
> A := Matrix([[1,1],[1/2,3/2]]);
> Mapcolor(A);
```
Like **Headtail**, **Mapcolor** uses a sample of equally-spaced unit input vectors v. But instead of drawing v, **Mapcolor** draws a colored circle at the head of v. The color is determined by the angle θ between v and $A\,v$, with the color nearer to red when θ is nearer to 0 degrees or 180 degrees. Since v is an eigenvector exactly when $\theta = 0$ or $\theta = 180$, the red regions on the color wheel indicate the approximate locations of the eigenvectors.

Here is a picture of the same color wheel, along with two unit eigenvectors of A. The regions diametrically opposite each eigenvector are also red, since the negative of an eigenvector is also an eigenvector.

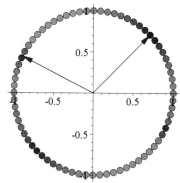

Now that you see how Mapcolor works on 2 by 2 matrices, let's apply it to a 3 by 3 matrix:
```
> B:=Matrix([[6,-2,-1],[-2,6,-1],[-1,-1,5]]);
> Mapcolor(B);
```
As in the 2 by 2 case, **Mapcolor** generates a large sample of unit input vectors v, calculates the angle between v and $B\,v$, and chooses a color corresponding to that angle, with the color nearer to red when the angle is nearer to 0 degrees or 180 degrees. This color is drawn on the surface of the unit sphere around the head of the input vector v.

Move the colored sphere around and observe the six red regions. What does this tell you? Execute the

next line to see how three eigenvectors for this matrix fit into the **Mapcolor** display.
```
> display({Mapcolor(B),Drawvec3d
   (<1,1,-2>,<1,1,1>,<-1,1,0>)},axes=none);
```
The **Mapcolor** command has an optional argument that lets you control the color contrast. The default value is 40, and increasing this value heightens the contrast. Try it:
```
> Mapcolor(B,resolution=100);
```
Note: The **Mapcolor** command does not detect eigenvectors associated with a zero eigenvalue.

The Problem: Below is a list of three 3 by 3 matrix transformations followed by three pictures created by the command **Mappictures**. Your task is to match each picture to the appropriate matrix. Explain your choices by relating the picture to what you know about the eigenspaces of the matching matrix. You can test your answers by constructing the three matrices and applying **Mapcolor** to each.

A is a 3 by 3 matrix that reflects vectors across a plane through the origin.
B is a 3 by 3 matrix that rotates all vectors by an acute angle about a given line through the origin.
C is a 3 by 3 matrix that scales the x, y, and z components of every vector by the factors 2, 4, and 8.
```
> Mappictures(1);
> Mappictures(2);
> Mappictures(3);
```
Maple note: Delete the three above pictures before you save this file; otherwise, the file may become too large for your system.

 Student Workspace

Linear Algebra Modules Project
Chapter 6, Module 2

The Characteristic Polynomial

 Purpose of this module

The purpose of this module is to introduce the characteristic polynomial of a square matrix; the roots of this polynomial are the eigenvalues of the matrix. Also included are explorations designed to develop more insights about eigenvalues and eigenvectors.

 Prerequisites

Eigenvalues and eigenvectors; determinants; linear independence, basis, and dimension; null space and rank of a matrix.

 Commands used in this module

```
> restart; with(LinearAlgebra): with(plots): with(Lamp):
  UseHardwareFloats := false: Digits := 6:
```

Tutorial

Section 1: Definition and Properties of the Characteristic Polynomial

In this section, we derive a polynomial associated with a matrix whose roots are the eigenvalues of the matrix. This polynomial is derived by linking the following key facts:

- λ is an eigenvalue of a square matrix A if and only if the matrix $A - \lambda I$ is singular. (This is Theorem 3 in Module 1.)

- A square matrix M is singular if and only if $\text{Det}(M) = 0$. (This is Theorem 14 in Chapter 3.)

Putting these together, we conclude:

Theorem 4: λ is an eigenvalue of the square matrix A if and only if $\text{Det}(A - \lambda I) = 0$.

=====================

Example 1A: Use Theorem 4 to find the eigenvalues of the matrix A below.

```
> A := Matrix([[1,2],[2,1]]);
```

$$A := \begin{bmatrix} 1 & 2 \\ 2 & 1 \end{bmatrix}$$

Solution: We compute the expression $\text{Det}(A - \lambda I)$, where λ is treated as an unknown:

```
> I := Idmat(2):
  M := Expand(A-lambda*I);
```

$$M := \begin{bmatrix} -\lambda + 1 & 2 \\ 2 & -\lambda + 1 \end{bmatrix}$$

Maple note: Maple will not multiply a matrix by a symbolic scalar such as λ (because it does not know whether the symbol stands for a scalar or a matrix); the **Expand** command forces the scalar multiplication.

```
> Det(M);
```

$$\lambda^2 - 2\lambda - 3$$

Then we solve the quadratic polynomial equation $\text{Det}(A - \lambda I) = 0$ for λ:

$$\lambda^2 - 2\lambda - 3 = (\lambda - 3)(\lambda + 1) = 0$$

So the eigenvalues of A are $\lambda = 3$ and $\lambda = -1$.

===========================

Definition: The polynomial $p(\lambda) = \text{Det}(A - \lambda I)$ is called the ***characteristic polynomial*** for the matrix A. The equation $p(\lambda) = 0$ is called the ***characteristic equation*** of A.

We may now rephrase Theorem 4 as follows:

- The roots of the characteristic equation of A are the eigenvalues of A.

Exercise 1.1: By hand, find the characteristic polynomial of the matrix B below and use it to find the eigenvalues of B.

$$B = \begin{bmatrix} 1 & 3 \\ 2 & 2 \end{bmatrix}$$

 Student Workspace

 Answer 1.1

Find the characteristic polynomial:

$$p(\lambda) = \text{Det}(B - \lambda I) = \text{Det}\left(\begin{bmatrix} 1 - \lambda & 3 \\ 2 & 2 - \lambda \end{bmatrix}\right) = (1 - \lambda)(2 - \lambda) - 6 = \lambda^2 - 3\lambda - 4$$

Factor the polynomial and find its roots: $(\lambda - 4)(\lambda + 1) = 0$ has the solutions $\lambda = 4$ and $\lambda = -1$, which are therefore the eigenvalues of B.

Algebraic multiplicity of an eigenvalue

A theorem from elementary algebra says that r is a root of a polynomial equation $p(\lambda) = 0$ if and only if $\lambda - r$ is a factor of the polynomial $p(\lambda)$. So if we factor the characteristic polynomial $p(\lambda)$ in the form

$$p(\lambda) = a(\lambda - r_1)(\lambda - r_2) \ldots (\lambda - r_n)$$

then $r_1, r_2, \ldots, r_n$ are the roots of $p(\lambda)$ and hence are the eigenvalues of A.

Definition: The number of times that a factor $\lambda - r$ occurs in the characteristic polynomial is called the ***algebraic multiplicity*** of the eigenvalue r.

========================

Example 1B: Factor the characteristic polynomial in Example 1A to find the algebraic multiplicities of the two eigenvalues.

Solution:

```
> A := Matrix([[1,2],[2,1]]);
```

$$A := \begin{bmatrix} 1 & 2 \\ 2 & 1 \end{bmatrix}$$

```
> I := Idmat(2);
> p := Det(A-lambda*I);
```

$$I := \begin{bmatrix} 1 & 0 \\ 0 & 1 \end{bmatrix}$$

$$p := \lambda^2 - 2\lambda - 3$$

```
> factor(p);
```

$$(\lambda + 1)(\lambda - 3)$$

Since the factor $\lambda + 1$ occurs just once, the eigenvalue $\lambda = -1$ has algebraic multiplicity 1; similarly, $\lambda = 3$ has algebraic multiplicity 1. We can confirm our above results by using the command **Charpoly**, which computes the characteristic polynomial of a matrix:

```
> Charpoly(A,lambda);
```

$$\lambda^2 - 2\lambda - 3$$

We can also confirm the results of Example 1A by using the command **Evalues**, which computes all the eigenvalues at once. It also shows the algebraic multiplicity of the eigenvalues by repeating each eigenvalue the number of times it appears in the factorization of the characteristic polynomial:

```
> Evalues(A);
```

$$\begin{bmatrix} 3 \\ -1 \end{bmatrix}$$

Since each eigenvalue appears just once in this vector, each has algebraic multiplicity 1.

========================

Exercise 1.2: In answering the questions about the matrix B below, use the method of Example 1A; do not use **Charpoly** or **Evalues** except as a check on your answers.

```
> B := Matrix([[4,10,-10],[-5,-11,5],[-5,-5,-1]]);
```

$$B := \begin{bmatrix} 4 & 10 & -10 \\ -5 & -11 & 5 \\ -5 & -5 & -1 \end{bmatrix}$$

(a) Find the characteristic polynomial of B.
(b) Factor the characteristic polynomial, and use the factorization to determine the eigenvalues of B and their algebraic multiplicities.

+ Student Workspace

− Answer 1.2

```
> I := Idmat(3):
  poly := Det(B-lambda*I);
```

$$poly := -\lambda^3 - 8\lambda^2 + 12\lambda + 144$$

```
> factor(poly);
```

$$-(\lambda - 4)(\lambda + 6)^2$$

So $-(\lambda - 4)(\lambda + 6)^2$ is the factored form of the characteristic polynomial of B; the eigenvalues of B are therefore $\lambda = 4$ with algebraic multiplicity 1 and $\lambda = -6$ with algebraic multiplicity 2. We confirm by using **Evalues**:

```
> Evalues(B);
```

$$\begin{bmatrix} -6 \\ -6 \\ 4 \end{bmatrix}$$

=====================

Example 1C: Finding the eigenvalues of an upper triangular matrix is particularly easy, since the determinant of an upper triangular matrix is simply the product of its diagonal entries. Here, for example, is an upper-triangular matrix K and the matrix $K - \lambda I$, which is also upper triangular:

$$K = \begin{bmatrix} 2 & -3 & 1 \\ 0 & -4 & 5 \\ 0 & 0 & 7 \end{bmatrix} \qquad K - \lambda I = \begin{bmatrix} 2-\lambda & -3 & 1 \\ 0 & -4-\lambda & 5 \\ 0 & 0 & 7-\lambda \end{bmatrix}$$

So $Det(K - \lambda I) = (2 - \lambda)(-4 - \lambda)(7 - \lambda)$, which equals zero when λ is 2 or -4 or 7. So the eigenvalues of an upper-triangular matrix are its diagonal entries. We confirm:

```
> K := Matrix([[2,-3,1],[0,-4,5],[0,0,7]]):
> Evalues(K);
```

$$\begin{bmatrix} 7 \\ -4 \\ 2 \end{bmatrix}$$

=====================

Degree of the Characteristic Polynomial

As you may have observed by now, the characteristic polynomial of a 2 by 2 matrix has degree 2 and the characteristic polynomial of a 3 by 3 matrix has degree 3. More generally:

- The characteristic polynomial of an n by n matrix has degree n.

Since a polynomial of degree n has at most n roots, we conclude:

Theorem 2: An n by n matrix has at most n distinct eigenvalues.

Caution: The characteristic equation $p(\lambda) = 0$ is especially handy for finding eigenvalues of 2 by 2 matrices. But since the degree of $p(\lambda)$ gets larger as the matrices get larger, the equation $p(\lambda) = 0$ may become difficult or even impossible to solve for larger matrices. Maple can sometimes help here by solving the equation for us. Furthermore, if we and Maple are unable to compute the eigenvalues exactly, Maple can approximate them numerically. However, there are sometimes better ways to find eigenvalues and eigenvectors, as we saw in Section 3 of Module 1 and as we will see in the Section 3 of this module.

Section 2: Evectors Command

The powerful command **Evectors(A)** produces the eigenvalues of A, the algebraic multiplicity of each eigenvalue, and a basis for each eigenspace. However, some care is required to interpret the output correctly.

========================

Example 2A: Interpret the results of the command **Evectors(A)** below.

```
> A := Matrix([[3,0,0,1],[1,3,0,1],[1,0,3,1],[1,0,0,3]]);
```

$$A := \begin{bmatrix} 3 & 0 & 0 & 1 \\ 1 & 3 & 0 & 1 \\ 1 & 0 & 3 & 1 \\ 1 & 0 & 0 & 3 \end{bmatrix}$$

```
> h,P := Evectors(A);
```

$$h, P := \begin{bmatrix} 4 \\ 3 \\ 3 \\ 2 \end{bmatrix}, \begin{bmatrix} 1 & 0 & 0 & -1 \\ 2 & 1 & 0 & 0 \\ 2 & 0 & 1 & 0 \\ 1 & 0 & 0 & 1 \end{bmatrix}$$

Solution: The **Evectors** command produces a sequence of two outputs. The first is the vector of eigenvalues (just as in the **Evalues** command), and the second is a matrix whose columns contain eigenvectors that provide a basis for each of the corresponding eigenspaces. In the above command, we have given the name h to the vector of eigenvalues and the name P to the matrix of eigenvectors.

Specifically, in the above example, the vector h tells us that the eigenvalues of A are $\lambda = 4$ with algebraic multiplicity 1, $\lambda = 3$ with algebraic multiplicity 2, and $\lambda = 2$ with algebraic multiplicity 1. The matrix P

tells us that the following sets of vectors are bases for the three eigenspaces:

$$\lambda = 4: \left[\begin{bmatrix} 1 \\ 2 \\ 2 \\ 1 \end{bmatrix}\right] \quad \lambda = 3: \left[\begin{bmatrix} 0 \\ 1 \\ 0 \\ 0 \end{bmatrix}, \begin{bmatrix} 0 \\ 0 \\ 1 \\ 0 \end{bmatrix}\right] \quad \lambda = 2: \left[\begin{bmatrix} -1 \\ 0 \\ 0 \\ 1 \end{bmatrix}\right]$$

Thus the geometric multiplicities of these three eigenvalues (i.e., the dimensions of each of their eigenspaces) is the same as their algebraic multiplicities. We can check these answers by our earlier methods. The characteristic polynomial shows us the eigenvalues and their algebraic multiplicities. Then, for each eigenvalue λ, we find a basis for each eigenspace by finding a basis for the solution set of $(A - \lambda I)x = 0$:

```
> factor(Charpoly(A,lambda));
```

$$(\lambda - 2)(\lambda - 4)(-3 + \lambda)^2$$

```
> I := Idmat(4):
  Nullbasis(A-4*I);
```

$$\left[\begin{bmatrix} 1 \\ 2 \\ 2 \\ 1 \end{bmatrix}\right]$$

```
> Nullbasis(A-3*I);
```

$$\left[\begin{bmatrix} 0 \\ 1 \\ 0 \\ 0 \end{bmatrix}, \begin{bmatrix} 0 \\ 0 \\ 1 \\ 0 \end{bmatrix}\right]$$

```
> Nullbasis(A-2*I);
```

$$\left[\begin{bmatrix} -1 \\ 0 \\ 0 \\ 1 \end{bmatrix}\right]$$

======================

As you will see in the next exercise, the geometric multiplicity of an eigenvalue can be less than its algebraic multiplicity.

Exercise 2.1: Use the `Evectors` command to find the eigenvalues of the matrix B below. Also determine the algebraic and geometric multiplicities of each eigenvalue.

```
> B := Matrix([[-2,0,-5,-4],[1,2,0,1],[-5,0,-3,-5],[4,0,5,6]]);
```

$$B := \begin{bmatrix} -2 & 0 & -5 & -4 \\ 1 & 2 & 0 & 1 \\ -5 & 0 & -3 & -5 \\ 4 & 0 & 5 & 6 \end{bmatrix}$$

Maple note: You will discover that the matrix of eigenvectors in this example has some columns of asterisks (*). This indicates that the geometric multiplicity of the associated eigenvalue is less than its algebraic multiplicity. Specifically, for each eigenvalue λ, the number of columns of asterisks is the difference between the algebraic and geometric multiplicities of λ.

 Student Workspace

 Answer 2.1

> Evectors(B);

$$\begin{bmatrix} -3 \\ 2 \\ 2 \\ 2 \end{bmatrix}, \begin{bmatrix} -1 & 0 & -1 & * \\ 0 & 1 & 0 & * \\ -1 & 0 & 0 & * \\ 1 & 0 & 1 & * \end{bmatrix}$$

> Nullbasis(B+3*I);

$$\left[\begin{bmatrix} -1 \\ 0 \\ -1 \\ 1 \end{bmatrix} \right]$$

> Nullbasis(B-2*I);

$$\left[\begin{bmatrix} 0 \\ 1 \\ 0 \\ 0 \end{bmatrix}, \begin{bmatrix} -1 \\ 0 \\ 0 \\ 1 \end{bmatrix} \right]$$

The eigenvalue $\lambda = -3$ has algebraic multiplicity 1 and geometric multiplicity 1; the eigenvalue $\lambda = 2$ has algebraic multiplicity 3 and geometric multiplicity 2. Bases for the two eigenspaces are

$$\lambda = -3: \left[\begin{bmatrix} -1 \\ 0 \\ -1 \\ 1 \end{bmatrix} \right] \quad \lambda = 2: \left[\begin{bmatrix} 0 \\ 1 \\ 0 \\ 0 \end{bmatrix}, \begin{bmatrix} -1 \\ 0 \\ 0 \\ 1 \end{bmatrix} \right]$$

Although the geometric multiplicity of an eigenvalue can be smaller than its algebraic multiplicity, it can never be larger:

Theorem 5: The geometric multiplicity of an eigenvalue is always less than or equal to its algebraic multiplicity.

Furthermore, since the sum of the algebraic multiplicities of the eigenvalues of a matrix is the degree of the characteristic polynomial and since that degree is the size of the matrix, we conclude:

Theorem 6: The sum of the geometric multiplicities of the eigenvalues of an n by n matrix is at most n.

Section 3: Explorations

In this section, we employ yet another method for finding eigenvalues and eigenvectors: we look for patterns. Our basic tool will be Theorem 3 from Module 1:

Theorem 3: A number λ is an eigenvalue of a square matrix A if and only if the matrix $A - \lambda I$ is singular.

Note that each of the following properties is equivalent to the property that $A - \lambda I$ is singular (see Theorem 6 of Chapter 3):

- the columns of $A - \lambda I$ are linearly dependent.
- the rows of $A - \lambda I$ are linearly dependent.
- the rank of $A - \lambda I$ is less than n (where A is n by n).
- the null space of $A - \lambda I$ is nonzero.

So in looking for eigenvalues, we may look for values of λ for which one of these properties holds.

======================

Example 3A: Consider the matrix A below and the matrix $A - \lambda I$:

```
> A := Matrix([[4, 4, 0], [4, 4, 0], [1, 2, 1]]);
```

$$A := \begin{bmatrix} 4 & 4 & 0 \\ 4 & 4 & 0 \\ 1 & 2 & 1 \end{bmatrix}$$

```
> I := Idmat(3):
  M := Expand(A-lambda*I);
```

$$M := \begin{bmatrix} -\lambda + 4 & 4 & 0 \\ 4 & -\lambda + 4 & 0 \\ 1 & 2 & -\lambda + 1 \end{bmatrix}$$

Observe that when $\lambda = 1$, the last column of $A - \lambda I$ is zero, and hence the columns of $A - 1 I$ are linearly dependent (see below). Therefore $\lambda = 1$ is an eigenvalue of A.

```
> subs(lambda=1,M);
```

$$\begin{bmatrix} 3 & 4 & 0 \\ 4 & 3 & 0 \\ 1 & 2 & 0 \end{bmatrix}$$

======================

Exercise 3.1: Find one of the remaining two eigenvalues of the matrix A above by observing those values of λ that make the columns or rows of A linearly dependent.

Student Workspace

Answer 3.1

Let's substitute $\lambda = 0$ in the formula for $M = A - \lambda I$:
```
> subs(lambda=0,M);
```
$$\begin{bmatrix} 4 & 4 & 0 \\ 4 & 4 & 0 \\ 1 & 2 & 1 \end{bmatrix}$$

The first two rows of $A - 0I$ are equal, and therefore the rows of $A - 0I$ are linearly dependent; hence $\lambda = 0$ is an eigenvalue of A.

The other eigenvalue is a little less obvious. Substitute $\lambda = 8$ in the formula for $M = A - \lambda I$:
```
> subs(lambda=8,M);
```
$$\begin{bmatrix} -4 & 4 & 0 \\ 4 & -4 & 0 \\ 1 & 2 & -7 \end{bmatrix}$$

The first two rows of $A - 8I$ are negatives of one another, and therefore the rows of $A - 8I$ are linearly dependent; hence $\lambda = 8$ is an eigenvalue of A.

According to Theorem 2, these are all the eigenvalues of the matrix A, since a 3 by 3 matrix has at most 3 distinct eigenvalues.

Sometimes we can find the geometric multiplicity of the eigenvalues of a matrix by a simple observation rather than by a calculation. This observation is based on Theorem 9 of Chapter 5, which says:

- The dimension of the null space of an n by n matrix of rank r is $n - r$.

======================

Example 3B: Use the above fact about the dimension of a null space to find the geometric multiplicity of the eigenvalue $\lambda = 2$ of the matrix A below.
```
> A := Matrix([[2,0,0,1],[1,2,0,1],[1,0,2,1],[1,0,0,2]]);
```
$$A := \begin{bmatrix} 2 & 0 & 0 & 1 \\ 1 & 2 & 0 & 1 \\ 1 & 0 & 2 & 1 \\ 1 & 0 & 0 & 2 \end{bmatrix}$$
```
> I := Idmat(4):
  M := Expand(A-lambda*I);
```
$$M := \begin{bmatrix} -\lambda+2 & 0 & 0 & 1 \\ 1 & -\lambda+2 & 0 & 1 \\ 1 & 0 & -\lambda+2 & 1 \\ 1 & 0 & 0 & -\lambda+2 \end{bmatrix}$$

Solution:
```
> subs(lambda=2,M);
```

$$\begin{bmatrix} 0 & 0 & 0 & 1 \\ 1 & 0 & 0 & 1 \\ 1 & 0 & 0 & 1 \\ 1 & 0 & 0 & 0 \end{bmatrix}$$

The matrix $A - 2I$ has two zero columns, and therefore is singular; so $\lambda = 2$ is an eigenvalue of A. Furthermore, since the first and fourth columns of $A - 2I$ are linearly independent, the rank of $A - 2I$ must be 2. Therefore, by the above remark, the eigenvalue $\lambda = 2$ has geometric multiplicity $4 - 2 = 2$. Calculation of a null space basis for $A - 2I$ confirms this conclusion:

> `Nullbasis(A-2*I);`

$$\left[\begin{bmatrix} 0 \\ 1 \\ 0 \\ 0 \end{bmatrix}, \begin{bmatrix} 0 \\ 0 \\ 1 \\ 0 \end{bmatrix} \right]$$

===================

Let's formalize these observations as a theorem: Since the geometric multiplicity of an eigenvalue λ of a matrix A is the dimension of the null space of $A - \lambda I$, we have:

Theorem 7: The geometric multiplicity of the eigenvalue λ of an n by n matrix A is $n - r$, where r is the rank of $A - \lambda I$.

Exercise 3.2: Find another eigenvalue of the matrix A above, and find its geometric multiplicity.

> `A := Matrix([[2,0,0,1],[1,2,0,1],[1,0,2,1],[1,0,0,2]]):`
> `I := Idmat(4):`
> `M := Expand(A-lambda*I);`

$$M := \begin{bmatrix} -\lambda + 2 & 0 & 0 & 1 \\ 1 & -\lambda + 2 & 0 & 1 \\ 1 & 0 & -\lambda + 2 & 1 \\ 1 & 0 & 0 & -\lambda + 2 \end{bmatrix}$$

 Student Workspace

Answer 3.2

> `subs(lambda=1,M);`

$$\begin{bmatrix} 1 & 0 & 0 & 1 \\ 1 & 1 & 0 & 1 \\ 1 & 0 & 1 & 1 \\ 1 & 0 & 0 & 1 \end{bmatrix}$$

If $\lambda = 1$, the first and fourth columns of the resulting matrix $A - I$ are identical; so $A - I$ is singular and therefore $\lambda = 1$ is an eigenvalue. Since the first three columns of A are linearly independent, the matrix $A - I$ has rank 3; therefore, by Theorem 7, the eigenvalue $\lambda = 1$ has geometric multiplicity $4 - 3 = 1$. Calculation of a null space basis for $A - I$ confirms this result:

```
> Nullbasis(A-1*I);
```

$$\begin{bmatrix} -1 \\ 0 \\ 0 \\ 1 \end{bmatrix}$$

Problems

Problem 1: Practice using the characteristic polynomial

For the matrices A and B below, do all of the following using the method of Example 1A; do not use **Charpoly** or **Evalues** except as a check.
(a) Find the characteristic polynomial and factor it.
(b) Find the eigenvalues and their algebraic multiplicities.
(c) Use the **Nullbasis** command to find a basis for each eigenspace and the geometric multiplicity of each eigenvalue.

```
> A := Matrix([[4,-1,1],[-1,4,-1],[1,-1,4]]);
> B := Matrix([[-1,-1,2,-3],[4,3,0,6],[0,0,5,9],[0,0,-1,-1]]);
```

Student Workspace

Problem 2: Eigenvalues depending on a variable

For what values of the variable k does the matrix A below have
(a) three distinct eigenvalues?
(b) two distinct eigenvalues?
(c) only one eigenvalue?

```
> A := Matrix([[1,k,-1],[1,0,1],[-1,-2,1]]);
```

Student Workspace

Problem 3: Shear matrices

Let S be the horizontal shear matrix:
$$S = \begin{bmatrix} 1 & k \\ 0 & 1 \end{bmatrix}$$

```
> S := Xshearmat(k):
```
(a) (By hand) Find the eigenvalues of S from the characteristic polynomial of S. Also find the corresponding eigenvectors. (Assume that $k \neq 0$.)
(b) Give a geometric interpretation of the eigenvalues of S and their corresponding eigenvectors. In particular, explain why the geometric multiplicity is less than the algebraic multiplicity.

Student Workspace

Problem 4: Matrices of ones

(a) Use the commands below to construct a square matrix J of all 1's. What are the eigenvalues of J and their geometric multiplicities? (Suggestions: Examine the matrix $J - \lambda I$ to look for patterns that suggest values of λ for which $J - \lambda I$ is singular, and keep in mind that the eigenvalues can depend on the size of the matrix. If that doesn't help enough, use the **Evalues** and **Nullbasis** commands.)

(b) For every n by n matrix of all 1's, explain why one of the eigenvalues must always have an eigenspace of dimension $n - 1$. (Hint: What can you say about the rank of the matrix $A - \lambda I$ for this value of λ?)

```
> n := 4:    # change the value of n to experiment.
  J := Matrix(n,n,1);
```

+ Student Workspace

Problem 5: Eigenvalues of the transpose

For each of the matrices below, apply the **Evectors** command to both the matrix and its transpose.

(a) Verify that, for each matrix A, A and A^T have the same eigenvalues. Do the eigenvalues of A and A^T have the same algebraic multiplicity? same geometric multiplicity? same eigenvectors?

(b) Prove that your conclusions are correct for all square matrices. (Hint: You will find the following theorems useful: $\text{Det}(A) = \text{Det}(A^T)$, which is Theorem 13 in Chapter 3, and dimension of row space of A equals dimension of column space of A, which is Theorem 8 in Chapter 5.)

```
> A1 := Matrix([[4,-1,1],[-1,4,-1],[1,-1,4]]);
> A2 := Matrix([[-1,-1,2,-3],[4,3,0,6],[0,0,5,9],
     [0,0,-1,-1]]);
> A3 := Matrix([[2,b,c],[0,-1,e],[0,0,3]]);
```

+ Student Workspace

Problem 6: Characteristic polynomial applied to the matrix

We say that a square matrix A satisfies a polynomial equation such as $t^3 + 4t^2 + 5t + 2 = 0$ if the equation holds when t is replaced by A; that is,

$$A^3 + 4A^2 + 5A + 2I = O$$

(Not only is the number t is replaced by the matrix A but the number 2 is replaced by the matrix $2I$ and the number 0 by the zero matrix.)

(a) For each of the matrices below, use **Charpoly** to compute the characteristic polynomial and check that the matrix satisfies its own characteristic equation.

(b) (By hand) For an arbitrary upper-triangular matrix with diagonal entries $d_1, d_2, ..., d_n$, write the characteristic polynomial in factored form (like Maple produces for the matrix A_3 below). Explain why such a matrix must satisfy its own characteristic equation. (In fact, the Cayley-Hamilton theorem says

that <u>every</u> square matrix satisfies its own characteristic equation.)
```
> A1 := Matrix([[4,-1,1],[-1,4,-1],[1,-1,4]]):
> A2 := Matrix([[-1,-1,2,-3],[4,3,0,6],[0,0,5,9],
    [0,0,-1,-1]]):
> A3 := Matrix([[a,b,c],[0,d,e],[0,0,f]]):
```
 Student Workspace

 Problem 7: Letter N matrices

A "letter N matrix" is a square matrix of 0's and 1's, where the 1's form the shape of the letter N. Use the commands below to construct and study letter N matrices of various sizes. (See the worked example in the closed section below.)

(a) What are the eigenvalues and their geometric multiplicities for the n by n letter N matrix?

(b) For every n by n letter N matrix, explain why one of the eigenvalues must always have an eigenspace of dimension $n - 2$.

(c) State a basis for the eigenspace associated with each of the eigenvalues other than the eigenvalue in part (b).

```
> n := 4:    # change the value of n to experiment.
> N := LetterN(n);
```

Example for Problems 7 and 8

A "letter S matrix" is a square matrix of 0's and 1's, where the 1's form the shape of the letter S. With the help of the **Evectors** command, find the eigenvalues of the n by n "letter S" matrices below; also find their geometric and algebraic multiplicities.

Solution: For several values of n, we examine the matrix $S - \lambda I$ to look for a pattern, and we apply the **Evalues** command to check that pattern. Then we explain why this pattern holds for every n.

```
> S := LetterS(3);
> Expand(S-lambda*Idmat(3));
  Evalues(S);
> S := LetterS(4):
  Expand(S-lambda*Idmat(4));
  Evalues(S);
> S := LetterS(5):
  Expand(S-lambda*Idmat(5));
  Evalues(S);
```

The pattern appears to be:

0 and 2 are eigenvalues with algebraic multiplicity 2;

1 is an eigenvalue with algebraic multiplicity $n - 2$.

The following explanation verifies that this pattern does indeed hold for every n. The first and last columns of S are identical, which means that 0 is an eigenvalue of S. This is the only linear dependence relation for the columns of S, which means that S has rank $n - 1$ and hence 0 has geometric multiplicity 1. Similarly, the first and last columns of $S - 2I$ are negatives of one another, and there are no other linear dependence relations for the columns of $S - 2I$. The middle $n - 2$ rows of the matrix $S - I$ are all zero, which means that $S - I$ has rank 2 and hence 1 has geometric multiplicity $n - 2$.

Furthermore, since the sum of all the geometric multiplicities is $1 + 1 + (n - 2) = n$, all the algebraic multiplicities are equal to the geometric multiplicities by Theorems 5 and 6.

 Student Workspace

 Problem 8: Letter L matrices

A "letter L matrix" is a square matrix of 0's and 1's, where the 1's form the shape of the letter L. Use the commands below to construct and study letter L matrices of various sizes. (See the worked example in Problem 7.)

(a) What are the eigenvalues and their geometric multiplicities for the n by n letter L matrix?

(b) For every n by n letter L matrix, explain why the eigenvalues have the geometric multiplicities claimed in (a).

```
> n := 4;    # change the value of n to experiment.
> L := LetterL(n);
```

 Student Workspace

Linear Algebra Modules Project
Chapter 6, Module 3

Eigenvector Bases and Discrete Dynamical Systems

Purpose of this module

The purpose of this module is to introduce bases that consist of eigenvectors of a matrix. We then apply such bases to the study of sequences x_k defined by recursion formulas of the form $x_{k+1} = A\, x_k$.

Prerequisites

Eigenvalues, eigenvectors, and eigenspaces; Module 4 of Chapter 3, "Markov Chains."

Commands used in this module

```
> restart; with(LinearAlgebra): with(plots): with(Lamp):
  UseHardwareFloats := false: Digits := 6:
```

Tutorial

Section 1: Eigenvector Bases for R^n

If an n by n matrix A has n linearly independent eigenvectors, we can use these vectors to form an *eigenvector basis* of R^n. The following strategy will produce a basis of eigenvectors of A, if there is one: Find a basis for each eigenspace of A, and take the collection of all these basis vectors. Note that this strategy will produce n eigenvectors if and only if the sum of the geometric multiplicities of A is exactly n (see Theorem 6). That is, it will produce n eigenvectors if and only if the geometric multiplicity of every eigenvalue equals its algebraic multiplicity (see Theorem 5).

The following theorem guarantees that this strategy will produce an eigenvector basis, if there is one.

> **Theorem 8:** Suppose A is a square matrix and $\{v_1, ..., v_m\}$ is a set of eigenvectors of A constructed as follows: Select a basis for each eigenspace of A and take all these basis vectors collectively. Then $\{v_1, ..., v_m\}$ is linearly independent.

====================

Example 1A: For the matrix A below, find an eigenvector basis of R^4.

```
> A := Matrix([[3,0,0,1],[1,3,0,1],[1,0,3,1],[1,0,0,3]]);
```

Module 6.3 Eigenvector Bases and Discrete Dynamical Systems

$$A := \begin{bmatrix} 3 & 0 & 0 & 1 \\ 1 & 3 & 0 & 1 \\ 1 & 0 & 3 & 1 \\ 1 & 0 & 0 & 3 \end{bmatrix}$$

Solution: We apply the command **Evectors(A)**:

```
> h,P := Evectors(A);
```

$$h, P := \begin{bmatrix} 4 \\ 3 \\ 3 \\ 2 \end{bmatrix}, \begin{bmatrix} 1 & 0 & 0 & -1 \\ 2 & 1 & 0 & 0 \\ 2 & 0 & 1 & 0 \\ 1 & 0 & 0 & 1 \end{bmatrix}$$

The collection of columns of the matrix P, $\{v_1, v_2, v_3, v_4\}$, is an eigenvector basis of R^4. We confirm that it is a basis of R^4 by showing that it is linearly independent. Solving the equation $x_1 v_1 + x_2 v_2 + x_3 v_3 + x_4 v_4 = 0$ is equivalent to solving $P x = 0$:

```
> Matsolve(P,zero,free=x);
```

$$\begin{bmatrix} 0 \\ 0 \\ 0 \\ 0 \end{bmatrix}$$

Since the only solution is the trivial solution, $\{v_1, v_2, v_3, v_4\}$ is linearly independent.

==========================

As we saw in Section 3 of Module 2 of Chapter 5, a basis for a subspace gives us a new coordinate system with its own sets of coordinates. For example, the 2 by 2 matrix B below has eigenvalues 36 and 18 with linearly independent eigenvectors $u = \begin{bmatrix} 4 \\ 3 \end{bmatrix}$ and $v = \begin{bmatrix} -2 \\ 3 \end{bmatrix}$ respectively. Therefore the set $\boldsymbol{B} = \{u, v\}$ is an eigenvector basis for R^2:

```
> B := Matrix([[30,8],[9,24]]);
```

$$B := \begin{bmatrix} 30 & 8 \\ 9 & 24 \end{bmatrix}$$

```
> u := Vector([4,3]);
  B.u;
```

$$u := \begin{bmatrix} 4 \\ 3 \end{bmatrix}$$

$$\begin{bmatrix} 144 \\ 108 \end{bmatrix}$$

```
> v := Vector([-2,3]);
  B.v;
```

$$v := \begin{bmatrix} -2 \\ 3 \end{bmatrix}, \quad B v := \begin{bmatrix} -36 \\ 54 \end{bmatrix}$$

Chapter 6 Eigenvalues and Eigenvectors

==========================

Example 1B: Express the vector $y = \begin{bmatrix} 6 \\ 9 \end{bmatrix}$ as a linear combination of the eigenvectors u and v.

Solution: We are looking for a_1 and a_2 such that $a_1 u + a_2 v = y$. This can be rewritten as the matrix-vector equation $P\,a = y$, where $P = [u\ v]$ and $a = \begin{bmatrix} a_1 \\ a_2 \end{bmatrix}$. Since P is invertible (why?), we can solve for a: $a = P^{(-1)} y$. We calculate a_1 and a_2 below.

```
> y := Vector([6,9]);
  P := Matrix([u,v]);
```

$$y := \begin{bmatrix} 6 \\ 9 \end{bmatrix},\ P := \begin{bmatrix} 4 & -2 \\ 3 & 3 \end{bmatrix}$$

```
> a := P^(-1).y;
```

$$a := \begin{bmatrix} 2 \\ 1 \end{bmatrix}$$

Therefore $a_1 = 2$ and $a_2 = 1$. Let's check that these values of a_1 and a_2 do yield the given vector y:

```
> a[1]*u+a[2]*v;
```

$$\begin{bmatrix} 6 \\ 9 \end{bmatrix}$$

We say that the point $y = (6, 9)$ has *coordinates* $(2, 1)$ *relative to the basis* **B**. So y has standard coordinates $(6, 9)$ and **B**-*coordinates* $(2, 1)$.

==========================

The next command produces a picture of both coordinate systems, the standard xy grid and the uv coordinate grid. Note that the point with **B**-coordinates $(2, 1)$ (in a blue circle) is the point $y = (6, 9)$.

```
> Basisgrid(u,v,[2,1]);
```

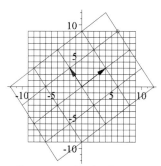

Summary: If y is a point in standard coordinates and x is that same point in **B**-coordinates, we can switch between the two by the conversion formulas

$$x = P^{(-1)} y \quad \text{and} \quad y = P\,x \qquad [1]$$

where P is the matrix whose columns are the vectors in the basis **B**.

Exercise 1.1: (a) For the 3 by 3 matrix C below, find an eigenvector basis B of R^3.
(b) Find the B-coordinates of the point with standard coordinates $(2, 6, -5)$ by using formulas [1].
(c) Find the standard coordinates of the point with B-coordinates $(1, 2, -3)$ by using [1].

```
> C := Matrix([[21,-9,-3],[57,-25,-9],[-3,3,3]]);
```

$$C := \begin{bmatrix} 21 & -9 & -3 \\ 57 & -25 & -9 \\ -3 & 3 & 3 \end{bmatrix}$$

 Student Workspace

Answer 1.1

(a) The columns of the matrix P of eigenvectors form an eigenvector basis B:

```
> h,P := Evectors(C);
```

$$h, P := \begin{bmatrix} -3 \\ 2 \\ 0 \end{bmatrix}, \begin{bmatrix} -1 & 0 & -1 \\ -3 & -1 & -3 \\ & 3 & 2 \\ 1 & 1 & 1 \end{bmatrix}$$

For (b) we compute $P^{(-1)} y$, where y is the given point in standard coordinates; for (c) we compute $P x$, where x is the given point in B-coordinates:

```
> y := Vector([2,6,-5]);
  P^(-1).y;
```

$$y := \begin{bmatrix} 2 \\ 6 \\ -5 \end{bmatrix}, \begin{bmatrix} 1 \\ 0 \\ -6 \end{bmatrix}$$

```
> x := Vector([1,2,-3]);
  P.x;
```

$$x := \begin{bmatrix} 1 \\ 2 \\ -3 \end{bmatrix}, \begin{bmatrix} \frac{1}{2} \\ \frac{5}{6} \\ 0 \end{bmatrix}$$

So $(1, 0, -6)$ are the B-coordinates of the point whose standard coordinates are $(2, 6, -5)$. And $(\frac{1}{2}, \frac{5}{6}, 0)$ are the standard coordinates of the point whose B-coordinates are $(1, 2, -3)$.

Section 2: Eigenvector Decomposition

Next, we continue the development begun in Module 4 of Chapter 3, "Markov Chains -- An Application." Here we will use eigenvector bases to get a deeper understanding of the types of sequences we studied

344   Chapter 6 Eigenvalues and Eigenvectors

there, and we will apply our methods to a greater variety of matrices.

Let's review the general form of the problem we studied in the Markov chains module. Suppose x_0 is a vector in R^n and A is an n by n matrix. Then we can define a sequence of vectors $\{x_k\}$ by the recursion formula

$$x_{k+1} = A\, x_k \qquad [2]$$

That is, x_1 is defined by $x_1 = A\, x_0$; then x_2 is defined in terms of x_1 by $x_2 = A\, x_1$; etc. Note that x_2 can be computed directly from A and x_0: $x_2 = A\, x_1 = A\, A\, x_0 = A^2\, x_0$. Continuing this process for x_3, x_4, and so on, we arrive at:

$$x_k = A^k\, x_0 \qquad [3]$$

The basic example we studied in the "Markov Chains" module was the urban versus rural population problem about Lampland:

- (1) Each year 6% of the urban population moves to rural areas and the remaining 94% stay in urban areas.

- (2) Each year 9% of the rural population moves to urban areas and the remaining 91% stay in rural areas.

- (3) At the beginning of our study, the Lampland population is 45% urban and 55% rural.

We then set up a 2 by 2 migration matrix M and a sequence of state vectors, $x_0, x_1, x_2, \ldots$, that were related by the recursion formula

$$x_{k+1} = M\, x_k$$

where:

```
> x0 := Vector([.45,.55]);
  M := Matrix([[.94,.09],[.06,.91]]);
```

$$x0 := \begin{bmatrix} .45 \\ .55 \end{bmatrix}, \quad M := \begin{bmatrix} .94 & .09 \\ .06 & .91 \end{bmatrix}$$

Formula [3] suggests that we can study the powers A^k in order to get information about $\{x_k\}$. The key to this study will involve an eigenvector basis associated with the matrix A. We show how to use an eigenvector basis to compute the iterates x_k for the Lampland migration matrix M:

```
> h,P := Evectors(M);
```

$$h, P := \begin{bmatrix} 1 \\ \dfrac{17}{20} \end{bmatrix}, \begin{bmatrix} \dfrac{3}{2} & -1 \\ 1 & 1 \end{bmatrix}$$

So $v_1 = \begin{bmatrix} 1.5 \\ 1 \end{bmatrix}$ is an eigenvector of M with eigenvalue $\lambda_1 = 1$, and $v_2 = \begin{bmatrix} -1 \\ 1 \end{bmatrix}$ is an eigenvector with eigenvalue $\lambda_2 = 17/20$. Since v_1 and v_2 are linearly independent, $\{v_1, v_2\}$ is an eigenvector basis of R^2. So we can express the initial state x_0 as a linear combination of v_1 and v_2:

$$x_0 = a_1 v_1 + a_2 v_2$$

From this and the fact that $x_k = M^k x_0$ (formula [3]), we can find x_k:

$$x_k = M^k x_0 = M^k (a_1 v_1 + a_2 v_2) = a_1 M^k v_1 + a_2 M^k v_2$$

Now recall the fact that if v is an eigenvector of a matrix A with eigenvalue λ so that $A v = \lambda v$, then $A^k v = \lambda^k v$ for every positive integer k (Problem 7 in Module 1). Therefore

$$x_k = a_1 \lambda_1^k v_1 + a_2 \lambda_2^k v_2$$

With little effort, the above derivation can be modified to prove the following eigenvector decomposition theorem for any square matrix that has an associated eigenvector basis:

Theorem 9: Let A be an n by n matrix with n linearly independent eigenvectors $v_1, ..., v_n$ and associated eigenvalues $\lambda_1, ..., \lambda_n$. Then, for any vector x_0 in R^n, we have the eigenvector decomposition

$$A^k x_0 = a_1 \lambda_1^k v_1 + ... + a_n \lambda_n^k v_n \qquad [4]$$

where the scalars $a_1, ..., a_n$ are determined by the equation $x_0 = a_1 v_1 + ... + a_n v_n$.

Note also that the equation $x_0 = a_1 v_1 + ... + a_n v_n$ can be rewritten as

$$x_0 = P a \qquad [5]$$

where P is the matrix whose columns are the eigenvectors $v_1, ..., v_n$ and a is the column vector with components $a_1, ..., a_n$. Since $\{v_1, ..., v_n\}$ is linearly independent, P is invertible and hence equation [5] has the unique solution $a = P^{(-1)} x_0$.

======================

Example 2A: Use the above eigenvector decomposition to find the steady-state vector for the Lampland urban/rural population problem and to plot the urban and rural populations as functions of the time parameter k.

Solution: First we list again the eigenvalues and associated eigenvectors of the migration matrix M:

```
> v1 := Column(P,1);
  v2 := Column(P,2);
```

$$v1 := \begin{bmatrix} \frac{3}{2} \\ 1 \end{bmatrix}, \quad v2 := \begin{bmatrix} -1 \\ 1 \end{bmatrix}$$

```
> h;
```

$$\begin{bmatrix} 1 \\ \frac{17}{20} \end{bmatrix}$$

Next we find the weights a_1, a_2 in the equation $a_1 v_1 + a_2 v_2 = x_0$ by solving $P a = x_0$ for a, where $P = [v_1, v_2]$ and $a = \begin{bmatrix} a_1 \\ a_2 \end{bmatrix}$:

```
> a := P^(-1).x0;
```

$$a := \begin{bmatrix} .400000 \\ .150000 \end{bmatrix}$$

Thus $a_1 = .4$ and $a_2 = .15$, and so the eigenvector decomposition formula [4] is:

$$x_k = .4 \; 1^k \begin{bmatrix} 1.5 \\ 1 \end{bmatrix} + .15 \left(\frac{17}{20}\right)^k \begin{bmatrix} -1 \\ 1 \end{bmatrix}$$

Notice that as k becomes large, the factor $\left(\dfrac{17}{20}\right)^k$ approaches 0, and thus the iterates x_k get closer and closer to the steady-state vector $.4 \begin{bmatrix} 1.5 \\ 1 \end{bmatrix} = \begin{bmatrix} .6 \\ .4 \end{bmatrix}$. This method for computing the iterates x_k provides insights that our earlier methods did not; it shows clearly how each eigenvector and eigenvalue affects the long-term behavior of x_k. Here is the same formula computed by Maple:

```
> xk := Expand(a[1]*1^k*v1+a[2]*(17/20)^k*v2);
```

$$xk := \begin{bmatrix} -.150000 \left(\dfrac{17}{20}\right)^k + .600000 \\ .150000 \left(\dfrac{17}{20}\right)^k + .400000 \end{bmatrix}$$

We next plot each of the components of x_k (as functions of k) by using the command **Componentplot**. This reveals the long-term steady state of the urban and rural populations as the two graphs approach the horizontal lines $y = .6$ and $y = .4$ asymptotically.

```
> Componentplot(M,x0,points=30);
```

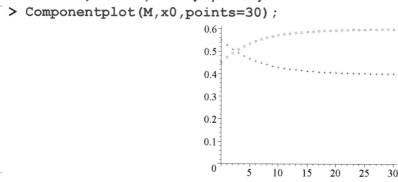

Module 6.3 Eigenvector Bases and Discrete Dynamical Systems

The eigenvector decomposition also shows why the limit of the iterates x_k is $\begin{bmatrix} .6 \\ .4 \end{bmatrix}$ for every initial state x_0. The eigenvector decomposition formula [4] for the migration matrix M has the form

$$x_k = a_1 \, 1^k \begin{bmatrix} 1.5 \\ 1 \end{bmatrix} + a_2 \left(\frac{17}{20}\right)^k \begin{bmatrix} -1 \\ 1 \end{bmatrix}$$

for every x_0. Since $a_1 v_1 + a_2 v_2 = x_0$ and v_1, v_2 are the given eigenvectors of A, only the coefficients a_1 and a_2 can change when x_0 changes. So as k becomes large, the iterates x_k get closer and closer to the limit $a_1 \begin{bmatrix} 1.5 \\ 1 \end{bmatrix}$. Furthermore, a_1 is always .4, and hence the steady-state limit is always $.4 \begin{bmatrix} 1.5 \\ 1 \end{bmatrix} = \begin{bmatrix} .6 \\ .4 \end{bmatrix}$. Here is why $a_1 = .4$: Since the components of x_0 add to 1 and are non-negative, the same is true for every x_k and hence also for the limit. Therefore, by solving $1.5 \, a_1 + a_1 = 1$, we get $a_1 = .4$.

To see the long-term behavior of the urban and rural populations under very different initial conditions, we change x_0, repeat the computation of x_k, and again plot the components. Note especially the limits, which are again .6 and .4, respectively.

```
> x0 := Vector([.95,.05]);
> a  := P^(-1).x0;
```

$$x0 := \begin{bmatrix} .95 \\ .05 \end{bmatrix}$$

$$a := \begin{bmatrix} .400000 \\ -.350000 \end{bmatrix}$$

```
> xk := Expand(a[1]*1^k*v1+a[2]*(17/20)^k*v2);
```

$$xk := \begin{bmatrix} .350000 \left(\dfrac{17}{20}\right)^k + .600000 \\ -.350000 \left(\dfrac{17}{20}\right)^k + .400000 \end{bmatrix}$$

```
> Componentplot(M,x0,points=30);
```

Chapter 6 Eigenvalues and Eigenvectors

Exercise 2.1: (a) For the stochastic matrix N and initial state x_0 below, find the eigenvector decomposition [4] for the iterates $x_k = N^k x_0$.
(b) From your answer to (a), find the steady-state limit of the iterates x_k. Observe that this limit is the same for every x_0.

```
> N := Matrix([[.9,.4],[.1,.6]]);
> x0 := Vector([.5,.5]);
```

 Student Workspace

Answer 2.1

```
> h,P := Evectors(N);
```

$$h, P := \begin{bmatrix} 1 \\ \frac{1}{2} \end{bmatrix}, \begin{bmatrix} 4 & -1 \\ 1 & 1 \end{bmatrix}$$

```
> a := P^(-1).x0;
```

$$a := \begin{bmatrix} .200000 \\ .300000 \end{bmatrix}$$

(a) Therefore $x_k = .2 \, 1^k \begin{bmatrix} 4 \\ 1 \end{bmatrix} + .3 \left(\frac{1}{2}\right)^k \begin{bmatrix} -1 \\ 1 \end{bmatrix}$.

(b) Since the limit of $\left(\frac{1}{2}\right)^k$ is 0, the limit of x_k is $.2 \begin{bmatrix} 4 \\ 1 \end{bmatrix} = \begin{bmatrix} .8 \\ .2 \end{bmatrix}$. Since, for every x_0, the limit has the form $a_1 \begin{bmatrix} 4 \\ 1 \end{bmatrix}$ and since the components of this vector must add to 1, the limit is always $\begin{bmatrix} .8 \\ .2 \end{bmatrix}$.

Section 3: Discrete Dynamical Systems and Their Trajectories

In this section, we give a new name to the problem we have been studying, and we consider a greater variety of matrices.

A *discrete dynamical system* is a sequence of vectors x_k related to one another by a square matrix A as follows:

$$x_{k+1} = A \, x_k, \quad \text{for } k = 0, 1, 2, \dots$$

If we think of k as representing time in some units, we can think of a dynamical system as describing the evolving behavior over time of the set of variables represented by the vectors x_k. We refer to this behavior as the "dynamics" of the "system" of vectors x_k.

========================

Example 3A: (a) For the matrix A and initial vector x_0 below, find the eigenvector decomposition [4] of the iterates $x_k = A^k x_0$.

Module 6.3 Eigenvector Bases and Discrete Dynamical Systems

(b) From this decomposition, determine the long-term behavior of the points x_k.

```
> A := Matrix([[.525,.15],[-.1875,.975]]);
> x0 := Vector([1.5,1]);
```

Solution: (a) First we find the eigenvalues and associated eigenvectors of A:

```
> h,P := Evectors(A);
```

$$h, P := \begin{bmatrix} \frac{9}{10} \\ \frac{3}{5} \end{bmatrix}, \begin{bmatrix} \frac{2}{5} & 2 \\ 1 & 1 \end{bmatrix}$$

Therefore we have the eigenvector decomposition

$$x_k = a_1 \left(\frac{9}{10}\right)^k \begin{bmatrix} .4 \\ 1 \end{bmatrix} + a_2 \left(\frac{3}{5}\right)^k \begin{bmatrix} 2 \\ 1 \end{bmatrix}$$

Since both $\left(\frac{9}{10}\right)^k$ and $\left(\frac{3}{5}\right)^k$ have limit 0, the iterates x_k tend toward the zero vector in the limit. To complete part (a), here's the computation of the coefficients a_1 and a_2:

```
> a := P^(-1).x0;
```

$$a := \begin{bmatrix} .312500 \\ .687500 \end{bmatrix}$$

==========================

As in Section 1, we could plot the components of x_k as functions of k. However, in this section we will exploit a different kind of picture, one that displays better the "dynamics" of the system. We will plot the "trajectory" of the vectors x_k, which is a plot of the points $x_0, x_1, x_2, \ldots$. Such a trajectory is much like the plot of a parametrized curve $r(t) = (x(t), y(t))$ in which we think of t as time; here we can think of k as time.

The command **Trajectory(A,x0,points=n)** plots the trajectory of the vectors x_k for k from 0 to n as an animation:

```
> Trajectory(A,x0,points=20);
```

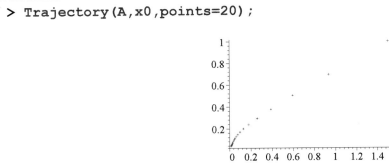

Chapter 6 Eigenvalues and Eigenvectors

==========================

Example 3B: For the matrix A and initial vector x_0 in Example 3A, use the eigenvector decomposition to explain the behavior we saw above of the trajectory of the points x_k.

Solution: Since $x_k = a_1 \left(\dfrac{9}{10}\right)^k \begin{bmatrix} .4 \\ 1 \end{bmatrix} + a_2 \left(\dfrac{3}{5}\right)^k \begin{bmatrix} 2 \\ 1 \end{bmatrix}$, and since both eigenvalues are less than one in absolute value, the points x_k will converge to the origin for every x_0. We sometimes say that the points x_k are "attracted to the origin." Furthermore, since $\dfrac{3}{5} < \dfrac{9}{10}$, the points x_k will be close to $a_2 \left(\dfrac{9}{10}\right)^k \begin{bmatrix} .4 \\ 1 \end{bmatrix}$ for large k, which tells us that x_k approaches the origin along a line with direction vector $\begin{bmatrix} .4 \\ 1 \end{bmatrix}$. The following picture, which includes the two eigenvectors, shows this behavior more clearly:

```
> v1,v2 := Column(P,1..2);
```

$$v1, v2 := \begin{bmatrix} \dfrac{2}{5} \\ 1 \end{bmatrix}, \begin{bmatrix} 2 \\ 1 \end{bmatrix}$$

```
> p1 := Trajectory(A,x0,points=20):
  p2 := Drawvec(v1,v2,headcolor=blue,headlength=.15):
  display([p1,p2]);
```

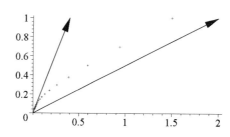

==========================

Exercise 3.1: Carry out an analysis in the style of Example 3B for the matrix B below. That is, find the eigenvector decomposition for x_k, and use it to explain the behavior of the trajectory of x_k. Use the initial vector x_0 given below.

```
> B := Matrix([[.9,-.12],[-.5,.8]]);
> x0 := Vector([1,3]);
```

+ Student Workspace

Answer 3.1

```
> h,P := Evectors(B);
```

$$h, P := \begin{bmatrix} \dfrac{11}{10} \\ \dfrac{3}{5} \end{bmatrix}, \begin{bmatrix} \dfrac{-3}{5} & \dfrac{2}{5} \\ 1 & 1 \end{bmatrix}$$

```
> v1,v2 := Column(P,1..2);
```

$$v1, v2 := \begin{bmatrix} \dfrac{-3}{5} \\ 1 \end{bmatrix}, \begin{bmatrix} \dfrac{2}{5} \\ 1 \end{bmatrix}$$

```
> p1 := Trajectory(B,x0,points=30):
  p2 := Drawvec(v1,v2, headcolor=blue,headlength=.2):
  display({p1,p2});
```

Since the larger eigenvalue of B is $11/10$, the points x_k eventually move away from the origin and get arbitrarily far away. An associated eigenvector is $\begin{bmatrix} -.6 \\ 1 \end{bmatrix}$, which tells us that eventually the trajectory follows the line through the origin with direction vector $\begin{bmatrix} -.6 \\ 1 \end{bmatrix}$. We sometimes say that "the origin is a saddle point" for this dynamical system, since the shape of the trajectory is similar to level curves near a saddle point of a function of two variables.

The "saddle point" shape of the trajectory for the above matrix B is typical when one of the eigenvalues is larger than 1 and the other is between 0 and 1. At first the points are attracted toward the origin along the eigenvector associated with the smaller eigenvalue; then they are "repelled" from the origin along the eigenvector associated with the larger eigenvalue:

```
> p1 := Trajectory(B,[seq([cos(t*Pi/6),sin(t*Pi/6)],t=0..12)],
    points=6,lines=on):
  p2 := Drawvec(v1,v2,headcolor=blue):
  display({p1,p2});
```

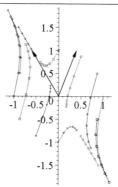

However, note what happens below when the initial vector x_0 is an eigenvector associated with the smaller eigenvalue. Then change x_0 to be an eigenvector associated with the larger eigenvalue.

```
> x0 := Vector([.4,1]):
> Trajectory(B,x0,points=30);
```

Exercise 3.2: Carry out an analysis in the style of Example 3B for the stochastic matrix C below. That is, find the eigenvector decomposition for x_k, and use it to explain the behavior of the trajectory of x_k. Use the initial vector x_0 given below.

```
> x0 := Vector([.1,.9]);
  C := Matrix([[.9,.2],[.1,.8]]);
```

+ Student Workspace

− Answer 3.2

```
> h,P := Evectors(C);
```

$$h, P := \begin{bmatrix} 1 \\ \frac{7}{10} \end{bmatrix}, \begin{bmatrix} 2 & -1 \\ 1 & 1 \end{bmatrix}$$

```
> v1,v2 := Column(P,1..2);
```

$$v1, v2 := \begin{bmatrix} 2 \\ 1 \end{bmatrix}, \begin{bmatrix} -1 \\ 1 \end{bmatrix}$$

```
> p1 := Trajectory(C,x0,points=20):
  p2 := Drawvec(v1,v2,headcolor=blue,headlength=.2):
  display({p1,p2});
```

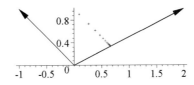

Thus $x_k = a_1 \, 1^k \begin{bmatrix} 2 \\ 1 \end{bmatrix} + a_2 \, .7^k \begin{bmatrix} -1 \\ 1 \end{bmatrix}$, where the values of a_1 and a_2 depend on x_0. Since C is a stochastic matrix, the larger eigenvalue is 1. An associated eigenvector is $\begin{bmatrix} 2 \\ 1 \end{bmatrix}$, which tells us that the trajectory moves toward a point on the line through the origin with that direction vector. Furthermore, since the components of the state vectors for a Markov chain always add to 1, all the points x_k lie on the line $x + y = 1$. What is the intersection point of the two lines that every x_k converges to? Why is the other eigenvector parallel to the line $x + y = 1$?

====================

Example 3C: Here is a system whose dynamics are more complicated:
```
> A := Matrix([[.8,.5],[-.1,1.0]]);
```
$$A := \begin{bmatrix} .8 & .5 \\ -.1 & 1.0 \end{bmatrix}$$
```
> Trajectory(A,<1,1>,points=40);
```

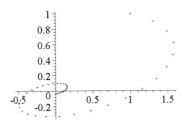

We sometimes say that the points are "attracted to the origin along a spiral." As we will see in a later module, the spiral behavior is caused by the presence of complex eigenvalues.

====================

Section 4: Predator-Prey Model : Foxes and Rabbits

In this section we consider a dynamical systems model that simulates the interactions between two populations, a population of foxes (the predators) and a population of rabbits (the prey). We denote the fox and rabbit populations at time k by $x_k = \begin{bmatrix} F_k \\ R_k \end{bmatrix}$, where k is in years and where F_k is the number of foxes after k years and R_k is the number of rabbits (in hundreds) after k years. Also we suppose that:

$$F_{k+1} = .6 F_k + .4 R_k$$
$$R_{k+1} = -.125 F_k + 1.2 R_k$$

The term $.6 F_k$ in the first equation says that in the absence of rabbits for food only 60% of the foxes would survive each year. On the other hand, the presence of rabbits increases the number of foxes, which is represented by the term $.4 R_k$. Looking at the second equation, we see the term $1.2 R_k$, which represents a 20% growth of rabbits per year if there were no foxes. Finally the effect of the foxes on the rabbits is represented by the term $-.125 F_k$; thus, the greater the number of foxes, the more the rabbit population will decrease in a given year.

Here is the transition matrix M for the fox and rabbit populations:

```
> M := Matrix([[.6,.4],[-.125,1.2]]);
```

$$M := \begin{bmatrix} .6 & .4 \\ -.125 & 1.2 \end{bmatrix}$$

Let's find the eigenvalues and eigenvectors of M and use them to find an eigenvector decomposition of the iterates x_k:

```
> h,P := Evectors(M);
```

$$h, P := \begin{bmatrix} \frac{11}{10} \\ \frac{7}{10} \end{bmatrix}, \begin{bmatrix} 4 & 4 \\ 5 & \\ 1 & 1 \end{bmatrix}$$

Therefore

$$x_k = a_1 \left(\frac{11}{10}\right)^k \begin{bmatrix} .8 \\ 1 \end{bmatrix} + a_2 \left(\frac{7}{10}\right)^k \begin{bmatrix} 4 \\ 1 \end{bmatrix}$$

Since one eigenvalue is larger than 1 and the other is smaller than 1, the origin is a saddle point for this dynamical system. More specifically, since x_k is approximately equal to $a_1 \left(\frac{11}{10}\right)^k \begin{bmatrix} .8 \\ 1 \end{bmatrix}$ for large values of k, both the fox and rabbit populations grow at a yearly rate of 10% (except in the special case when $a_1 = 0$), and the populations are in the ratio of 8 foxes to every 1000 rabbits. Note, moreover, that these conclusions are independent of the initial population x_0. Let's look at the trajectory and the component

plots in which we start with 50 foxes and 1500 rabbits:
```
> x0 := Vector([50,15]);
> Trajectory(M,x0,points=30);
```

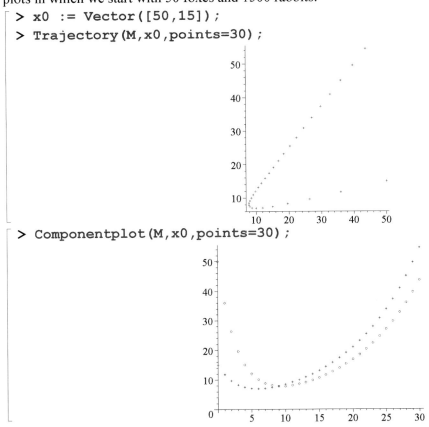

```
> Componentplot(M,x0,points=30);
```

Exercise 4.1: In the above plots, observe that the fox population decreases dramatically at first, before both populations eventually get larger and larger. Note the contrasting situation below where we start with 5 foxes and 1300 rabbits. Give a plausible explanation for why these different initial conditions lead to such different trajectories.

```
> v1,v2 := Column(P,1..2);
```

$$v1, v2 := \begin{bmatrix} \frac{4}{5} \\ 1 \end{bmatrix}, \begin{bmatrix} 4 \\ 1 \end{bmatrix}$$

```
> x0 := Vector([5,10]):
  p1 := Trajectory(M,x0,points=20):
  p2 := Drawvec(20*v1,20*v2,headcolor=blue,
    headlength=6):
  display({p1,p2});
```

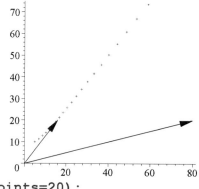

```
> Componentplot(M,x0,points=20);
```

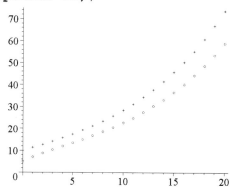

 Student Workspace

 Answer 4.1

When we start with 50 foxes and 1500 rabbits, there are not enough rabbits to sustain the fox population; so at first the fox population decreases until there are few enough to be sustained. When we start with 5 foxes and 1000 rabbits, there are plenty of rabbits to sustain the fox population.

Problems

Problem 1: Coordinates relative to an eigenvector basis

(a) For the 4 by 4 matrix N below, find an eigenvector basis B of R^4.
(b) Find the B-coordinates of the point whose standard coordinates are (4, 3, 1, 0).
(c) Find the standard coordinates of the point whose B-coordinates are (4, 0, 4, −3).

```
> N := Matrix([[3,0,0,1],[1,2,0,1],[1,0,2,1],[1,0,0,3]]);
```

Student Workspace

Module 6.3 Eigenvector Bases and Discrete Dynamical Systems   357

Problem 2: Eigenvector decomposition for 3 by 3 matrices

(a) Write out the eigenvector decomposition [4] for $x_k = A^k x_0$, where A is the 3 by 3 matrix below, and x_0 is the vector below. (Use the **Evectors** command.)

```
> A := Matrix([[3,2,2],[1,4,1],[ -2,-4,-1]]);
> x0 := Vector([4,2,-5]);
```

(b) (By hand) Derive the eigenvector decomposition formula [4] for the case of a general 3 by 3 matrix.

+ Student Workspace

Problem 3: More eigenvector decompositions

Answer the following for each of the matrices below.

(a) Find the eigenvector decomposition [4] of the iterates $x_k = M^k x_0$. Use the **Evectors** command, but leave the scalar weights a_1, a_2, etc. as letters.

(b) Using this decomposition, describe the long-term behavior of the points x_k. (Do they approach a limit? Do they eventually follow some straight line? Do points move toward or away from the origin, or is the origin a saddle point?)

```
> A := Matrix([[1,-.5],[1,2.5]]);
> B := Matrix([[.8,-.2,-.2],[-.2,2,.8],[.2,-1.4,-.2]]);
> C := Matrix([[.85,.15,.05],[.05,.75,.05],[.1,.1,.9]]);
> E := Matrix([[-1,1.2,-2],[.8,0,1],[1.6,-1.2,2.6]]);
```

+ Student Workspace

Problem 4: More foxes and rabbits

This problem is a follow-up to the fox versus rabbit model presented in Section 3.

(a) Determine the long-term behavior of the model when the (2, 2) entry of the matrix M is changed from 1.2 to 1.05. (Assume that a degraded environment lowers the fecundity of the rabbits.) Base your analysis on an eigenvector decomposition of x_k.

(b) Determine a value of the (2, 2) entry that leads to constant levels of the fox and rabbit populations, so that eventually neither population is changing. What is the ratio of the sizes of the populations in this case?

+ Student Workspace

Problem 5: Leslie population model

We consider an animal population whose size and age distribution are determined by characteristics of the female. (We assume there will always be enough males to propagate the species.) The maximal lifespan of the Lesser Horned Beetle (LHB, for short) is 4 years. Of the newborn female LHB's, 38% live to one year; of the one-year old females, 30% live to two; of the two-year old females, 25% live to three. Furthermore, on average, each one-year-old female produces 2 female offspring and each two-year-old female produces 2 female offspring.

If $x_k = \langle a_k, b_k, c_k, d_k \rangle$ denotes the number of female LHB's in each of the four age categories after k years, we have the equations

$$a_{k+1} = 2 b_k + 2 c_k$$
$$b_{k+1} = .38 a_k$$
$$c_{k+1} = .3 b_k$$
$$d_{k+1} = .25 c_k$$

(a) From the above system of equations, find the transition matrix M such that $x_{k+1} = M x_k$. (M is called a *Leslie* matrix.) Hint: Check that your matrix M satisfies the equation

$$\begin{bmatrix} a_{k+1} \\ b_{k+1} \\ c_{k+1} \\ d_{k+1} \end{bmatrix} = M \begin{bmatrix} a_k \\ b_k \\ c_k \\ d_k \end{bmatrix}$$

(b) Find the eigenvector decomposition [4] of the iterates x_k. (Use the **Evectors** command.)
(c) Using this eigenvector decomposition, describe the long-term behavior of x_k.
(d) Using the initial state vector $x_0 = \langle 4, 3, 2, 1 \rangle$, find an explicit formula for the x_k's. Also, use **Componentplot** to plot all four components. Label the graphs to indicate which gives the number of newborns, the number of one-year olds, the number of two-year olds, and the number of three-year olds.
+ Student Workspace

Problem 6: Three contrasting populations

The general Leslie matrix (see Problem 5) for describing a population by age groups has the form

$$\begin{bmatrix} a_1 & a_2 & a_3 & \cdots & a_{n-1} & a_n \\ p_1 & 0 & 0 & \cdots & 0 & 0 \\ 0 & p_2 & 0 & \cdots & 0 & 0 \\ \cdot & \cdot & \cdot & \cdots & 0 & 0 \\ \cdot & \cdot & \cdot & \cdots & 0 & 0 \\ 0 & 0 & 0 & \cdots & p_{n-1} & 0 \end{bmatrix}$$

Here a_j is the average number of offspring per individual in age group j, and p_j is the probability that an individual in age group j survives one time period to become a member of the next age group, $j + 1$. The tables below give the values of these birth and death parameters, a_j and p_j, for three populations of human females (*Demographic Yearbook, 1981*). The time period is ten years.

$$\begin{bmatrix} \text{Venezuela} \\ \text{Age group in years} \quad a_j \quad p_j \\ \begin{array}{ccc} 0\text{-}10 & 0 & .953 \\ 10\text{-}20 & .259 & .994 \\ 20\text{-}30 & 1.099 & .990 \\ 30\text{-}40 & .699 & .983 \\ 40\text{-}50 & .154 & .963 \\ 50\text{-}60 & 0 & .915 \\ 60\text{-}70 & 0 & \end{array} \end{bmatrix} \begin{bmatrix} \text{Netherlands} \\ \text{Age group in years} \quad a_j \quad p_j \\ \begin{array}{ccc} 0\text{-}10 & 0 & .990 \\ 10\text{-}20 & .022 & .997 \\ 20\text{-}30 & .532 & .996 \\ 30\text{-}40 & .209 & .993 \\ 40\text{-}50 & .009 & .980 \\ 50\text{-}60 & 0 & .952 \\ 60\text{-}70 & 0 & \end{array} \end{bmatrix} \begin{bmatrix} \text{Bulgaria} \\ \text{Age group in years} \quad a_j \quad p_j \\ \begin{array}{ccc} 0\text{-}10 & 0 & .978 \\ 10\text{-}20 & .196 & .996 \\ 20\text{-}30 & .687 & .994 \\ 30\text{-}40 & .107 & .990 \\ 40\text{-}50 & .006 & .976 \\ 50\text{-}60 & 0 & .938 \\ 60\text{-}70 & 0 & \end{array} \end{bmatrix}$$

(a) Use the vectors below and the commands **Diagmat** and **Copyinto** to construct the three corresponding Leslie matrices.

```
> aV := Vector([0,.259,1.099,.699,.154,0,0]):
> pV := Vector([.953,.994,.990,.983,.963,.915]):
> aN := Vector([0,.022,.532,.209,.009,0,0]):
  pN := Vector([.990,.997,.996,.993,.980,.952]):
> aB := Vector([0,.196,.687,.107,.006,0,0]):
  pB := Vector([.978,.996,.994,.990,.976,.938]):
```

(b) Use the **Evectors** command to find the eigenvalues and eigenvectors of each of the matrices.

(c) From the information in (b), describe the long-term behavior of each age group for each of the three populations. In particular, determine the long-term rate of growth of the populations, the long-term distribution of the age groups within each of the three populations, and describe the differences between the three populations.

(d) Use **Componentplot** to plot all seven components for each of the three populations.

[+] Student Workspace

Linear Algebra Modules Project
Chapter 6, Module 4

Diagonalization and Similarity

Purpose of this module

The purpose of this module is to study the process by which the eigenvectors of a matrix can be used to diagonalize a matrix. We then study the more general concept of similar matrices and investigate its geometric meaning.

Prerequisites

Eigenvalues and eigenvectors; eigenvector basis.

Commands used in this module

```
> restart: with(LinearAlgebra): with(plots): with(Lamp):
  UseHardwareFloats := false: Digits := 6:
```

Tutorial

Section 1: Diagonalizing a Matrix

We have seen that the eigenvalues and eigenvectors of a matrix provide us with valuable insights into how a matrix acts geometrically as a transformation of vectors. In this section we study how a matrix can be "diagonalized" and then factored into a product of matrices that reveals all the eigenvector and eigenvalue information about the matrix in a compact and useful form.

==================

Example 1A: (How to diagonalize a matrix) For the matrix A defined below, we first calculate its eigenvalues and eigenvectors:

```
> A := Matrix([[1,4],[2,3]]);
```

$$A := \begin{bmatrix} 1 & 4 \\ 2 & 3 \end{bmatrix}$$

```
> h,P := Evectors(A);
```

$$h, P := \begin{bmatrix} 5 \\ -1 \end{bmatrix}, \begin{bmatrix} 1 & -2 \\ 1 & 1 \end{bmatrix}$$

So A has eigenvalues $\lambda_1 = 5$ and $\lambda_2 = -1$ with eigenvectors $v_1 = \begin{bmatrix} 1 \\ 1 \end{bmatrix}$ and $v_2 = \begin{bmatrix} -2 \\ 1 \end{bmatrix}$, respectively.

Consider now the matrix P, the whose columns are the eigenvectors v_1 and v_2:

> P;

$$\begin{bmatrix} 1 & -2 \\ 1 & 1 \end{bmatrix}$$

Since these eigenvectors are linearly independent, P has an inverse, and so we can calculate the product $P^{(-1)} A P$:

> K := P^(-1).A.P;

$$K := \begin{bmatrix} 5 & 0 \\ 0 & -1 \end{bmatrix}$$

K is the diagonal matrix we are looking for. Notice that K is not only diagonal, but the diagonal entries are in fact the eigenvalues of A. Furthermore, the diagonal entries appear in the same order as their corresponding eigenvectors appear as columns in P.

========================

The following theorem summarizes the method of Example 1A for diagonalizing a square matrix:

Theorem 10: If A is an n by n matrix with n linearly independent eigenvectors, then A can be diagonalized. Specifically, if P is a matrix whose columns are n linearly independent eigenvectors of A, then $P^{(-1)} A P = K$, where K is the diagonal matrix whose diagonal entries are the corresponding eigenvalues.

We will refer to P, the matrix of eigenvectors, as a "diagonalizing matrix" for A.

========================

Example 1B: Diagonalize the matrix A below by the method of Example 1A and Theorem 10.

> A := Matrix([[1,-2,4,2],[-2,1,4,2],[0,-2,5,2],[-2,-2,4,5]]);

$$A := \begin{bmatrix} 1 & -2 & 4 & 2 \\ -2 & 1 & 4 & 2 \\ 0 & -2 & 5 & 2 \\ -2 & -2 & 4 & 5 \end{bmatrix}$$

Solution: The **Evectors** command gives us the eigenvalues and eigenvectors; then we denote the columns of the eigenvector matrix by v_1, v_2, v_3, v_4:

> h,P := Evectors(A);

$$h, P := \begin{bmatrix} 5 \\ 3 \\ 3 \\ 1 \end{bmatrix}, \begin{bmatrix} 1 & 1 & 0 & 1 \\ 1 & 1 & 1 & 1 \\ 1 & 1 & 0 & 0 \\ 1 & 0 & 1 & 1 \end{bmatrix}$$

> v1,v2,v3,v4 := Column(P,1..4);

$$v1, v2, v3, v4 := \begin{bmatrix} 1 \\ 1 \\ 1 \\ 1 \end{bmatrix}, \begin{bmatrix} 1 \\ 1 \\ 1 \\ 0 \end{bmatrix}, \begin{bmatrix} 0 \\ 1 \\ 0 \\ 1 \end{bmatrix}, \begin{bmatrix} 1 \\ 1 \\ 0 \\ 1 \end{bmatrix}$$

Chapter 6 Eigenvalues and Eigenvectors

We note that the eigenvalue 5 has an eigenspace basis $\{v_1\}$, the eigenvalue 3 has an eigenspace basis $\{v_2, v_3\}$, and the eigenvalue 1 has an eigenspace basis $\{v_4\}$. Taking the four eigenspace basis vectors v_1, v_2, v_3, v_4 together will give us four linearly independent eigenvectors, as we confirm next:

```
> Matsolve(P,zero);
```

$$\begin{bmatrix} 0 \\ 0 \\ 0 \\ 0 \end{bmatrix}$$

Since $P x = 0$ has only the zero solution, the columns of P are linearly independent and hence P is invertible. Next we compute the diagonal matrix $K = P^{(-1)} A P$. Note that the diagonal entries of K are indeed the eigenvalues produced by the **Evectors** command and that they appear in K in the same order as their corresponding eigenvectors in P.

```
> K := P^(-1).A.P;
```

$$K := \begin{bmatrix} 5 & 0 & 0 & 0 \\ 0 & 3 & 0 & 0 \\ 0 & 0 & 3 & 0 \\ 0 & 0 & 0 & 1 \end{bmatrix}$$

========================

Exercise 1.1: The matrix Q below has the columns of P above but in a different order. Predict what the diagonal matrix $Q^{(-1)} A Q$ will be before doing any computation.

```
> Q := Matrix([v2,v3,v4,v1]);
```

$$Q := \begin{bmatrix} 1 & 0 & 1 & 1 \\ 1 & 1 & 1 & 1 \\ 1 & 0 & 0 & 1 \\ 0 & 1 & 1 & 1 \end{bmatrix}$$

 Student Workspace

 Answer 1.1

If we reorder the columns of P, their corresponding eigenvalues in K will be reordered to match the order of the eigenvectors:

```
> K := Q^(-1).A.Q;
```

$$K := \begin{bmatrix} 3 & 0 & 0 & 0 \\ 0 & 3 & 0 & 0 \\ 0 & 0 & 1 & 0 \\ 0 & 0 & 0 & 5 \end{bmatrix}$$

For example, the eigenvalue 3 appears as the first and second entries on the diagonal of K, since the first and second columns of Q are eigenvectors with eigenvalue 3.

Exercise 1.2: Below is a four-step proof of Theorem 10 for 2 by 2 matrices. Give a brief justification for each step. We are given a 2 by 2 matrix A with linearly independent eigenvectors v_1, v_2 and corresponding eigenvalues λ_1, λ_2. Let $P = [\,v_1,\, v_2\,]$ and $K = \begin{bmatrix} \lambda_1 & 0 \\ 0 & \lambda_2 \end{bmatrix}$. Then

(1) $A\,P = A\,[\,v_1,\, v_2\,] = [\,A\,v_1,\, A\,v_2\,]$

(2) $P\,K = [\,v_1,\, v_2\,] \begin{bmatrix} \lambda_1 & 0 \\ 0 & \lambda_2 \end{bmatrix} = [\,\lambda_1\,v_1,\, \lambda_2\,v_2\,]$

(3) $A\,P = P\,K$

(4) $P^{(-1)}\,A\,P = K$

+ Student Workspace

− Answer 1.2

(1) Definition of P; definition of matrix multiplication.
(2) Definition of P and K; definition of matrix multiplication.
(3) $A\,v_1 = \lambda_1\,v_1$ and $A\,v_2 = \lambda_2\,v_2$, by the definition of eigenvalue and eigenvector. So, from steps (1) and (2), $A\,P$ and $P\,K$ are the same matrix.
(4) Multiply the equation in step (3) on the left by $P^{(-1)}$.

Diagonalizing the Powers of a Matrix

Suppose we have diagonalized a matrix A in the usual way:

$$P^{(-1)}\,A\,P = K$$

If we multiply this equation on the left by P and on the right by $P^{(-1)}$, we get a factorization of A into a product of three matrices:

$$A = P\,K\,P^{(-1)}$$

From this factorization of A, we can derive a similar factorization of A^2:

$$A^2 = A\,A = P\,K\,P^{(-1)}\,P\,K\,P^{(-1)} = P\,K\,I\,K\,P^{(-1)} = P\,K^2\,P^{(-1)}$$

It is not surprising that A and A^2 have the same diagonalizing matrix P, since A and A^2 have the same eigenvectors (Problem 7 in Module 1). Similarly, for any positive integer power n, we have a factorization of A^n:

$$A^n = P\,K^n\,P^{(-1)} \qquad [1]$$

Exercise 1.3: For the matrix A in Example 1A (repeated below), use formula [1] to compute A^5.

```
> A := Matrix([[1,4],[2,3]]):
```

+ Student Workspace

Answer 1.3

```
> h,P := Evectors(A);
```

$$h, P := \begin{bmatrix} 5 \\ -1 \end{bmatrix}, \begin{bmatrix} 1 & -2 \\ 1 & 1 \end{bmatrix}$$

```
> K := P^(-1).A.P;
```

$$K := \begin{bmatrix} 5 & 0 \\ 0 & -1 \end{bmatrix}$$

Since $P^{(-1)} A P = K$, we can compute A^5 from $A^5 = P K^5 P^{(-1)}$:

```
> P.(K^5).P^(-1);
> A^5;
```

$$\begin{bmatrix} 1041 & 2084 \\ 1042 & 2083 \end{bmatrix}, \begin{bmatrix} 1041 & 2084 \\ 1042 & 2083 \end{bmatrix}$$

In the next example we use the above formula for A^n to study the powers of a Markov matrix.

===================

Example 1C: Find a formula for the nth power of the Markov matrix M below. Also find the limit of M^n as n goes to infinity.

```
> M := Matrix([[.94,.09],[.06,.91]]);
```

$$M := \begin{bmatrix} .94 & .09 \\ .06 & .91 \end{bmatrix}$$

```
> h,P := Evectors(M);
```

$$h, P := \begin{bmatrix} 1 \\ \frac{17}{20} \end{bmatrix}, \begin{bmatrix} \frac{3}{2} & -1 \\ 1 & 1 \end{bmatrix}$$

So M has eigenvalues 1 and 17/20 with eigenvectors $v = \begin{bmatrix} 1.5 \\ 1 \end{bmatrix}$ and $w = \begin{bmatrix} -1 \\ 1 \end{bmatrix}$ respectively. The eigenvector v is a steady state vector for the Markov matrix M, since $M v = 1\, v = v$. To see how this steady-state vector arises directly from the powers of M, we diagonalize M:

```
> K := P^(-1).M.P;
```

$$K := \begin{bmatrix} 1.00000 & 0. \\ 0. & .850000 \end{bmatrix}$$

Using the fact that the nth power of a diagonal matrix is found by simply taking the nth powers of the diagonal entries and using formula [1], we have the following formula for M^n:

$$M^n = P K^n P^{(-1)} = P \begin{bmatrix} 1^n & 0 \\ 0 & \left(\frac{17}{20}\right)^n \end{bmatrix} P^{(-1)}$$

Since the limit of K^n, as n goes to infinity, is $\begin{bmatrix} 1 & 0 \\ 0 & 0 \end{bmatrix}$, the limit of M^n is $Q = P \begin{bmatrix} 1 & 0 \\ 0 & 0 \end{bmatrix} P^{(-1)}$. Thus:

```
> L := Matrix([[1,0],[0,0]]);
> Q := P.L.P^(-1);
```

$$L := \begin{bmatrix} 1 & 0 \\ 0 & 0 \end{bmatrix}, \quad Q := \begin{bmatrix} \frac{3}{5} & \frac{3}{5} \\ \frac{2}{5} & \frac{2}{5} \end{bmatrix}$$

In particular, note that the columns of this limit matrix are identical and that they are a multiple (2/5) of the eigenvector v. Furthermore, if x_0 is any initial vector, then the limit of $M^n x_0$ is $Q x_0$, which is also a multiple of v:

```
> x0 := Vector([a,b]);
  Q.x0;
```

$$x0 := \begin{bmatrix} a \\ b \end{bmatrix}$$

$$\begin{bmatrix} \frac{3}{5}a + \frac{3}{5}b \\ \frac{2}{5}a + \frac{2}{5}b \end{bmatrix}$$

```
> Expand((a+b)*(2/5)*Vector([3/2,1]));
```

$$\begin{bmatrix} \frac{3}{5}a + \frac{3}{5}b \\ \frac{2}{5}a + \frac{2}{5}b \end{bmatrix}$$

Hence the limit of the state vectors, $x_n = M^n x_0$, is always an eigenvector with eigenvalue 1 and hence a steady-state vector.

Section 2: Similar Matrices

In Section 1 we found that we can diagonalize an n by n matrix, provided we can find n linearly independent eigenvectors for the matrix. If the matrix does not have n linearly independent eigenvectors, we cannot diagonalize it. However, even in this case we can carry out a process like diagonalization that has useful consequences.

If we have a matrix A and any invertible matrix P, we can calculate the product $P^{(-1)} A P$. The resulting matrix may not be a diagonal matrix, but it does have an interesting relationship to A.

Definition: Two n by n matrices A and B are **similar** if there is an invertible matrix P such that:

$$P^{(-1)} A P = B, \text{ or equivalently, } A = P B P^{(-1)}$$

Chapter 6 Eigenvalues and Eigenvectors

In particular, when we diagonalize a matrix A by finding a matrix P such that $P^{(-1)} A P = K$ is diagonal, we say that A is *similar* to the diagonal matrix K.

We now show that all 2 by 2 reflection matrices are similar to one another. Let R be a matrix that reflects vectors in R^2 across a line L that passes through the origin. Recall that R has an eigenvalue 1 whose associated eigenvectors are parallel to L and an eigenvalue -1 whose associated eigenvectors are perpendicular to L. Since every 2 by 2 reflection matrix has these eigenvalues, they must all be similar to the diagonal matrix with diagonal entries 1 and -1; hence they are all similar to one another (see Problem 5). This suggests that all 2 by 2 reflection matrices are much alike. The following example illustrates this by examining a typical reflection matrix.

======================

Example 2A: Let R be the reflection matrix that reflects vectors across the line L that makes an angle of $\pi/3$ radians with the x axis.

```
> R := Reflectmat(Pi/3);
```

$$R := \begin{bmatrix} \frac{-1}{2} & \frac{1}{2}\sqrt{3} \\ \frac{1}{2}\sqrt{3} & \frac{1}{2} \end{bmatrix}$$

Two handy eigenvectors for R are u, a unit vector parallel to L, and v, a unit vector perpendicular to L:

$$u = \begin{bmatrix} \cos\left(\frac{\pi}{3}\right) \\ \sin\left(\frac{\pi}{3}\right) \end{bmatrix} \quad \text{and} \quad v = \begin{bmatrix} -\sin\left(\frac{\pi}{3}\right) \\ \cos\left(\frac{\pi}{3}\right) \end{bmatrix}$$

See the figure below with u and v on the unit circle.

```
> u := Vector([cos(Pi/3),sin(Pi/3)]);
  v := Vector([-sin(Pi/3),cos(Pi/3)]);
```

$$u := \begin{bmatrix} \frac{1}{2} \\ \frac{1}{2}\sqrt{3} \end{bmatrix}, \quad v := \begin{bmatrix} -\frac{1}{2}\sqrt{3} \\ \frac{1}{2} \end{bmatrix}$$

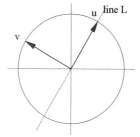

Module 6.4 Diagonalization and Similarity

Let's check that a matrix with columns u and v does in fact diagonalize R.

```
> Q := Matrix([u,v]);
```

$$Q := \begin{bmatrix} \frac{1}{2} & -\frac{1}{2}\sqrt{3} \\ \frac{1}{2}\sqrt{3} & \frac{1}{2} \end{bmatrix}$$

```
> K := Q^(-1).R.Q;
```

$$K := \begin{bmatrix} 1 & 0 \\ 0 & -1 \end{bmatrix}$$

Now let's take a closer look at the matrices K and Q from a geometric point of view. First we recognize that K is simply the reflection across the x axis. And Q turns out to be the rotation matrix that rotates vectors about the origin by $\frac{\pi}{3}$ radians.

```
> Reflectmat(0);
```

$$\begin{bmatrix} 1 & 0 \\ 0 & -1 \end{bmatrix}$$

```
> Rotatemat(Pi/3);
```

$$\begin{bmatrix} \frac{1}{2} & -\frac{1}{2}\sqrt{3} \\ \frac{1}{2}\sqrt{3} & \frac{1}{2} \end{bmatrix}$$

So this rotation matrix is a diagonalizing matrix for the reflection R, and the formula $K = Q^{(-1)} R Q$ shows that R is similar to K, the reflection across the x axis. Moreover, we can readily interpret the product $Q^{(-1)} R Q$ geometrically (see figure below). Let v_1 be any vector in the plane, and let's apply the matrices of the product $Q^{(-1)} R Q$ successively to v_1; that is, let $v_2 = Q v_1$, and $v_3 = R v_2 = R Q v_1$, and $v_4 = Q^{(-1)} v_3 = Q^{(-1)} R Q v_1$. We then have the following picture:

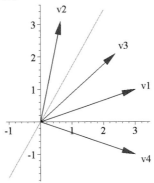

Thus, the equation $K v_1 = Q^{(-1)} R Q v_1$ says that if we rotate a vector v_1 counter-clockwise by 60 degrees to v_2, then reflect v_2 across the 60-degree line to v_3, and then rotate v_3 back by 60 degrees to v_4, the overall result will be a reflection across the x axis, which is exactly what $K v_1$ is.

Of course there is nothing special about the specific reflection in this example. We can diagonalize any 2 by 2 reflection matrix in the same way, using a corresponding rotation matrix. We will use Maple to check the general case. (The command `map(simplify,K)` tells Maple to apply its trigonometric identities to simplify every component of K.)

```
> R := Reflectmat(theta);
```
$$R := \begin{bmatrix} \cos(2\theta) & \sin(2\theta) \\ \sin(2\theta) & -\cos(2\theta) \end{bmatrix}$$

```
> Q := Rotatemat(theta);
```
$$Q := \begin{bmatrix} \cos(\theta) & -\sin(\theta) \\ \sin(\theta) & \cos(\theta) \end{bmatrix}$$

```
> K := Q^(-1).R.Q:
> map(simplify,K);
```
$$\begin{bmatrix} 1 & 0 \\ 0 & -1 \end{bmatrix}$$

================================

Example 2A is intended to suggest that similar matrices are much alike. As we will see, when two matrices are similar, in the sense of the above definition, they have many of the same properties. The following theorem describes one such property:

Theorem 11: If A and B are similar matrices, they have the same characteristic polynomial and hence the same eigenvalues with the same algebraic multiplicities.

Proof: We will use the following properties of determinants:

$$\text{Det}(MN) = \text{Det}(M)\text{Det}(N) \quad \text{and} \quad \text{Det}(P^{(-1)}) = \frac{1}{\text{Det}(P)}$$

Writing B as $P^{(-1)} A P$ and I as $P^{(-1)} I P$, we have:

$$\text{Det}(B - \lambda I) = \text{Det}(P^{(-1)} A P - \lambda P^{(-1)} I P)$$
$$= \text{Det}(P^{(-1)} (A - \lambda I) P)$$
$$= \text{Det}(P^{(-1)}) \text{Det}(A - \lambda I) \text{Det}(P)$$
$$= \text{Det}(A - \lambda I)$$

Therefore A and B have the same characteristic polynomial.

Example 2B: Given any square matrix A, we can construct a similar matrix B by choosing any invertible n by n matrix for P, and defining B to be $P^{(-1)} A P$. Below we construct B in this way and check that A and B have the same characteristic polynomial:

```
> A := Matrix([[3,-4],[1,-2]]);
```

$$A := \begin{bmatrix} 3 & -4 \\ 1 & -2 \end{bmatrix}$$

```
> P := Matrix([[2,3],[1,2]]);
```

$$P := \begin{bmatrix} 2 & 3 \\ 1 & 2 \end{bmatrix}$$

```
> B := P^(-1).A.P;
```

$$B := \begin{bmatrix} 4 & 5 \\ -2 & -3 \end{bmatrix}$$

```
> Charpoly(A,lambda);
> Charpoly(B,lambda);
```

$$-2 - \lambda + \lambda^2, \quad -2 - \lambda + \lambda^2$$

Do similar matrices also have the same eigenvectors? Execute the next input region and take a close look at the result. Notice that while the eigenvalues and their algebraic multiplicities are the same, their eigenvectors are quite different.

```
> Evectors(A);
```

$$\begin{bmatrix} 2 \\ -1 \end{bmatrix}, \begin{bmatrix} 4 & 1 \\ 1 & 1 \end{bmatrix}$$

```
> Evectors(B);
```

$$\begin{bmatrix} 2 \\ -1 \end{bmatrix}, \begin{bmatrix} \frac{-5}{2} & -1 \\ 1 & 1 \end{bmatrix}$$

Exercise 2.1: (a) Check that when each of the above eigenvectors of A is multiplied by $P^{(-1)}$, the result is an eigenvector of B with the same eigenvalue.
(b) (By hand) Show in general that if u is an eigenvector of a matrix A with eigenvalue λ, then the vector $P^{(-1)} u$ is an eigenvector of $P^{(-1)} A P$ with the same eigenvalue.

+ Student Workspace

− Answer 2.1

(a) We apply $P^{(-1)}$ to each of the eigenvectors of A computed above, and we note that the results are (multiples of) the eigenvectors of B computed above:

```
> u := Vector([4,1]):
```

```
        v := Vector([1,1]):
>  P^(-1).u;
   P^(-1).v;
```

$$\begin{bmatrix} 5 \\ -2 \end{bmatrix}$$

$$\begin{bmatrix} -1 \\ 1 \end{bmatrix}$$

(b) Since $A u = \lambda u$, we have that

$$P^{(-1)} A P P^{(-1)} u = P^{(-1)} A u = P^{(-1)} \lambda u = \lambda P^{(-1)} u$$

Exercise 2.1(b) is the key to proving the following theorem:

Theorem 12: If A and B are similar matrices, their eigenvalues have the same geometric multiplicities.

Exercise 2.2: The matrices A and B below are similar, and $\lambda = 3$ is an eigenvalue for both. Check that $\lambda = 3$ has the same geometric multiplicity for A and B.

```
>  A := Matrix([[5,-4,2],[4,-5,4],[2,-4,5]]):
>  P := Matrix([[-1,0,0],[-1,0,-1],[1,1,-1]]):
>  B := P^(-1).A.P:
```

+ Student Workspace

− Answer 2.2

```
>  I := Idmat(3):
>  Nullbasis(A-3*I);
```

$$\begin{bmatrix} \begin{bmatrix} 2 \\ 1 \\ 0 \end{bmatrix}, \begin{bmatrix} -1 \\ 0 \\ 1 \end{bmatrix} \end{bmatrix}$$

```
>  Nullbasis(B-3*I);
```

$$\begin{bmatrix} \begin{bmatrix} \frac{-1}{2} \\ 1 \\ 0 \end{bmatrix}, \begin{bmatrix} \frac{-1}{2} \\ 0 \\ 1 \end{bmatrix} \end{bmatrix}$$

Since both null spaces have dimension 2, the eigenvalue $\lambda = 3$ has geometric multiplicity 2 for both A and B.

− Section 3: Geometric Meaning of Similarity and Diagonalization

In this section we will think of matrices as matrix transformations. That is, if A is an n by n matrix, we will study the behavior of the function T defined by

$$T(x) = A x, \text{ where } x \text{ ranges over all vectors in } R^n.$$

We will see that by diagonalizing the matrix A we can understand the behavior of the matrix transformation T geometrically. To see how to interpret a matrix transformation geometrically, we first consider diagonal matrix transformations.

========================

Example 3A: Describe the geometric behavior of the diagonal matrix K below.

```
> K := Diagmat([2,-1]);
```

$$K := \begin{bmatrix} 2 & 0 \\ 0 & -1 \end{bmatrix}$$

Solution: As a function, K takes vectors on the x axis to twice themselves and reverses the direction of vectors on the y axis:

```
> K.Vector([x,0]);
  K.Vector([0,y]);
```

$$\begin{bmatrix} 2x \\ 0 \end{bmatrix}, \begin{bmatrix} 0 \\ -y \end{bmatrix}$$

So the behavior of K as a function is easy to describe when K is applied to vectors along either axis. (Notice, of course, that the x axis is an eigenspace of K with associated eigenvalue 2, and the y axis is an eigenspace of K with eigenvalue -1.) Furthermore, once we know how K behaves when applied to vectors on either axis, we can also describe how K behaves when applied to other vectors, since all vectors in R^2 can be decomposed into their x and y components. For example, if K is applied to the vector $\begin{bmatrix} -1 \\ 3 \end{bmatrix}$, K doubles the x component to -2 and reverses the sign of the y component to -3:

```
> K.Vector([-1,3]);
```

$$\begin{bmatrix} -2 \\ -3 \end{bmatrix}$$

More geometrically, we can say that K stretches figures by a factor of 2 in the x direction and reflects figures across the x axis (i.e., in the direction of the y axis). For example, here's a picture of K applied to the House figure:

```
> Transform(K,House);
```

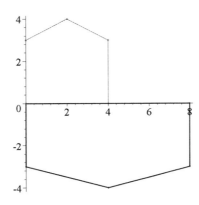

========================

The eigenvectors and eigenvalues of the diagonal matrix K gave us the insights that allowed us to describe the behavior of K as a geometric transformation. The matrix A below is not diagonal, but it can be diagonalized. Its eigenvectors and eigenvalues will also be the key to describing its behavior as a geometric transformation.

```
> A := Matrix([[1,-2],[-1,0]]);
```

$$A := \begin{bmatrix} 1 & -2 \\ -1 & 0 \end{bmatrix}$$

```
> h,P := Evectors(A);
```

$$h, P := \begin{bmatrix} 2 \\ -1 \end{bmatrix}, \begin{bmatrix} -2 & 1 \\ 1 & 1 \end{bmatrix}$$

So the eigenvalues of A are 2 and -1 with associated eigenvectors u and v below:

```
> u,v := Column(P,1..2);
```

$$u, v := \begin{bmatrix} -2 \\ 1 \end{bmatrix}, \begin{bmatrix} 1 \\ 1 \end{bmatrix}$$

```
> K := P^(-1).A.P;
```

$$K := \begin{bmatrix} 2 & 0 \\ 0 & -1 \end{bmatrix}$$

The columns of the matrix P are linearly independent eigenvectors of A, and K is the diagonal matrix similar to A that P produces. We saw in Example 3A how to describe the behavior of K, and now we want to use that knowledge to describe the behavior of A. Since u is an eigenvector of A with eigenvalue 2, A takes vectors in the eigenspace Span$\{u\}$ to twice themselves. And, since v is an eigenvector of A with eigenvalue -1, A reverses the direction of vectors in the eigenspace Span$\{v\}$. Furthermore, we can describe what A does to every vector in R^2 by decomposing vectors into their u and v components.

======================

Example 3B: Describe what A does to the vector $y = \begin{bmatrix} 5 \\ 2 \end{bmatrix}$ by decomposing y into its u and v components, where u and v are the above eigenvectors of A.

Solution: To solve $a\,u + b\,v = y$ for a and b, we write the equation in the form $[u, v]\begin{bmatrix} a \\ b \end{bmatrix} = y$ or more succinctly as $P\,x = y$, where $P = [u, v]$ and $x = \begin{bmatrix} a \\ b \end{bmatrix}$. Then we solve for x:

```
> y := Vector([5,2]);
```

$$y := \begin{bmatrix} 5 \\ 2 \end{bmatrix}$$

```
> x := P^(-1).y;
```

$$x := \begin{bmatrix} -1 \\ 3 \end{bmatrix}$$

Thus $a = -1$ and $b = 3$. So A doubles the u component of y from -1 to -2 and reverses the sign of the v component of y from 3 to -3. Let's check this result:

> -2*u-3*v, A.y;

$$\begin{bmatrix} 1 \\ -5 \end{bmatrix} \begin{bmatrix} 1 \\ -5 \end{bmatrix}$$

Since $A y = P K P^{(-1)} y = P K x$, we can also compute $A y$ as follows:
> P.K.x;

$$\begin{bmatrix} 1 \\ -5 \end{bmatrix}$$

In a sense, therefore, A behaves very much like the similar diagonal matrix K, and its behavior is almost as simple to describe. The diagonal matrix K behaves especially simply when applied to vectors on the x and y axes, while A behaves especially simply when applied to vectors on the u and v axes.

========================

Let's compare the behavior of K and A visually by applying both to the Bug figure:
> Transform(K,Bug);

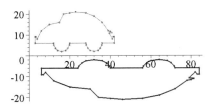

> Transform(A,Bug);

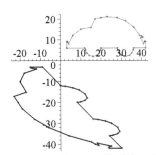

Although the results are not identical, you can see the similarities. Both matrices reflect the bug in one direction and stretch it in another direction. We can even argue that, in a sense, A and K represent the same function in two different coordinate systems, the standard system and the uv system. To see this, let's look at a smaller version of the house figure together with the standard basis vectors, e_1 and e_2:
> shed := .4*House:
> p1 := Drawmatrix(shed,figcolor=black):
> p2 := Drawvec([1,0],[0,1],headcolor=black,headscale=.2):
> display([p1,p2],view=[-3..2,-.2..3]);

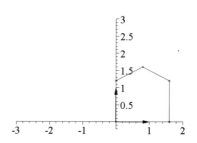

Next we construct the shed figure in *uv* coordinates (calling it *uv_shed*), draw *uv_shed* together with the eigenvector basis vectors *u* and *v*, and draw shed and *uv_shed* together. Note especially that the coordinates of *uv_shed* relative to the basis $\{u, v\}$ are the same as the coordinates of the shed figure relative to the basis $\{e_1, e_2\}$.

```
> uv_shed := P.shed:
> p3 := Drawmatrix(uv_shed):
> p4 := Drawvec([u,label="u"],[v,label="v"]):
> display([p3,p4]);
```

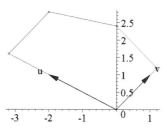

```
> display([p1,p2,p3,p4]);
```

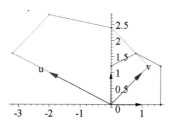

Finally, we apply the original matrix *A* to *uv_shed*, and observe that *A* behaves just as *K* did; that is, *A* reflects across the *u* axis in the direction of the *v* axis and stretches by a factor of 2 in the direction of the *u* axis:

```
> p5 := Transform(A,uv_shed):
> display([p3,p4,p5]);
```

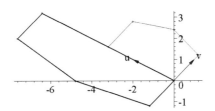

Exercise 3.1: (a) Use the diagonalization process to show that the matrices R and K below are similar.

```
> R := Reflectmat(Pi/4);
> K := Reflectmat(0);
```

$$R := \begin{bmatrix} 0 & 1 \\ 1 & 0 \end{bmatrix}$$

$$K := \begin{bmatrix} 1 & 0 \\ 0 & -1 \end{bmatrix}$$

(b) Use geometry to explain how R and K can be thought of as essentially the same matrix transformation.

+ Student Workspace

− Answer 3.1

```
> h,P := Evectors(R);
```

$$h, P := \begin{bmatrix} 1 \\ -1 \end{bmatrix}, \begin{bmatrix} 1 & -1 \\ 1 & 1 \end{bmatrix}$$

```
> P^(-1).R.P, K;
```

$$\begin{bmatrix} 1 & 0 \\ 0 & -1 \end{bmatrix}, \begin{bmatrix} 1 & 0 \\ 0 & -1 \end{bmatrix}$$

Alternatively, we could use the following rotation matrix instead of P:

```
> Q := Rotatemat(Pi/4);
> Q^(-1).R.Q;
```

$$Q := \begin{bmatrix} \frac{1\sqrt{2}}{2} & -\frac{1\sqrt{2}}{2} \\ \frac{1\sqrt{2}}{2} & \frac{1\sqrt{2}}{2} \end{bmatrix}, \begin{bmatrix} 1 & 0 \\ 0 & -1 \end{bmatrix}$$

(b) R is the same reflection matrix as K, if you think of the coordinate system as having been rotated $\pi/4$ radians (45 degrees).

Problems

Problem 1: Compatible products under diagonalization

(a) Diagonalize the matrix A below; that is, find a diagonalizing matrix P and a diagonal matrix K so that $P^{(-1)} A P = K$.

(b) Check that the diagonalizing matrix P you found in (a) is also a diagonalizing matrix for the matrix B below. What are the corresponding eigenvalues of B?

(c) Check that P is also a diagonalizing matrix for the product $A\,B$. What are the corresponding eigenvalues of $A\,B$?

(d) (By hand) Show that if an invertible matrix P diagonalizes two matrices A and B, then P also diagonalizes the product $A\,B$. How are the eigenvalues of $A\,B$ related to those of A and B?

```
> A := Matrix([[20,3,19],[-9,0,-9],[-23,-3,-22]]);
> B := Matrix([[23,-15,24],[-6,4,-6],[-21,15,-22]]);
```

Student Workspace

Problem 2: Markov matrices, powers, and stable vectors

Let M be the Markov matrix below.

(a) Use a diagonalization of M to calculate M^5 and M^{10}.

(b) (By hand) Write down a formula for M^n. Then find the exact limit of M^n as n goes to infinity.

(c) (By hand) Use your result in (b) to explain the following. If x_0 is any vector in R^2 whose components add to 1, then the limit of $M^n x_0$ is always $\begin{bmatrix} \dfrac{3}{7} \\ \dfrac{4}{7} \end{bmatrix}$.

```
> M := Matrix([[1/5,3/5],[4/5,2/5]]);
```

Student Workspace

Problem 3: Square roots of matrices

Recall that a positive real number r has two square roots -- one positive and one negative. For example, both 2 and -2 are square roots of 4. Matrices typically have more square roots, because we have more places to choose signs plus or minus.

(a) (By hand) The 2 by 2 diagonal matrix `Diagmat([1,4])` has four diagonal square roots (i.e., diagonal matrices whose square is the given matrix). Find them. (Hint: One of the diagonal square roots is `Diagmat([-1,2])`.) For each of these square roots, state its eigenvalues.

(b) (By hand) Write down the eight diagonal square roots of `Diagmat([1,4,9])`.

(c) Find a square root of the matrix A below. That is, find a matrix C such that $C^2 = A$ (Hint: Diagonalize A and use any of the square roots of its similar diagonal matrix K to find C.)

```
> A := Matrix([[13,18],[-6,-8]]);
```

$$A := \begin{bmatrix} 13 & 18 \\ -6 & -8 \end{bmatrix}$$

(d) Apply the idea in (c) to find a square root of the projection matrix **Projectmat(Pi/4)**. By considering the nature of the eigenvalues of this matrix, explain why this matrix has fewer distinct square roots than the matrix in (c).

(e) By the same reasoning as in (a), the 2 by 2 identity matrix I_2 has four diagonal square roots. But the fact that the eigenvalue 1 for I_2 has geometric multiplicity 2 turns out to mean that I_2 has infinitely many choices of square root, and these are related in interesting ways by similarity. Choose any diagonal square root T of I_2, such as **Diagmat([-1,1])**, and any nonsingular matrix S, such as $\begin{bmatrix} 1 & 2 \\ 3 & 4 \end{bmatrix}$. Check that $S^{(-1)} T S$ is a square root of I_2. Find at least two nondiagonal square roots of I_2 by this method.

(f) (By hand) Prove by matrix algebra that every matrix of the form $S^{(-1)} T S$, where $T^2 = I_n$ and S is nonsingular, is a square root of I_n (I_n is the n by n identity matrix).

+ Student Workspace

Problem 4: Construct a matrix from its eigenvalues and eigenvectors

(a) Construct a 2 by 2 matrix A that has the eigenvectors $\begin{bmatrix} 2 \\ 3 \end{bmatrix}$ and $\begin{bmatrix} 1 \\ 2 \end{bmatrix}$ with associated eigenvalues 4 and 1, respectively. Hint: Use $P^{(-1)} A P = K$, where K is diagonal.

(b) (By hand) Suppose you are given n linearly independent vectors $v_1, ..., v_n$ and n scalars $t_1, ..., t_n$. Describe in general how to construct an n by n matrix A with eigenvectors $v_1, ..., v_n$ and corresponding eigenvalues $t_1, ..., t_n$.

+ Student Workspace

Problem 5: Similar to similar implies similar

(By hand) If A, B, C are n by n matrices such that A and B are similar and B and C are similar, prove that A and C are similar. **Caution:** The invertible matrix that relates A and B is not necessarily the same matrix that relates B and C.

+ Student Workspace

Problem 6: Geometric behavior of a matrix transformation

(a) Describe the geometric behavior of the diagonal matrix transformation K below.
(b) Describe the geometric behavior of the matrix transformation A below. (See the comparable discussion in Section 3.)
(c) For the vector y below, find the coordinates of y relative to the eigenvector basis you found in (b). Use these coordinates to describe what A does to the vector y.

```
> K := Diagmat([0,-2,-2]);
> A := Matrix([[-3,1,-2],[1,-3,2],[2,-2,2]]);
> y := Vector([2,2,-1]);
```

Student Workspace

Problem 7: Projection matrices

The matrix A below projects vectors onto the line making an angle of θ radians with the x axis:
```
> A := Projectmat(theta);
```

(a) Use geometry to explain why two of the eigenvectors for A are $u = \begin{bmatrix} \cos(\theta) \\ \sin(\theta) \end{bmatrix}$ and $v = \begin{bmatrix} -\sin(\theta) \\ \cos(\theta) \end{bmatrix}$. What are the associated eigenvalues for u and v?

(b) Construct the matrix $P = [u, v]$ to diagonalize A. What familiar matrix transformation is P?
(c) Use the general form of A (i.e., `Projectmat(theta)`) and P to find the diagonal matrix K. In order to get Maple to simplify the result, use `map(simplify,K)`.
(d) What familiar geometric transformation is K?
(e) Are all projection matrices similar? Use your results from parts (a)-(d) to support your answer.
(f) Give a geometric explanation for why the matrix product $P K P^{(-1)}$ equals A. Hint: Apply $P^{(-1)}$ to an arbitrary vector v. What does $P^{(-1)}$ do to v geometrically? Then apply K. Then apply P.

Student Workspace

Problem 8: Shear matrices

(a) Find the eigenvalues of the horizontal shear matrix $H = \begin{bmatrix} 1 & 1 \\ 0 & 1 \end{bmatrix}$, and find their algebraic and geometric multiplicities. Explain why this matrix cannot be diagonalized.

(b) Show that every horizontal shear matrix, $S = \begin{bmatrix} 1 & k \\ 0 & 1 \end{bmatrix}$ where $k \neq 0$, is similar to the special shear matrix H in (a). (Hint: Use $P = \begin{bmatrix} 1 & 0 \\ 0 & \frac{1}{k} \end{bmatrix}$ as the diagonalizing matrix.) Explain geometrically why

$P^{(-1)}SP = H$, especially describing the effects of P and $P^{(-1)}$.

(c) Show that the vertical shear matrix $V = \begin{bmatrix} 1 & 0 \\ 1 & 1 \end{bmatrix}$ is similar to the horizontal shear matrix H in (a). (Hint: To change the coordinates, use a certain reflection matrix for P.)

(d) Find the eigenvalues of $A = \begin{bmatrix} 3 & -1 \\ 4 & -1 \end{bmatrix}$, and find their algebraic and geometric multiplicities.

(e) Show that the matrix A in (d) is similar to the horizontal shear matrix H. Hint: The equations

$$A \begin{bmatrix} 1 \\ 2 \end{bmatrix} = 1 \begin{bmatrix} 1 \\ 2 \end{bmatrix} + 0 \begin{bmatrix} 1 \\ 1 \end{bmatrix}, \quad A \begin{bmatrix} 1 \\ 1 \end{bmatrix} = 1 \begin{bmatrix} 1 \\ 2 \end{bmatrix} + 1 \begin{bmatrix} 1 \\ 1 \end{bmatrix}$$

show that the behavior of A on the two vectors $\begin{bmatrix} 1 \\ 2 \end{bmatrix}$ and $\begin{bmatrix} 1 \\ 1 \end{bmatrix}$ is similar to the behavior of H on the two standard basis vectors $\begin{bmatrix} 1 \\ 0 \end{bmatrix}$ and $\begin{bmatrix} 0 \\ 1 \end{bmatrix}$.

+ Student Workspace

Linear Algebra Modules Project
Chapter 6, Module 5

Complex Eigenvalues and Eigenvectors

Purpose of this module

The purpose of this module is to introduce complex eigenvalues and eigenvectors, to interpret them geometrically, and to use them in analyzing certain discrete dynamical systems.

Prerequisites

Addition and multiplication of complex numbers; eigenvalues and eigenvectors; similarity; and Module 3, "Eigenvector Bases and Discrete Dynamical Systems."

Commands used in this module

```
> restart; with(LinearAlgebra): with(plots): with(Lamp):
  UseHardwareFloats := false: Digits := 6:
```

Tutorial

Section 1: Complex Numbers and Complex Vectors

A complex number z has the form $z = a + b\,i$, where a and b are real numbers and $i = \sqrt{-1}$.

Definition: If $z = a + b\,i$, then a is called the **real part** of z and b is called the **imaginary part** of z

===================

Example 1A: The complex number $z = 3 + 2\,i$ has the real part 3 and imaginary part 2. Here are Maple's commands **Re(z)** and **Im(z)** for extracting the real and imaginary parts of z, respectively:

```
> z := 3+2*i;
```
$$z := 3 + 2\,i$$

```
> Re(z), Im(z);
```
$$3, 2$$

Maple note: The default symbol for $\sqrt{-1}$ in Maple is I. The Lamp library overrides this default and sets the symbol for $\sqrt{-1}$ to the more familiar i. However, when you save a worksheet, Maple changes all occurrences of the complex number i in the output back to I. To avoid confusion, remove all output before you save this worksheet.

===================

Definition: If $z = a + b\,i$ then the ***conjugate*** of z, denoted $\overline{(z)}$, is the complex number $a - b\,i$.

```
> conjugate(z);
```
$$3 - 2\,i$$

Rules of algebra for the complex conjugate: If z and w are complex numbers then

- $$\overline{(z + w)} = \overline{(z)} + \overline{(w)}$$

- $$\overline{(z\,w)} = \overline{(z)}\,\overline{(w)}$$

Exercise 1.1: (a) Verify $\overline{(z\,w)} = \overline{(z)}\,\overline{(w)}$ by direct calculation. (Let $z = a + b\,i$ and $w = c + d\,i$.)
(b) Simplify $\overline{(\overline{(z)})}$. That is, what do you get when you take the conjugate of the conjugate of z?
(c) For which complex numbers z is it the case that $z = \overline{(z)}$?

+ Student Workspace

− Answer 1.1

(a) $z\,w = (a + b\,i)(c + d\,i) = a\,c - b\,d + (a\,d + b\,c)\,i$, and so $\overline{(z\,w)} = a\,c - b\,d - (a\,d + b\,c)\,i$.

$\overline{(z)}\,\overline{(w)} = (a - b\,i)(c - d\,i) = a\,c - b\,d - (a\,d + b\,c)\,i = \overline{(z\,w)}$.

(b) $\overline{(\overline{(z)})} = \overline{(a - b\,i)} = a + b\,i = z$.

(c) $z = \overline{(z)}$ if and only if z is real.

Definition: If $z = a + b\,i$ then the ***absolute value*** (or ***magnitude***) of z, denoted $|z|$, is $\sqrt{a^2 + b^2}$.

Rules of algebra for the absolute value: If z and w are complex numbers, then

- $$|z\,w| = |z|\,|w|$$

- $$|\overline{(z)}| = |z|$$

For example, using the complex number $z = 3 + 2\,i$ defined earlier, we have

```
> abs(z);
  abs(conjugate(z));
```
$$\sqrt{13},\ \sqrt{13}$$

Geometry of Complex Numbers

We can picture the complex number $z = a + b\,i$ as the point in the plane with coordinates (a, b). Note that the real part of z is then the x coordinate and the imaginary part is the y coordinate. Also, the conjugate of z is the reflection of z across the x axis, and the absolute value of z is the distance between (a, b) and the origin.

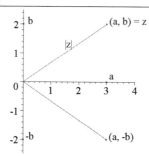

Recall that any point (a, b) in the plane has polar coordinates (r, θ), where r is the distance between the origin and the point (a, b), and θ is the angle measured counterclockwise from the positive x axis to the ray going from the origin through the point (a, b). So we can write $(a, b) = (r \cos(\theta), r \sin(\theta))$. We can therefore write complex numbers in *polar form*:

$$z = a + b\,i = r\cos(\theta) + r\sin(\theta)\,i = r(\cos(\theta) + \sin(\theta)\,i)$$

Note that $r = |z|$. The polar angle θ is usually referred to as the *argument* of z.

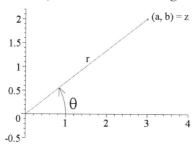

====================

Example 1B: Write the complex number $-1 + i$ in polar form.

Solution: The complex number $-1 + i$ can be pictured as the point (-1, 1). So $r = \sqrt{2}$ and $\theta = \dfrac{3\pi}{4}$.

Therefore $-1 + i = \sqrt{2}\left(\cos\left(\dfrac{3\pi}{4}\right) + \sin\left(\dfrac{3\pi}{4}\right)i\right)$. Maple provides more than one way to compute the magnitude and argument of a complex number:

> `convert(-1+i,polar);`

$$\mathrm{polar}\left(\sqrt{2}, \frac{3}{4}\pi\right)$$

> `abs(-1+i), argument(-1+i);`

$$\sqrt{2}, \frac{3}{4}\pi$$

====================

Exercise 1.2: (By hand) Write the complex numbers $2i$ and $2 - 2i$ in polar form.

 Student Workspace

 Answer 1.2

The point $(0, 2)$ is 2 units from the origin and is rotated 90 degrees counterclockwise from the positive x axis. So $r = 2$ and $\theta = \dfrac{\pi}{2}$, and therefore $2i = 2\cos\left(\dfrac{\pi}{2}\right) + 2\sin\left(\dfrac{\pi}{2}\right)i$.

The point $(2, -2)$ is $\sqrt{8}$ units from the origin and is rotated 45 degrees clockwise from the positive x axis. So $r = \sqrt{8} = 2\sqrt{2}$ and $\theta = -\dfrac{\pi}{4}$, and therefore

$$2 - 2i = 2\sqrt{2}\cos\left(-\dfrac{\pi}{4}\right) + 2\sqrt{2}\sin\left(-\dfrac{\pi}{4}\right)i$$

Computational note: Although we were able to find the polar coordinates of $2i$ and $2 - 2i$ using only geometry, we will sometimes encounter angles that are not familiar multiples of π. In this case, we can find the magnitude and argument of a complex number $a + ib$ from the formulas

$$|a + ib| = \sqrt{a^2 + b^2} \quad \text{and} \quad \text{argument}(a + ib) = \arctan\left(\dfrac{b}{a}\right) \quad (\text{provided } a \neq 0)$$

Complex Vectors

We now expand our concept of a vector by allowing n-tuples of <u>complex</u> numbers. For example, here are two vectors that are 3-tuples of complex numbers:

$$v = \begin{bmatrix} 3 - i \\ 9i \\ -4 + 3i \end{bmatrix} \quad w = \begin{bmatrix} 2 + 4i \\ 5 - 9i \\ 7 \end{bmatrix}$$

We add such vectors componentwise and multiply them by complex scalars componentwise. We denote the set of all n-tuples of complex numbers by C^n, much as we denoted the set of all n-tuples of real numbers by R^n.

==================

Example 1C: (a) Add the vectors v and w above, and multiply w by the complex scalar $k = 1 + i$.
(b) Decompose w in the form $w = x + iy$, where x and y are vectors with real components. (x is called the *real part* of w and y is called the *imaginary part* of w.)

Chapter 6 Eigenvalues and Eigenvectors

Solution: (a) $v + w = \begin{bmatrix} 5+3i \\ 5 \\ 3+3i \end{bmatrix}$ and $kw = \begin{bmatrix} (1+i)(2+4i) \\ (1+i)(5-9i) \\ (1+i)7 \end{bmatrix} = \begin{bmatrix} -2+6i \\ 14-4i \\ 7+7i \end{bmatrix}$

(b) $w = \begin{bmatrix} 2 \\ 5 \\ 7 \end{bmatrix} + i \begin{bmatrix} 4 \\ -9 \\ 0 \end{bmatrix}$

In Maple:

```
> v := Vector([3-i,5+9*i,-4+3*i]);
  w := Vector([2+4*i,5-9*i,7]);
```

$$v := \begin{bmatrix} 3-i \\ 5+9i \\ -4+3i \end{bmatrix}, \quad w := \begin{bmatrix} 2+4i \\ 5-9i \\ 7 \end{bmatrix}$$

```
> v+w;
```

$$\begin{bmatrix} 5+3i \\ 10 \\ 3+3i \end{bmatrix}$$

```
> k := 1+i;
  k*w;
```

$$k := 1+i$$

$$\begin{bmatrix} -2+6i \\ 14-4i \\ 7+7i \end{bmatrix}$$

```
> map(Re,w), map(Im,w);
```

$$\begin{bmatrix} 2 \\ 5 \\ 7 \end{bmatrix}, \begin{bmatrix} 4 \\ -9 \\ 0 \end{bmatrix}$$

Maple note: To apply a Maple command (such as **Re** and **Im** above) to each component of a vector or matrix, use the **map** command as above.

======================

Definition: If $w = x + iy$, where x and y are real n-tuples, then the ***conjugate*** of w, denoted $\overline{w}$, is the complex vector $x - iy$.

```
> map(conjugate,w);
```

$$\begin{bmatrix} 2-4i \\ 5+9i \\ 7 \end{bmatrix}$$

The following rules of algebra can be deduced from the corresponding rules for the conjugate of a complex number: If v and w are complex n-tuples and k is a complex scalar, then

- $\overline{(v+w)} = \overline{(v)} + \overline{(w)}$
- $\overline{(kw)} = \overline{(k)}\,\overline{(w)}$

Exercise 1.3: Find the complex conjugate and the real and imaginary parts of the vector v below:

$$v = \begin{bmatrix} 3-i \\ 9i \\ -4+3i \end{bmatrix}$$

+ Student Workspace

− Answer 1.3

$$\overline{(v)} = \begin{bmatrix} 3+i \\ -9i \\ -4-3i \end{bmatrix} \text{ and } \text{Re}(v) = \begin{bmatrix} 3 \\ 0 \\ -4 \end{bmatrix} \text{ and } \text{Im}(v) = \begin{bmatrix} -1 \\ 9 \\ 3 \end{bmatrix}.$$

− **Section 2: Introduction to Complex Eigenvalues and Eigenvectors**

In our discussion of complex eigenvalues and eigenvectors, we will use many of the concepts and methods of linear algebra that we have developed in earlier modules. There is no obstacle to doing so, since, from the beginning of this course, we could just as easily have used complex numbers in place of real numbers. That is, the coefficients of a linear system can be complex, and the solutions can be n-tuples of complex numbers. Gaussian elimination works just as it did for real numbers. Also the entries of a matrix can be complex numbers, and matrix algebra then works just as it did for real matrices. The concepts of subspace, span, linear independence, basis, and dimension are defined for C^n just as they are for R^n, but the scalars are complex numbers instead of real numbers.

====================

Example 2A: Find the eigenvalues and eigenvectors of the rotation matrix which rotates vectors 90 degrees counterclockwise:

$$\begin{bmatrix} 0 & -1 \\ 1 & 0 \end{bmatrix}$$

Solution: First, let's recall how this matrix behaves geometrically as a transformation. As in Module 1, we can look at the picture that shows equally-spaced input vectors (red shaft, unit length, tail at origin) with their corresponding output vectors (thick blue shaft, tail at head of its input vector):

```
> A := Rotatemat(Pi/2);
```

$$A := \begin{bmatrix} 0 & -1 \\ 1 & 0 \end{bmatrix}$$

```
> Headtail(A);
```

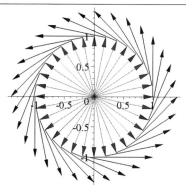

Since A rotates every nonzero vector by 90 degrees, there can be no real eigenvectors. However, let's see what happens when we try to find eigenvalues and eigenvectors by our usual techniques. First we calculate the characteristic polynomial of A:

$$\text{Det}(A - \lambda I) = \text{Det}\left(\begin{bmatrix} 0 - \lambda & -1 \\ 1 & 0 - \lambda \end{bmatrix}\right) = \lambda^2 + 1$$

Thus the characteristic equation is $\lambda^2 + 1 = 0$. This equation has no real roots, which confirms that A has no real eigenvalues. However, the characteristic equation does have two complex roots: Since $\lambda^2 = -1$, we have $\lambda_1 = i$ and $\lambda_2 = -i$, which are <u>complex</u> eigenvalues. To find the corresponding eigenvectors, we solve the matrix equation $(A - \lambda I) x = 0$ for each eigenvalue λ:

If $\lambda = i$, the augmented matrix corresponding to $(A - \lambda I) x = 0$ is:

$$\begin{bmatrix} -i & -1 & 0 \\ 1 & -i & 0 \end{bmatrix}$$

As a first step in reducing this matrix, let's interchange the two rows to get a 1 in the first pivot position:

$$\begin{bmatrix} 1 & -i & 0 \\ -i & -1 & 0 \end{bmatrix}$$

Now to get a zero below that 1, we add i times row 1 to row 2:

$$\begin{bmatrix} 1 & -i & 0 \\ 0 & 0 & 0 \end{bmatrix}$$

This last row operation results in a zero row since $i(-i) = 1$. Now that the matrix is in echelon form, we can apply back substitution. We see that x_2 is a free variable and $x_1 = i x_2$. Thus the solutions of $(A - \lambda I) x = 0$ are $\begin{bmatrix} i x_2 \\ x_2 \end{bmatrix} = x_2 \begin{bmatrix} i \\ 1 \end{bmatrix}$. So the complex vector $v = \begin{bmatrix} i \\ 1 \end{bmatrix}$ is an eigenvector associated with the complex eigenvalue $\lambda = i$. To confirm this conclusion, let's check the definition of eigenvalue and eigenvector; that is, let's check that $A v = \lambda v$:

```
> v := Vector([i,1]);
```
$$v := \begin{bmatrix} i \\ 1 \end{bmatrix}$$

```
> A.v;
```
$$\begin{bmatrix} -1 \\ i \end{bmatrix}$$

```
> i*v;
```
$$\begin{bmatrix} -1 \\ i \end{bmatrix}$$

====================

Exercise 2.1: (By hand) Use the method of Example 2A to find an eigenvector associated with the other eigenvalue, $\lambda = -i$.

 Student Workspace

− Answer 2.1

The augmented matrix for the homogeneous sysem $(A - \lambda I) x = 0$, where $\lambda = -i$, is $\begin{bmatrix} i & -1 & 0 \\ 1 & i & 0 \end{bmatrix}$ which reduces to $\begin{bmatrix} 1 & i & 0 \\ 0 & 0 & 0 \end{bmatrix}$. Thus the solutions of $(A - \lambda I) x = 0$ are $\begin{bmatrix} -i x_2 \\ x_2 \end{bmatrix} = x_2 \begin{bmatrix} -i \\ 1 \end{bmatrix}$. So an eigenvector associated with the eigenvalue $\lambda = -i$ is $\begin{bmatrix} -i \\ 1 \end{bmatrix}$.

The **Evalues** and **Evectors** commands can find complex eigenvalues and eigenvectors:

```
> Evalues(A);
```
$$\begin{bmatrix} i \\ -i \end{bmatrix}$$

```
> Evectors(A);
```
$$\begin{bmatrix} i \\ -i \end{bmatrix}, \begin{bmatrix} i & -i \\ 1 & 1 \end{bmatrix}$$

Note: When eigenvectors are complex, recognizing two eigenvectors as multiples of one another is not always easy, since the scalar can be complex. For example, the following vectors are eigenvectors corresponding to the eigenvalue $\lambda = i$ for the rotation matrix A above: $\begin{bmatrix} i \\ 1 \end{bmatrix}, \begin{bmatrix} 1 \\ -i \end{bmatrix}, \begin{bmatrix} 2+3i \\ 3-2i \end{bmatrix}$. (Multiply the first vector by $-i$ and $3 - 2i$ to get the second and third vectors, respectively.)

Observe the following about the eigenvalues and eigenvectors of the rotation matrix A. The eigenvalues i

and $-i$ are complex conjugates of one another, and so are the corresponding eigenvectors $\begin{bmatrix} i \\ 1 \end{bmatrix}$ and $\begin{bmatrix} -i \\ 1 \end{bmatrix}$. The next theorem says that complex eigenvalues <u>always</u> occur in conjugate pairs and so do their corresponding eigenvectors.

> **Theorem 13:** Suppose A is an a square matrix with real entries. If A has a complex eigenvector w and associated eigenvalue λ, then $\overline{(w)}$ is also an eigenvector of A and its associated eigenvalue is $\overline{(\lambda)}$. That is, if $A\,w = \lambda\,w$, then $A\,\overline{(w)} = \overline{(\lambda)}\,\overline{(w)}$.

Exercise 2.2: Prove Theorem 13. Hint: Write $A\,w$ as a linear combination of the columns of A,

$$A\,w = \sum_{i=1}^{n} w_i\,A_i;$$ then apply the rules of algebra for complex conjugates, and use the fact that $\overline{(A_i)} = A_i$ (since A is real).

 Student Workspace

 Answer 2.2

> We are given that $A\,w = \lambda\,w$, which can be written $\sum_{i=1}^{n} w_i\,A_i = \lambda\,w$. Taking the complex conjugate of both sides of this equation and using the addition rule for complex conjugates, we get
>
> $$\sum_{i=1}^{n} \overline{(w_i\,A_i)} = \overline{(\lambda\,w)}.$$ Using the product rule for complex conjugates and the fact that A is real, the preceding equation can be written as $\sum_{i=1}^{n} \overline{(w_i)}\,A_i = \overline{(\lambda)}\,\overline{(w)}$ and hence as $A\,\overline{(w)} = \overline{(\lambda)}\,\overline{(w)}$.

 Section 3: Scale-Rotations

In this section and the next, we attempt to determine which real matrices have complex eigenvalues and eigenvectors. For example, we will see that all real matrices of the form

$$\begin{bmatrix} a & -b \\ b & a \end{bmatrix} \qquad [1]$$

have complex eigenvalues and eigenvectors (unless $b = 0$). These matrices are the so-called "scale-rotations," a name whose significance will soon be apparent. In the next section, we will show that every 2 by 2 real matrix with complex eigenvalues and eigenvectors is similar to a scale-rotation.

Module 6.5 Complex Eigenvalues and Eigenvectors

Example 3A: Find the eigenvalues and eigenvectors of any scale-rotation $K = \begin{bmatrix} a & -b \\ b & a \end{bmatrix}$, $b \neq 0$.

Solution:

```
> K := Matrix([[a,-b],[b,a]]);
```
$$K := \begin{bmatrix} a & -b \\ b & a \end{bmatrix}$$

```
> Evectors(K);
```
$$\begin{bmatrix} a+ib \\ a-ib \end{bmatrix}, \begin{bmatrix} i & -i \\ 1 & 1 \end{bmatrix}$$

So the eigenvalues of K are $a + b\,i$ and $a - b\,i$ with corresponding eigenvectors $\begin{bmatrix} i \\ 1 \end{bmatrix}$ and $\begin{bmatrix} -i \\ 1 \end{bmatrix}$, respectively. In particular, the eigenvalues depend on the entries of K in a very simple way, and the eigenvectors do not depend at all on the particular values of these entries.

Exercise 3.1: (a) Use Example 3A to find the eigenvalues and eigenvectors for the matrix A below.
(b) Find the polar coordinates for the complex eigenvalues you found in (a).

```
> A := Matrix([[1,-1],[1,1]]);
```
$$A := \begin{bmatrix} 1 & -1 \\ 1 & 1 \end{bmatrix}$$

+ Student Workspace

− Answer 3.1

(a) By Example 3A, the eigenvalues of A are $1 + i$ and $1 - i$ with corresponding eigenvectors $\begin{bmatrix} i \\ 1 \end{bmatrix}$ and $\begin{bmatrix} -i \\ 1 \end{bmatrix}$, respectively. Let's check:

```
> Evectors(A);
```
$$\begin{bmatrix} 1+i \\ 1-i \end{bmatrix}, \begin{bmatrix} i & -i \\ 1 & 1 \end{bmatrix}$$

(b) The eigenvalue $1 + i$ has polar coordinates $(r, \theta) = \left(\sqrt{2}, \dfrac{\pi}{4}\right)$; its conjugate, $1 - i$, has polar coordinates $(r, \theta) = \left(\sqrt{2}, -\dfrac{\pi}{4}\right)$.

Example 3B: We will describe the above matrix A geometrically, as a matrix transformation, and we will find that its geometric behavior is closely related to the polar coordinates of its complex eigenvalues.

We begin by applying **Headtail** to A:

> `Headtail(A);`

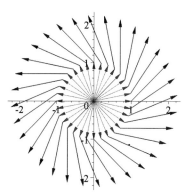

It appears that A both rotates vectors counterclockwise and stretches them. Let's assume this is correct; in other words, let's assume that A is the product of a dilation and a rotation (hence a "scale-rotation")

$$A = \begin{bmatrix} r & 0 \\ 0 & r \end{bmatrix} \begin{bmatrix} \cos(\theta) & -\sin(\theta) \\ \sin(\theta) & \cos(\theta) \end{bmatrix}$$

for some suitable values of r and θ.

We can see the effect of this transformation even more clearly by observing the effect that A has on the unit square:

> `unitsquare := Matrix([[0,1,1,0,0],[0,0,1,1,0]]);`
> `Transform(A,unitsquare);`

$$unitsquare := \begin{bmatrix} 0 & 1 & 1 & 0 & 0 \\ 0 & 0 & 1 & 1 & 0 \end{bmatrix}$$

From this picture we see that the rotation angle is $\theta = \dfrac{\pi}{4}$ and the stretch factor is $r = \sqrt{2}$ (the length of the diagonal of the unit square). Let's check:

> `S := Diagmat([sqrt(2),sqrt(2)]);`

$$S := \begin{bmatrix} \sqrt{2} & 0 \\ 0 & \sqrt{2} \end{bmatrix}$$

```
> R := Rotatemat(Pi/4);
```

$$R := \begin{bmatrix} \frac{1}{2}\sqrt{2} & -\frac{1}{2}\sqrt{2} \\ \frac{1}{2}\sqrt{2} & \frac{1}{2}\sqrt{2} \end{bmatrix}$$

```
> S.R, A;
```

$$\begin{bmatrix} 1 & -1 \\ 1 & 1 \end{bmatrix}, \begin{bmatrix} 1 & -1 \\ 1 & 1 \end{bmatrix}$$

So A rotates vectors counterclockwise $\pi/4$ radians and stretches them by a factor of $\sqrt{2}$. These same two numbers appear in the polar coordinates of the eigenvalues of the matrix A: $(r, \theta) = \left(\sqrt{2}, \frac{\pi}{4}\right)$ and $(r, \theta) = \left(\sqrt{2}, -\frac{\pi}{4}\right)$ (see Exercise 3.1). As we will soon learn, this is not a coincidence.

═══════════════════════════

We now investigate the general case. Note that the product of a dilation matrix and a rotation matrix

$$\begin{bmatrix} r & 0 \\ 0 & r \end{bmatrix} \begin{bmatrix} \cos(\theta) & -\sin(\theta) \\ \sin(\theta) & \cos(\theta) \end{bmatrix} = \begin{bmatrix} r\cos(\theta) & -r\sin(\theta) \\ r\sin(\theta) & r\cos(\theta) \end{bmatrix}$$

always has the form of a scale-rotation, where $a = r\cos(\theta)$ and $b = r\sin(\theta)$ (see the matrices [1]). Conversely, given any matrix of the form [1], we can factor it into a product of a dilation and a rotation by writing the point (a, b) in polar form:

$$a = r\cos(\theta) \quad \text{and} \quad b = r\sin(\theta)$$

Then

$$\begin{bmatrix} a & -b \\ b & a \end{bmatrix} = \begin{bmatrix} r\cos(\theta) & -r\sin(\theta) \\ r\sin(\theta) & r\cos(\theta) \end{bmatrix} = r\begin{bmatrix} \cos(\theta) & -\sin(\theta) \\ \sin(\theta) & \cos(\theta) \end{bmatrix} = \begin{bmatrix} r & 0 \\ 0 & r \end{bmatrix}\begin{bmatrix} \cos(\theta) & -\sin(\theta) \\ \sin(\theta) & \cos(\theta) \end{bmatrix}$$

We summarize this discussion and the result of Example 3A:

Theorem 14: If $K = \begin{bmatrix} a & -b \\ b & a \end{bmatrix}$ is any scale-rotation with $b \neq 0$, then its eigenvalues are $a + bi$ and $a - bi$ with corresponding eigenvectors $\begin{bmatrix} i \\ 1 \end{bmatrix}$ and $\begin{bmatrix} -i \\ 1 \end{bmatrix}$, respectively. Furthermore, we can factor the scale-rotation K as a product of a dilation and a rotation

$$K = \begin{bmatrix} r & 0 \\ 0 & r \end{bmatrix} \begin{bmatrix} \cos(\theta) & -\sin(\theta) \\ \sin(\theta) & \cos(\theta) \end{bmatrix}$$

where the numbers r and θ come from the polar coordinates of $a + bi = r\cos(\theta) + r\sin(\theta)i$.

Exercise 3.2: Use Theorem 14 to (a) write down the eigenvalues of the scale-rotation $\begin{bmatrix} \sqrt{3} & -1 \\ 1 & \sqrt{3} \end{bmatrix}$, and (b) write this matrix as the product of a dilation and a rotation.

 Student Workspace

Answer 3.2

(a) The eigenvalues are $a + bi = \sqrt{3} + i$ and $a - bi = \sqrt{3} - i$.

(b) The polar coordinates for $\sqrt{3} + i$ are $(r, \theta) = \left(2, \dfrac{\pi}{6}\right)$. So

$$\begin{bmatrix} \sqrt{3} & -1 \\ 1 & \sqrt{3} \end{bmatrix} = \begin{bmatrix} 2 & 0 \\ 0 & 2 \end{bmatrix} \begin{bmatrix} \cos\left(\dfrac{\pi}{6}\right) & -\sin\left(\dfrac{\pi}{6}\right) \\ \sin\left(\dfrac{\pi}{6}\right) & \cos\left(\dfrac{\pi}{6}\right) \end{bmatrix}.$$

Let's check:
```
> Diagmat([2,2]).Rotatemat(Pi/6);
```
$$\begin{bmatrix} \sqrt{3} & -1 \\ 1 & \sqrt{3} \end{bmatrix}$$

Discrete Dynamical Systems

Recall from Module 3 that a discrete dynamical system is a sequence of vectors $\{x_n\}$ related to one another by a square matrix A as follows:

$$x_{n+1} = A x_n \quad \text{for } n = 0, 1, 2, \ldots$$

We can compute the iterates x_n directly from the initial vector x_0 by the formula

$$x_n = A^n x_0$$

By examining the trajectories of these iterates, we will gain further geometric insights into the behavior of matrices with complex eigenvalues and eigenvectors.

====================

Example 3C: For the scale-rotation K below and any initial vector x_0, describe the path of the iterates x_n.
```
> K := Matrix([[1,-1/4],[1/4,1]]);
```
$$K := \begin{bmatrix} 1 & -\dfrac{1}{4} \\ \dfrac{1}{4} & 1 \end{bmatrix}$$

Solution: First let's apply the **Trajectory** command with a typical initial vector:
```
> x0 := Vector([1,0]):
> Trajectory(K,x0,points=40);
```

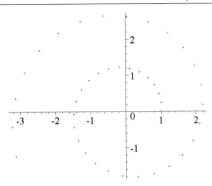

So the iterates x_n spiral counterclockwise around the origin and move away from the origin. In fact, we could have predicted this behavior by using what we know about scale-rotations. By Theorem 14, one eigenvalue of K is $1 + .25\,i$ whose polar coordinates are

```
> r := abs(1+.25*i);
  theta := argument(1+.25*i);
```

$$r := 1.03078$$
$$\theta := .244979$$

Thus K rotates vectors counterclockwise by about .24 radians (roughly 14 degrees) and stretches them by a factor of about 1.03, which is what we saw in the **Trajectory** animation.

==========================

Similarly, we can predict the trajectory of any discrete dynamical system $x_{n+1} = K x_n$ when the matrix K is a scale-rotation, $K = r \begin{bmatrix} \cos(\theta) & -\sin(\theta) \\ \sin(\theta) & \cos(\theta) \end{bmatrix}$:

- If $r < 1$ the iterates go to zero as n gets large (we say they are *attracted to the origin*);

- if $1 < r$ the iterates get arbitrarily large in absolute value as n gets large (we say they are *repelled by the origin*);

- if $r = 1$, the iterates all lie on a circle.

Furthermore, the sign and magnitude of the angle θ tells us which way the iterates are rotating about the origin and how much they rotate by at each step. In particular, the iterates make their first complete rotation about the origin when $n = \dfrac{2\pi}{\theta}$ (rounded to the nearest integer). The computation below shows that, in the above example, the x_n's make one complete rotation about the origin after 26 iterations:

```
> evalf(2*Pi/theta);
```

$$25.6478$$

Exercise 3.3: For the scale-rotation L below, compute r and θ, and use them to predict the behavior of the iterates $x_n = L^n x_0$.

Chapter 6 Eigenvalues and Eigenvectors

```
> x0 := Vector([1,0]):
  L := Matrix([[4/5,-3/5],[3/5,4/5]]);
```

$$L := \begin{bmatrix} \dfrac{4}{5} & \dfrac{-3}{5} \\ \dfrac{3}{5} & \dfrac{4}{5} \end{bmatrix}$$

+ Student Workspace

− Answer 3.3

The point $(\dfrac{4}{5}, \dfrac{3}{5})$ has polar coordinates $(r, \theta) = \left(1, \arctan\left(\dfrac{3}{4}\right)\right)$. Let's compute θ and $\dfrac{2\pi}{\theta}$ numerically:

```
> theta := arctan(.75);
  evalf(2*Pi/theta);
```

$$\theta := .643501$$

$$9.76406$$

Since $r = 1$, the iterates travel along a circle of radius 1 centered at the origin. Also, since $\theta > 0$, the iterates move counterclockwise, and the above computation shows that by the 10th iteration they complete one full rotation. Let's check our assertions:

```
> Trajectory(L,x0,points=10);
```

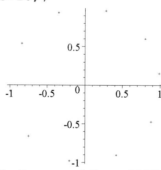

We can also give an explicit formula for the iterates as follows. If K is a scale-rotation, then we can write K as a positive scalar r times a rotation matrix:

$$K = r \begin{bmatrix} \cos(\theta) & -\sin(\theta) \\ \sin(\theta) & \cos(\theta) \end{bmatrix}$$

Next, we use the fact that the nth power of a rotation matrix with angle θ is simply the rotation through the angle $n\theta$; therefore

$$x_n = K^n x_0 = r^n \begin{bmatrix} \cos(n\theta) & -\sin(n\theta) \\ \sin(n\theta) & \cos(n\theta) \end{bmatrix} x_0 \qquad [2]$$

Finally, if we write x_0 as $\begin{bmatrix} c \\ d \end{bmatrix}$ and multiply through by r^n, we get

$$x_n = \begin{bmatrix} r^n(c\cos(n\theta) - d\sin(n\theta)) \\ r^n(c\sin(n\theta) + d\cos(n\theta)) \end{bmatrix} \qquad [3]$$

We can use formula [2] or [3] as a replacement for the usual eigenvector decomposition of x_n. (Compare Theorem 9 in Module 3.)

====================

Example 3D: (a) For the matrix K and initial vector x_0 in Example 3C (repeated below), write out formula [3] explicitly. Use the result to predict the graphs of the components of the iterates x_n as functions of n. (b) Check the prediction by plotting these components.

```
> K := Matrix([[1,-1/4],[1/4,1]]);
  x0 := Vector([1,0]):
```

$$K := \begin{bmatrix} 1 & \frac{-1}{4} \\ \frac{1}{4} & 1 \end{bmatrix}$$

Solution: (a) In Example 3C, we found the following values for r and θ:

```
> r := abs(1+.25*i);
  theta := argument(1+.25*i);
```

$$r := 1.03078$$
$$\theta := .244979$$

Thus, since $c = 1$ and $d = 0$ (the components of x_0), formula [3] becomes:

```
> xn := Vector([r^n*cos(n*theta),r^n*sin(n*theta)]);
```

$$xn := \begin{bmatrix} 1.03078^n \cos(.244979\, n) \\ 1.03078^n \sin(.244979\, n) \end{bmatrix}$$

Thus, both components have sinusoidal graphs with amplitudes that become arbitrarily large as n increases since $r > 1$. Since we have an explicit formula for the components, we can plot these using the **plot** command.

```
> plot([xn[1],xn[2]],n=0..40,color=[red,blue]);
```

This plot is misleading, however, since it displays the graphs of the components as if they were defined at all real numbers, whereas they are only defined at the non-negative integers. The `Componentplot` command plots the components at the integers only:

```
> Componentplot(K,x0,points=40);
```

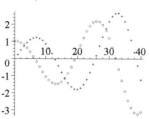

Section 4: Matrices Similar to Scale-Rotations

In this section we will show that every 2 by 2 real matrix with complex eigenvalues and eigenvectors is similar to a scale-rotation. First we consider a specific example of such a matrix.

Example 4A: The matrix A below is not a scale-rotation but it does have complex eigenvalues and eigenvectors. Describe the matrix A geometrically, as a matrix transformation, and relate its geometric behavior to the polar coordinates of its complex eigenvalues.

```
> A := Matrix([[5/4,-1/2],[1/4,3/4]]);
```

$$A := \begin{bmatrix} \frac{5}{4} & \frac{-1}{2} \\ \frac{1}{4} & \frac{3}{4} \end{bmatrix}$$

```
> h,Q := Evectors(A);
```

$$h, Q := \begin{bmatrix} 1 + \frac{1}{4}i \\ 1 - \frac{1}{4}i \end{bmatrix}, \begin{bmatrix} 1+i & 1-i \\ 1 & 1 \end{bmatrix}$$

Solution: We can try to interpret A geometrically by using the methods of Module 4. That is, we can diagonalize A using the above diagonalizing matrix Q:

```
> Q^(-1).A.Q;
```

$$\begin{bmatrix} 1 + \frac{1}{4}i & 0 \\ 0 & 1 - \frac{1}{4}i \end{bmatrix}$$

Although this diagonal matrix is similar to A, its complex entries do not help us interpret A geometrically. Furthermore, there is no diagonal matrix with real entries that is similar to A; for if there were, the real diagonal entries would be the eigenvalues of A, which we know are not real. However, the matrix A is similar to a scale-rotation, and we do know how to interpret scale-rotations geometrically (see Examples 3C and 3D):

```
> P := Matrix([[1,1],[1,0]]);
> K := P^(-1).A.P;
```

$$P := \begin{bmatrix} 1 & 1 \\ 1 & 0 \end{bmatrix}, \quad K := \begin{bmatrix} 1 & \frac{1}{4} \\ -\frac{1}{4} & 1 \end{bmatrix}$$

In the discussion immediately following this example, we will see where the matrix P comes from. First, however, let's compare the geometric behavior of the matrix A and the scale-rotation K. Specifically, let's compare a trajectory of the iterates $A^n x_0$ with a trajectory of the iterates $K^n x_0$:

```
> x0 := Vector([1,0]):
  Trajectory(K,x0,points=26);
```

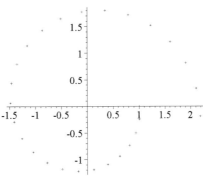

```
> Trajectory(A,x0,points=26);
```

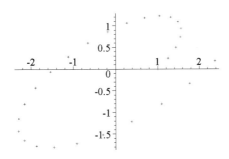

Note the following similarities in these trajectories. The iterates spiral away from the origin; they need 26 iterations for one complete revolution; and in one revolution they increase in length by a factor just a little bigger than 2.

======================

Now we show that every 2 by 2 real matrix with complex eigenvalues and eigenvectors is similar to a scale-rotation. At first, we allow A to be any n by n real matrix with complex eigenvalues and eigenvectors. Specifically, let $\lambda = a + b\,i$ be a complex eigenvalue of A with complex eigenvector $w = x + i\,y$, where a, b, x, and y are real. Then $A\,w = \lambda\,w$ can be written as

$$A(x + i\,y) = (a + b\,i)(x + i\,y)$$

Multiplying out, we get

$$A\,x + i\,A\,y = a\,x - b\,y + i\,(b\,x + a\,y)$$

Setting the real parts of the equation equal to one another and the imaginary parts equal to one another, we get two equations:

$$A\,x = a\,x - b\,y \quad \text{and} \quad A\,y = b\,x + a\,y$$

Next, let P be the matrix whose columns are x and y, the real and imaginary parts of the eigenvector w. Using $P = [x, y]$ and the above formulas for $A\,x$ and $A\,y$, we get:

$$A\,P = A\,[x, y] = [a\,x - b\,y, b\,x + a\,y] = [x, y]\begin{bmatrix} a & b \\ -b & a \end{bmatrix} = P\,K$$

If A is 2 by 2, then so is P. In fact, P is invertible (Problem 4), and so we can write this result as

$$P^{(-1)} A\,P = K, \quad \text{where } P = [x, y] \text{ and } K = \begin{bmatrix} a & b \\ -b & a \end{bmatrix} \quad [4]$$

Note that K is a scale-rotation, although the entry b in formula [4] is the negative of the value of b in formula [1] in Section 3.

Exercise 4.1: Find the matrices K and P in formula [4] for the matrix A in Example 4A (repeated below).
```
> A := Matrix([[5/4,-1/2],[1/4,3/4]]):
```
+ Student Workspace

− Answer 4.1

> Evectors(A);

$$\begin{bmatrix} 1 + \frac{1}{4}i \\ 1 - \frac{1}{4}i \end{bmatrix}, \begin{bmatrix} 1 + i & 1 - i \\ 1 & 1 \end{bmatrix}$$

So one eigenvalue of A is $\lambda = a + b\,i = 1 + \dfrac{i}{4}$, and a corresponding eigenvector is $w = x + i\,y = \begin{bmatrix} 1 + i \\ 1 \end{bmatrix} = \begin{bmatrix} 1 \\ 1 \end{bmatrix} + i\begin{bmatrix} 1 \\ 0 \end{bmatrix}$. Therefore, by formula [4], we have $P^{(-1)} A\,P = K$, where $P = \begin{bmatrix} 1 & 1 \\ 1 & 0 \end{bmatrix}$ and $K =$

$\begin{bmatrix} 1 & \frac{1}{4} \\ -\frac{1}{4} & 1 \end{bmatrix}$. Let's check:

```
> P := Matrix([[1,1],[1,0]]);
```
$$P := \begin{bmatrix} 1 & 1 \\ 1 & 0 \end{bmatrix}$$

```
> P^(-1).A.P;
```
$$\begin{bmatrix} 1 & \frac{1}{4} \\ \frac{-1}{4} & 1 \end{bmatrix}$$

To understand more deeply the trajectory of the iterates $x_n = A^n x_0$ and the plots of their components, we need formulas like the decompositions [2] and [3] that we derived for scale-rotations. Let A be any 2 by 2 matrix with a complex eigenvalue $\lambda = a + bi = r(\cos(\theta) + i\sin(\theta))$ and corresponding eigenvector $w = x + iy$. Then, with $P = [x, y]$ and $K = \begin{bmatrix} a & b \\ -b & a \end{bmatrix} = r\begin{bmatrix} \cos(\theta) & \sin(\theta) \\ -\sin(\theta) & \cos(\theta) \end{bmatrix}$, we have, by formula [4], $P^{(-1)} A P = K$. Therefore, by formula [2],

$$A^n = P K^n P^{(-1)} = r^n P \begin{bmatrix} \cos(n\theta) & \sin(n\theta) \\ -\sin(n\theta) & \cos(n\theta) \end{bmatrix} P^{(-1)}$$

Next we express the initial vector x_0 as a linear combination of x and y (the real and imaginary parts of the eigenvector w): $x_0 = c x + d y = P \begin{bmatrix} c \\ d \end{bmatrix}$. Then

$$x_n = A^n x_0 = r^n P \begin{bmatrix} \cos(n\theta) & \sin(n\theta) \\ -\sin(n\theta) & \cos(n\theta) \end{bmatrix} \begin{bmatrix} c \\ d \end{bmatrix} \qquad [5]$$

We can also expand [5] by writing P as $[x, y]$. We then have

$$x_n = r^n (c \cos(n\theta) + d \sin(n\theta)) x + r^n (-c \sin(n\theta) + d \cos(n\theta)) y \qquad [6]$$

Compare formulas [5] and [6] with the corresponding formulas [2] and [3] for scale-rotations. Also, note that since $P \begin{bmatrix} c \\ d \end{bmatrix} = x_0$, we can solve for c and d via $\begin{bmatrix} c \\ d \end{bmatrix} = P^{(-1)} x_0$.

As in Section 3, we see that there are three possibilities for the long-term behavior of the iterates x_n:

- If $r < 1$ the iterates go to zero as n gets large (we say they are *attracted to the origin*);

- if $1 < r$ the iterates get arbitrarily large in absolute value as n gets large (we say they are *repelled by the origin*);

- if $r = 1$, the iterates all lie on an ellipse.

Chapter 6 Eigenvalues and Eigenvectors

From the magnitude of the angle θ we can determine some information about the rotation of the iterates: They make their first complete rotation about the origin when $n = \dfrac{2\pi}{\theta}$ (rounded to the nearest integer). However, the sign of θ can be misleading, since the matrix P can reverse the direction of rotation.

======================

Example 4B: (a) Find the decomposition formula [6] for the matrix A in Example 4A and the initial vector x_0 below, and use it to predict the plots of the components of the iterates x_n as functions of n. (b) Check the prediction by using **Componentplot**.

```
> A := Matrix([[5/4,-1/2],[1/4,3/4]]):
> x0 := Vector([1,2]):
```

Solution: Recall that one eigenvalue of A is $\lambda = 1 + \dfrac{i}{4}$ and a corresponding eigenvector is $x + iy = \begin{bmatrix} 1 \\ 1 \end{bmatrix} + i \begin{bmatrix} 1 \\ 0 \end{bmatrix}$. Here is the computation of r, θ, x, y, c, and d, which are the values we need to write out formula [6]:

```
> r := abs(1+.25*i);
  theta := argument(1+.25*i);
```

$$r := 1.03078, \quad \theta := .244979$$

```
> x := Vector([1,1]);
  y := Vector([1,0]);
```

$$x := \begin{bmatrix} 1 \\ 1 \end{bmatrix}, \quad y := \begin{bmatrix} 1 \\ 0 \end{bmatrix}$$

```
> P := Matrix([x,y]):
  cd := P^(-1).x0;
  c := cd[1]:
  d := cd[2]:
```

$$cd := \begin{bmatrix} 2 \\ -1 \end{bmatrix}$$

Now we can put these numbers and vectors into formula [6]:

$$x_n = r^n (c \cos(n\theta) + d \sin(n\theta)) x + r^n (-c \sin(n\theta) + d \cos(n\theta)) y$$

```
> xn := map(expand,Expand(r^n*(c*cos(n*theta)+d*sin(n*theta))*x +
  r^n*(-c*sin(n*theta)+d*cos(n*theta))*y));
```

$$xn := \begin{bmatrix} 1.03078^n \cos(.244979\,n) - 3 \cdot 1.03078^n \sin(.244979\,n) \\ 2 \cdot 1.03078^n \cos(.244979\,n) - 1.03078^n \sin(.244979\,n) \end{bmatrix}$$

This is more complicated than the corresponding formula for the components of the x_n's in Example 3D, but the two have a lot in common:

$$xn := \begin{bmatrix} 1.03078^n \cos(.244979\, n) \\ 1.03078^n \sin(.244979\, n) \end{bmatrix}$$

Again, the graphs are sinusoidal, and, since r and θ are unchanged, the trigonometric functions have the same periods and amplitudes as before.

```
> Componentplot(A,x0,points=40);
```

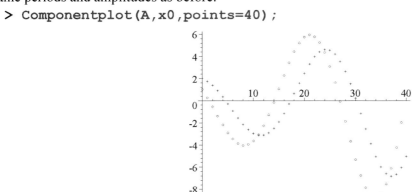

Exercise 4.2: (a) Find the polar coordinates of one of the eigenvalues of the matrix B below, and use them to predict the plots of the components of the iterates x_n as functions of n.
(b) Check your prediction by using **Componentplot** and the vector x_0 below.

```
> B := Matrix([[1,1/10],[-1/2,4/5]]):
  x0 := Vector([1,2]):
```

+ Student Workspace

 Answer 4.2

(a) We compute the eigenvalues of B and then find r, θ:

```
> Evectors(B);
```

$$\begin{bmatrix} \dfrac{9}{10}+\dfrac{1}{5}i \\ \dfrac{9}{10}-\dfrac{1}{5}i \end{bmatrix}, \begin{bmatrix} \dfrac{-1}{5}-\dfrac{2}{5}i & \dfrac{-1}{5}+\dfrac{2}{5}i \\ 1 & 1 \end{bmatrix}$$

```
> r := abs(.9+.2*i);
  theta := argument(.9+.2*i);
  evalf(2*Pi/theta);
```

$$r := .921954$$
$$\theta := .218669$$
$$28.7337$$

By formula [6], the plots will be sinusoidal. Since $r < 1$, their amplitudes will decay toward 0. The period of the cosine and sine terms will be the value of $\dfrac{2\pi}{\theta}$, which is approximately 29.

(b) Here is the plot of the two components of x_n as functions of n:

```
> Componentplot(B,x0,points=60);
```

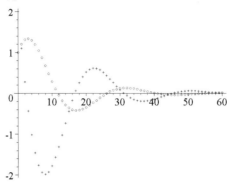

Section 5: A Leslie Model of Population

This application will give us an example of a 3 by 3 matrix with two complex eigenvalues and one real one, which will cause us to combine our above analysis of the iterates in the presence of complex eigenvalues with our earlier analysis of the iterates when the eigenvalues are real.

Consider an example of an animal population whose size is determined by characteristics of the female. (We assume there will always be enough males to propagate the species.) Suppose five-tenths of the newborns live to be one year old and four-tenths of the one-year-olds live to be two years old (and that none live beyond the age of three). Further, assume that on the average each one-year-old produces one female offspring and each two-year-old produces two female offspring. Then, if x_n is the 3-tuple that gives the number of females in each age group (newborn, one-year-old, and two-year-old), we have the discrete dynamical system $x_{n+1} = A\, x_n$, where A is given below:

```
> A := Matrix([[0,1,2],[.5,0,0],[0,.4,0]]);
```

$$A := \begin{bmatrix} 0 & 1 & 2 \\ .5 & 0 & 0 \\ 0 & .4 & 0 \end{bmatrix}$$

Such a system is called a *Leslie model* of populations by age group.

======================

Example 5A: Use the eigenvalues of A to determine the long-term trends for each of the above three age groups. Also, plot all three as functions of time, using the initial vector below.

```
> x0 := Vector([100,100,100]);
```

$$x0 := \begin{bmatrix} 100 \\ 100 \\ 100 \end{bmatrix}$$

Solution: We find the eigenvalues of A and use the magnitudes of the eigenvalues to give a rough description of the long-term trends:

```
> h := Evalues(A,method=numerical);
```

$$h := \begin{bmatrix} .957911 \\ -.478952 + .433804\,i \\ -.478952 - .433804\,i \end{bmatrix}$$

Maple note: The command **Evalues(A)** converts all entries in A to exact rationals before finding its eigenvalues. To instruct **Evalues** to skip this conversion and instead use an approximate method of solution, we added the option method=numerical.

```
> z := h[2];
```

$$z := -.478952 + .433804\,i$$

```
> r := abs(z);
  theta := argument(z);
  evalf(2*Pi/theta);
```

$$r := .646205$$
$$\theta := 2.40562$$
$$2.61187$$

The first component of the eigenvalues vector h is the real eigenvalue; the other two components are the complex eigenvalues. We found r and θ from the first of the complex eigenvalues, h_2.

Even though A has complex eigenvalues and eigenvectors, we can still write the eigenvector decomposition of x_n in the form (see Theorem 9 in Module 3)

$$x_n = a_1\,\lambda_1^{\,n}\,v_1 + a_2\,\lambda_2^{\,n}\,v_2 + a_3\,\lambda_3^{\,n}\,v_3$$

Since the real eigenvalue, λ_1, and both complex eigenvalues, λ_2 and λ_3, have absolute value less than 1, the iterates x_n tend to zero as n gets large. So, in the long run, all three age groups die out. Also, the period of the cosine and sine terms is about 2.6.

```
> Componentplot(A,x0,points=20,lines=on);
```

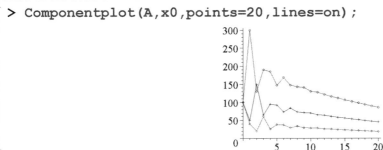

We see more than simply the fact that the three population age groups go to zero. We also see the effect of the trigonometric functions in the formula for the iterates: The populations fluctuate up and down and out of phase with one another for a while before settling down to a steady decrease. This is caused by the fact that $r < \lambda_1$, which tells us that the trigonometric terms (which have coefficient r^n) go to zero faster than the purely exponential terms (which have the larger coefficient λ_1^n).

=====================

Exercise 5.1: Change the (3, 2) entry in the Leslie matrix of Example 5A from 0.4 to 0.6. Find the eigenvalues of A, and from that information alone determine the long-term behavior of the three population age groups. Do they again all die out?

 Student Workspace

 Answer 5.1

```
> A := Matrix([[0,1,2],[.5,0,0],[0,.6,0]]);
```

$$A := \begin{bmatrix} 0 & 1 & 2 \\ .5 & 0 & 0 \\ 0 & .6 & 0 \end{bmatrix}$$

```
> h := Evalues(A,method=numerical);
```

$$h := \begin{bmatrix} 1.03820 \\ -.519105 + .555364\,i \\ -.519105 - .555364\,i \end{bmatrix}$$

```
> z := h[2];
  r := abs(z);
```

$$z := -.519105 + .555364\,i$$
$$r := .760197$$

So long as at least one eigenvalue is larger than one in absolute value, the iterates will not go to zero as n gets large. Here the real eigenvalue is larger than one. So none of the population groups die out. However, since the absolute value of the complex eigenvalue is less than 1, the trigonometric terms go to zero, which means that the graphs will eventually stop fluctuating up and down. Let's check:

```
> Componentplot(A,<100,100,100>,points=20,lines=on);
```

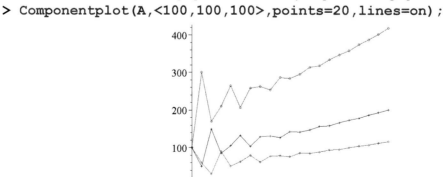

Problems

Problem 1: Find complex eigenvalues and eigenvectors

(a) For each of the matrices below, use the **Evectors** command to find the eigenvalues and a basis for each eigenspace. Write each eigenvector in the form $x + i y$, where x and y are real vectors.

(b) (By hand) Compute the characteristic polynomial for the matrix A below, and use it to find the eigenvalues of A. Also, solve the equation $A v = \lambda v$ for each eigenvalue λ, and check that your solutions include the eigenvectors in (a).

```
> A := Matrix([[1,5],[-2,3]]);
> B := Matrix([[-42,17,4],[-122,49,11],[22,-8,0]]);
```
+ Student Workspace

Problem 2: Rotation matrices

(By hand) For the general 2 by 2 rotation matrix $R = \begin{bmatrix} \cos(\theta) & -\sin(\theta) \\ \sin(\theta) & \cos(\theta) \end{bmatrix}$, compute the characteristic polynomial and use it to find the eigenvalues of R. (Assume θ is not a multiple of π.)

+ Student Workspace

Problem 3: Predict the trajectory

(a) For each of the matrices below, use formula [4] to find an invertible matrix P and a scale-rotation K so that $P K P^{(-1)}$ equals the given matrix.

(b) For each of the matrices below, find the polar coordinates of an eigenvalue and use them to predict the behavior of any trajectory of the iterates x_n. In particular, use r to predict whether the iterates are attracted to the origin or repelled by the origin or neither; use θ to predict the number of iterations required for one full revolution about the origin. (You may use the **Trajectory** command only to confirm your prediction.)

(c) For each of the matrices below, use r and θ to predict the behavior of the graphs of the components of the iterates x_n. (You may use the **Componentplot** command only to confirm your prediction.)

(d) For the matrix A and vector x_0, write out formula [6] for the iterates. (Formula [6] appears shortly before Example 4B.)

```
> A := Matrix([[1/5,1],[-2/5,3/5]]);
> B := Matrix([[1,-4/5],[4,-11/5]]);
> x0 := Vector([2,1]):
```
+ Student Workspace

Problem 4: Construct a matrix with complex eigenvalues

Construct a 2 by 2 matrix A with real entries whose eigenvalues are $2 + 3i$ and $2 - 3i$ with corresponding eigenvectors $\begin{bmatrix} 5 \\ 1 + 3i \end{bmatrix}$ and $\begin{bmatrix} 5 \\ 1 - 3i \end{bmatrix}$, respectively. (Hint: The matrix A cannot be a scale-rotation, since scale-rotations always have the eigenvectors $\begin{bmatrix} i \\ 1 \end{bmatrix}$ and $\begin{bmatrix} -i \\ 1 \end{bmatrix}$. Look instead for a matrix similar to a scale-rotation. Also, compare Problem 4 in Module 4.)

+ Student Workspace

Problem 5: Linear independence of the real and imaginary parts

Suppose A is a square matrix with real entries and that $x + iy$ (x and y real) is an eigenvector of A with a complex eigenvalue λ. Prove that $\{x, y\}$ is linearly independent. (Hint: If y is a multiple of x, show that x is itself a eigenvector of A with eigenvalue λ; then show that this conclusion is impossible by showing that a real vector cannot be an eigenvector associated with a complex eigenvalue. Similarly, if x is a multiple of y.)

Note: If $P = [x, y]$ is a 2 by 2 matrix whose columns are linearly independent, then P is invertible by Theorem 6 of Chapter 3. This proves an assertion in the derivation of formula [4] in Section 4.

+ Student Workspace

Problem 6: Dramatic changes in trajectories

The matrix A below is a Leslie matrix in which the (3, 2) entry is a parameter k. You will see that the trajectories of the iterates x_n can change dramatically when k changes.

```
> A := Matrix([[0,1,2],[.5,0,0],[0,k,0]]);
```

(a) Use Maple to confirm that the characteristic polynomial of A is

$$p(x) = -x^3 + .5\,x + k$$

(b) The animation below shows the graph of p as a function of x for $k = 0, .1, .2, ..., 1.0$. Run the animation, and determine the values of k (among the values $k = .0, .1, .2, ..., 1.0$) for which $p(x) = 0$ has three real roots and the values of k for which it has only one real root. Also determine the values of k for which all real roots of $p(x) = 0$ are less than 1 and the values of k for which one real root is larger than 1. (You may want to run the animation one frame at a time.)

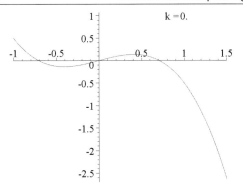

(c) Replace k by four different values that you found in (b), and apply the command **Trajectory3d(A,[x0],points=n)** for each of these values of k. Choose the first value of k so that the trajectory does not spiral but is attracted to the origin; choose the second value of k so that the trajectory spirals and is attracted to the origin; choose the third so that the trajectory stays on a plane but spirals in to a point other than the origin; and choose the fourth so that the trajectory spirals and is repelled by the origin. Use $x_0 = \langle 100, 100, 100 \rangle$.

(d) (Challenge) Find, by hand, the exact values of k for which the changes you observed in (b) take place.

+ Student Workspace

Problem 7: Spotted owl population dynamics

[Adapted from *Linear Algebra and Its Applications, 2nd ed.*, by David C. Lay; Addison-Wesley, 1996.] Recently, the population dynamics of the spotted owl have been studied in order to make decisions about the clear cutting of its habitat. We consider three components of the female population: juvenile females (denoted by j), subadult females (denoted by s), and adult females (denoted by a). The matrix A below is the transition matrix for the population vector $\langle j, s, a \rangle$.

```
> A := Matrix([[0,0,.33],[.18,0,0],[0,.71,.94]]);
```

That is, each year 0.33 juvenile females are produced on average by each adult female; 18% of the juvenile females survive to become subadult females; 71% of the subadult females survive to become adult females; and 94% of the adult females survive.

(a) Use the eigenvalues of A to predict what will happen to the three age groups in the long run.

(b) Plot the three components of x_n (i.e., the three age groups) as functions of n. Use $x_0 = \langle 100, 100, 100 \rangle$.

(c) Change the (2, 1) entry of A from 0.18 to 0.3. What changes in assumptions does this represent? What would the long-run consequences be for the three age groups under these conditions? In terms of the total female population of the spotted owl, what would the percentage of mature adult females be in the long run?

+ Student Workspace

Chapter 7: Orthogonality

Module 1. Orthogonal Vectors and Orthogonal Projections

Module 2. Orthogonal Bases

Module 3. Least-Squares Solutions

 Commands used in this chapter

Column(A,i); selects the ith column of the matrix A.
display([pict1, pict2]); displays together a group of previously defined pictures.
evalf(expr); evaluates the expression *expr* as a decimal approximation.
GramSchmidt([u1,u2,u3]); produces an orthogonal basis of the subspace spanned by $\{u_1, u_2, u_3\}$, using the Gram-Schmidt process.
map(simplify,M); applies the **simplify** command to every entry in the matrix or vector M.
Matrix([[a,b],[c,d]]); defines the matrix with rows $[a, b]$ and $[c, d]$.
plot(expr,x=a..b); plots a graph of a function of one variable.
pointplot([pt1,pt2,...,ptn]); plots the list of points $pt1, pt2, ..., ptn$, where each point is given as a list of two coordinates.
Row(A,i); selects the ith row of A.
Transpose(M); produces the transpose of the matrix M.
Vector([a,b,c]); defines the vector $\langle a, b, c \rangle$.

LAMP commands:
Dotprod(u,v); calculates the dot product of the vectors u and v.
Drawvec(u,[v,w]); draws the vector u with tail at the origin and the vector with tail at v and head at w. (Vector are in 2-space.)
Genmatrix([eqn1,eqn2],[x,y]); produces the augmented matrix for the linear system $[eqn1, eqn2]$ in the unknowns x and y.
Mag(u); calculates the magnitude (i.e., length) of the vector u.
Matsolve(A,b,free=t); solves the matrix-vector equation $Ax = b$ for x. The solution is expressed in terms of the free variable t.
Nullbasis(A); produces a basis for the null space of the matrix A.
Project(u,v); calculates the projection of the vector u onto the vector v.
Reflectmat(theta); produces the 2 by 2 matrix that reflects vectors across the line making the angle θ with the x axis.

`Residuals(f,data);` plots the points $[x, y]$ in *data* (a list of lists), the graph of the function f, and the residuals between the values of f and the data points.

`Rotatemat(theta);` produces the 2 by 2 matrix that rotates vectors by the angle θ.

`Vandermat(v,m,n);` produces the m by n matrix whose columns are the powers 0 through $n-1$ of the entries of the vector v.

Linear Algebra Modules Project
Chapter 7, Module 1

Orthogonal Vectors and Orthogonal Projections

 Purpose of this module

The purpose of this module is to introduce the following concepts and their basic properties: dot product, orthogonal vectors, orthogonal projection, magnitude, and distance in R^n. We use projections to solve a shortest distance problem that will be a key to an important area of applications in a later module.

 Prerequisites

Vector algebra, dot product of vectors in R^2 and R^3.

 Commands used in this module

```
> restart: with(LinearAlgebra): with(plots): with(Lamp):
  UseHardwareFloats := false: Digits := 6:
```

Tutorial

 Section 1: The Dot Product and Orthogonal Vectors in R^n

The definition of the dot product of two vectors in R^n is the natural generalization of the dot product for vectors in R^2 and R^3:

Definition: If $u = \langle u_1, ..., u_n \rangle$ and $v = \langle v_1, ..., v_n \rangle$ are vectors in R^n, then the **dot product** of u and v, denoted $u.v$, is the sum of the products of the corresponding components of u and v:

$$u.v = u_1 v_1 + \ldots + u_n v_n$$

====================

Example 1A : Let $u = \langle 2, 1, 4, -3 \rangle$ and $v = \langle 3, 2, 1, 3 \rangle$. Then

$$u.v = (2)(3) + (1)(2) + (4)(1) + (-3)(3) = 6 + 2 + 4 - 9 = 3$$

Using Maple:
```
> u := Vector([2,1,4,-3]);
  v := Vector([3,2,1,3]);
```

$$u := \begin{bmatrix} 2 \\ 1 \\ 4 \\ -3 \end{bmatrix}, \quad v := \begin{bmatrix} 3 \\ 2 \\ 1 \\ 3 \end{bmatrix}$$

```
> Dotprod(u,v);
                    3
```

Recall that in R^2 and R^3, two vectors are orthogonal (i.e., perpendicular) if and only if the dot product of the vectors equals 0. We define orthogonality in R^n similarly:

Definition: Vectors u and v in R^n are **orthogonal** if $u.v = 0$.

Note that the vectors u and v in Example 1A are not orthogonal since their dot product is 3, not 0.

Exercise 1.1: Change the last component of v (and no other components of v or u) so that $u.v = 0$.

+ Student Workspace

− Answer 1.1

```
> u := Vector([2,1,4,-3]):
> v := Vector([3,2,1,4]):
> Dotprod(u,v);
                    0
```

The dot product of two vectors is closely related to matrix multiplication. If we think of u and v as n by 1 matrices, then

$$u.v = u^T v$$

That is, $u.v$ equals the single entry in the 1 by 1 matrix product $u^T v$, as in:

$$[2 \; 1 \; 4 \; -3] \begin{bmatrix} 3 \\ 2 \\ 1 \\ 3 \end{bmatrix} = [3]$$

Length and Distance in R^n

Recall that for a vector $x = \langle a, b \rangle$ in R^2 we calculate the length of x, denoted $\|x\|$, by using the Pythagorean Theorem:

$$\|x\| = \sqrt{a^2 + b^2}$$

Alternatively, $\|x\|$ can be expressed in terms of a dot product, $\|x\| = \sqrt{x.x}$, since $x.x = a^2 + b^2$. We define the length of a vector in R^n similarly:

Definition: The **length** (or **magnitude**) of a vector $u = \langle u_1, ..., u_n \rangle$ in R^n is defined to be the square root of $u.u$:

$$\|u\| = \sqrt{u.u}$$

Note that if we square both sides of the above identity, we get $\|u\|^2 = u \cdot u$, a useful fact to remember.

The distance between any two vectors in R^n is defined as follows:

Definition: If u and v are vectors in R^n then the **distance between u and v** is the magnitude of the vector $u - v$, i.e., $\|u - v\|$.

Exercise 1.2: (By hand) Draw the vectors $u = \langle 1, -3 \rangle$ and $v = \langle 6, 4 \rangle$. Label the points P and Q corresponding to the heads of the vectors u and v. Now add the vector $u - v$ to your sketch, placing the tail of $u - v$ at the head of the vector v. Calculate the distance from P to Q in the following ways: (a) find the length of the vector $u - v$ using the dot product; (b) use the familiar distance formula $d = \sqrt{(x_1 - x_2)^2 + (y_1 - y_2)^2}$; and (c) use the command **Mag(u-v)**.

+ Student Workspace

− Answer 1.2

Here's a drawing:

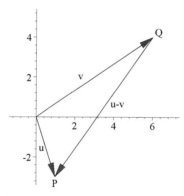

The length of $u - v$ by three methods:
```
> u := Vector([1,-3]):
  v := Vector([6,4]):
> sqrt(Dotprod(u-v,u-v));
  sqrt((1-6)^2 + (-3-4)^2);
  Mag(u-v);
```

$$\sqrt{74}, \ \sqrt{74}, \ \sqrt{74}$$

Rules of Algebra for the Dot Product

The dot product and length satisfy the following rules of algebra (where u, v, w denote vectors and c denotes a scalar):

$$u.v = v.u$$
$$u.(v + w) = u.v + u.w$$
$$(c\,u).v = u.(c\,v) = c\,(u.v)$$
$$||c\,u|| = |c|\,||u||$$

In fact, the second and third of these rules of algebra follow from the corresponding rules for matrix algebra, since the dot product can be treated as a matrix product.

The fundamental theorem relating orthogonality and length in R^2 is the Pythagorean Theorem. This theorem, in vector form, is also true in R^n:

Theorem 1: (The Pythagorean Theorem for R^n) If u and v are orthogonal vectors in R^n, then the square of $||u + v||$ equals the sum of the squares of $||u||$ and $||v||$; that is:

$$||u + v||^2 = ||u||^2 + ||v||^2$$

Exercise 1.3: (By hand) Draw the vectors $u = \langle 3, 1 \rangle$ and $v = \langle -2, 6 \rangle$. Check that these vectors are orthogonal. Add the vector $u + v$ to your sketch. Recall that for any vector x, $||x||^2 = x \cdot x$. Use this fact to check that $||u + v||^2 = ||u||^2 + ||v||^2$ for the given vectors.

+ Student Workspace

− Answer 1.3

```
> u := Vector([3,1]):
  v := Vector([-2,6]):
> Dotprod(u,v);
```
$$0$$

In the figure below, we have drawn an extra copy of the vector v with its tail at the head of u to make the right angle more noticable. The vector with the blue head is $u + v$.

```
> Drawvec([u,label="u"],[v,label="v"],
  [u+v,headcolor=blue],[u,u+v]);
```

```
> Dotprod(u+v,u+v);
```
$$50$$
```
> Dotprod(u,u) + Dotprod(v,v);
```
$$50$$

Exercise 1.4: Prove Theorem 1. Hint: Use the rules of algebra above to expand the right side of the identity

$$\|u+v\|^2 = (u+v).(u+v)$$

+ Student Workspace

− Answer 1.4

Since u and v are orthogonal, $u.v = 0$ and $v.u = 0$. So
$$\|u+v\|^2 = (u+v).(u+v) = u.u + v.u + u.v + v.v = u.u + v.v = \|u\|^2 + \|v\|^2$$

− Section 2: Projection of a Vector onto a Line

In this section we solve a geometry problem that will have significant applications. The problem is:

- Given a line through the origin and a point not on that line, find the shortest distance from the point to the line. Moreover, find the point on the line that yields this shortest distance.

======================

Example 2A: Given the point (1, 3) and the line $y = x/2$ (see figure below), find the point P on the line that is closest to (1, 3). That is, find P such that the distance from (1, 3) to P is shorter than the distance between (1, 3) and any other point on the line.

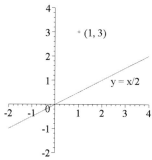

Solution: The point P will lie at the foot of the perpendicular drawn from (1, 3) to the given line. We will use vectors to find P. In the figure below, three vectors have been added to the picture. The vector $u = \langle 1, 3 \rangle$ has its head at our given point; the vector $v = \langle 4, 2 \rangle$ is a direction vector for the line $y = x/2$; and the blue vector with its head at P is the vector that we wish to calculate exactly. We will call this third vector p.

Also picture a fourth vector q going from the head of p (i.e., P) to the head of u (i.e., (1, 3)). Note that $q = u - p$ and also that q is orthogonal to v. Thus, $\|u - p\|$ gives us the shortest distance from (1, 3) to the line.

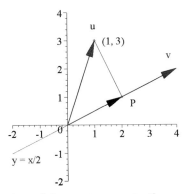

The blue vector p is called the *projection* of the vector u onto the vector v. Note that the projection vector p is parallel to the vector v and hence is a scalar multiple of v. Here is the standard formula for p:

$$p = \frac{u \cdot u}{v \cdot v} v \qquad [1]$$

We calculate p from formula [1]:
```
> u := Vector([1,3]):
  v := Vector([4,2]):
> p := Dotprod(u,v)/Dotprod(v,v)*v;
```
$$p := \begin{bmatrix} 2 \\ 1 \end{bmatrix}$$

So $P = (2, 1)$, and the distance from $(1, 3)$ to the line is $\|u - p\| = \sqrt{5}$:
```
> Mag(u-p);
```
$$\sqrt{5}$$

Note: Instead of choosing v to be $\langle 4, 2 \rangle$, we could have chosen v to be any nonzero vector along the line $y = x/2$.

========================

As a consequence of our method for finding the point on a line closest to a given point, we also obtain a vector $q = u - p$ that is orthogonal to the direction vector v and hence perpendicular to the line $y = x/2$:
```
> q := u-p;
  Dotprod(q,v);
```
$$q := \begin{bmatrix} -1 \\ 2 \end{bmatrix}$$
$$0$$

Exercise 2.1: For the vectors u and v below, find:
(a) the projection p of u onto v;
(b) the shortest distance from the head of u to the line containing v;

(c) a nonzero vector that is orthogonal to the line and lies in the plane containing both the line and the vector u.

```
> u := Vector([1,1,2]):
  v := Vector([-1,2,1]):
```

+ Student Workspace

− Answer 2.1

(a) The projection p is
```
> p := Dotprod(u,v)/Dotprod(v,v)*v;
```

$$p := \begin{bmatrix} -\frac{1}{2} \\ 1 \\ \frac{1}{2} \end{bmatrix}$$

(b) The distance is the length of $q = u - p$:
```
> q := u-p;
  Mag(q);
```

$$q := \begin{bmatrix} \frac{3}{2} \\ 0 \\ \frac{3}{2} \end{bmatrix}, \quad \frac{3\sqrt{2}}{2}$$

(c) The vector $q = u - p$ is orthogonal to v and lies in the plane containing u and p (i.e., u and v, since $p = k v$).
```
> Dotprod(q,v);
```
$$0$$

Exercise 2.2: Derive the formula $p = \dfrac{u \cdot u}{v \cdot v} v$ for the projection of u onto v by completing step (iii) of this argument:

(i) $p = k v$ since p is a multiple of v. (So our task is simply to find the scalar k.)
(ii) If p is the projection vector, then $u - p$ must be orthogonal to v. Therefore $(u - p).v = 0$, and hence $(u - k v).v = 0$.
(iii) Expand the final equation from step (ii) to solve for k.

+ Student Workspace

− Answer 2.2

$(u - k v).v = 0 \Rightarrow u.v - k v.v = 0 \Rightarrow u.v = k v.v \Rightarrow k = \dfrac{u \cdot u}{v \cdot v} \Rightarrow p = k v = \dfrac{u \cdot u}{v \cdot v} v.$

Maple note: For convenience we can use the command `Project(u,v)` to find the projection of any vector onto a second vector. For example, using the vectors *u* and *v* in Exercise 2.1, we have:

> `Dotprod(u,v)/Dotprod(v,v)*v;`

$$\begin{bmatrix} -\frac{1}{2} \\ 1 \\ \frac{1}{2} \end{bmatrix}$$

> `Project(u,v);`

$$\begin{bmatrix} -\frac{1}{2} \\ 1 \\ \frac{1}{2} \end{bmatrix}$$

Section 3: Projection of a Vector onto a Subspace

In Section 2 we saw how to find the shortest distance from a point to a line by projecting the point onto the line. In this section we introduce the more general idea of projection onto any subspace of R^n, which we will use to solve the following more general shortest distance problem:

- Given a subspace W of R^n and a point Q not in W, find the shortest distance from Q to W. Moreover, find the point in W that yields this shortest distance.

Before considering this general problem, we consider a more concrete version:

- Given a plane through the origin and a point not on that plane, find the shortest distance from the point to the plane. Moreover, find the point on the plane that yields this shortest distance.

========================

Example 3A: Let *u* be the vector given below, and let W be the plane spanned by the vectors v_1 and v_2 below. Find the point P on W closest to the point Q at the head of *u*. That is, find the projection vector *p* of *u* on W (so that P will be the head of *p*).

Note: This problem is much easier to solve if the spanning vectors for the plane are orthogonal; so we chose v_1 and v_2 that way.

> `u := Vector([-4,5,6]):`
> `v1 := Vector([-3,1,-1]):`
> `v2 := Vector([1,4,1]):`

> `Dotprod(v1,v2);`

$$0$$

Solution: The point P will lie at the foot of the perpendicular drawn from Q to the plane W. The figure below shows the two vectors v_1 and v_2 that span W, the vector u from the origin O to the point Q, and the vector p from O to P. The vector from P to Q (indicated by the vertical black line segment) is $u - p$.

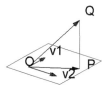

The key properties that p must have are therefore:

(i) p is in W and (ii) $u - p$ is orthogonal to every vector in W

To find the projection vector p, we cannot simply use the projection formula [1] in Section 2, since that gives us the projection onto a line, not a plane. Instead, formula [2] below does the job. Note that the formula depends on the vectors v_1 and v_2 being orthogonal.

- If W is a plane through the origin in R^3 and v_1, v_2 are orthogonal basis vectors for W, then the projection of any vector u onto W is the sum of the projections of u onto v_1 and u onto v_2:

$$p = \text{Project}(u, v_1) + \text{Project}(u, v_2) \qquad [2]$$

Calculate the projection vector using formula [2]:

```
> p := Project(u,v1)+Project(u,v2);
```

$$p := \begin{bmatrix} \dfrac{-16}{9} \\ \dfrac{53}{9} \\ \dfrac{2}{9} \end{bmatrix}$$

Thus $(-\dfrac{16}{9}, \dfrac{53}{9}, \dfrac{2}{9})$ is the point in W closest to the head of u.

Let's also check the defining properties (i) and (ii) of the projection p. In formula [2], the term Project(u, v_1) is a multiple of v_1 and hence is in W; similarly, Project(u, v_2) is in W. Therefore their sum p is in W, since W is a subspace. Furthermore, we can check that $u - p$ is orthogonal to both v_1 and v_2, which implies that $u - p$ is orthogonal to every vector in the span of v_1 and v_2, which is W:

```
> Dotprod(u-p,v1);
  Dotprod(u-p,v2);
```

$$0$$
$$0$$

Exercise 3.1: Verify formula [2] above for the projection p of a vector u onto a plane spanned by orthogonal vectors v_1, v_2. Since we have already explained why property (i) holds (i.e., why p lies in the plane), you need only verify property (ii) (i.e., show that $u - p$ is orthogonal to v_1 and v_2).

+ Student Workspace

− Answer 3.1

To show that $(u - p) \cdot (v_1) = 0$, first distribute the product:
$$(u - p) \cdot (v_1) = u \cdot (v_1) - p \cdot (v_1)$$
Substitute formula [2] for p and distribute again:
$$(u - p) \cdot (v_1) = u \cdot (v_1) - \frac{u \cdot (v_1)}{(v_1) \cdot (v_1)} (v_1) \cdot (v_1) - \frac{u \cdot (v_2)}{(v_2) \cdot (v_2)} (v_2) \cdot (v_1)$$
Simplify, using the fact that $(v_2) \cdot (v_1) = 0$:
$$(u - p) \cdot (v_1) = u \cdot (v_1) - u \cdot (v_1) = 0$$
Similarly, $(u - p) \cdot (v_2) = 0$.

Exercise 3.2: (a) Find the shortest distance from the head of u to the plane W in Example 3A. (b) Find a nonzero vector that is orthogonal to W.

+ Student Workspace

− Answer 3.2

(a) The shortest distance is the length of the vector $q = u - p$:
```
> Mag(u-p);
```
$$\frac{4}{3}\sqrt{22}$$

(b) The vector $q = u - p$ is orthogonal to v_1 and v_2 and hence to every vector in W.
```
> q := u-p;
```
$$q := \begin{bmatrix} \frac{-20}{9} \\ \frac{-8}{9} \\ \frac{52}{9} \end{bmatrix}$$

Now we return to the general problem posed at the beginning of the section, which requires us to define what we mean by a projection of a vector onto a subspace of R^n. The idea of the definition is explicit in Example 3A, where we required that the projection p of the vector u must lie in the plane W and must have the property that $u - p$ is orthogonal to every vector in the plane. Thus:

Definition: If u is any vector in R^n and W is any subspace of R^n, the **projection of u onto W** is the vector p that has the following two properties:

(i) p is in W and (ii) $u - p$ is orthogonal to every vector in W.

In the next module, we will see why the projection p defined above always exists and is uniquely determined by u and W; in the process, we will derive a general formula for p.

The following theorem tells us that the projection vector p solves the general problem posed at the beginning of this section.

Theorem 2: If u is any vector in R^n and W is any nonzero subspace of R^n, then the projection p of u onto W is the vector in W that is closest to u. That is, $\|u - p\| < \|u - w\|$ for every vector w in W ($w \neq p$).

Exercise 3.3: Prove Theorem 2 by completing this argument:
(i) Write the vector $u - w$ as the sum of two vectors: $u - w = (u - p) + (p - w)$.
(ii) Explain why the two vectors in the above sum, $u - p$ and $p - w$, are orthogonal to each other.
(iii) Apply Theorem 1 (Pythagorean Theorem) to the three vectors $u - w$, $u - p$, $p - w$, and deduce the conclusion of Theorem 2 from that formula.

 Student Workspace

Answer 3.3

(ii) Since p and w are in W, so is $p - w$. Since $u - p$ is orthogonal to every vector in W, $u - p$ and $p - w$ are orthogonal.
(iii) By Theorem 1, $\|u - w\|^2 = \|u - p\|^2 + \|p - w\|^2$. Since $w \neq p$, $0 < \|p - w\|^2$. Hence $\|u - p\|^2 < \|u - w\|^2$ and therefore $\|u - p\| < \|u - w\|$.

Problems

Problem 1: Distance and angle in the hypercube

The hypercube is the set of points (x_1, x_2, x_3, x_4) in R^4 such that x_1, x_2, x_3, x_4 all range between 0 and 1. So we can think of the hypercube as analogous to the solid unit cube in R^3.

(a) Find the length of longest diagonal of the hypercube, say from the vertex $(0, 0, 0, 0)$ to the vertex

(1, 1, 1, 1).
(b) Find the angle between this longest diagonal and any of the edges of the cube, say the edge from (0, 0, 0, 0) to (1, 0, 0, 0). Recall that the angle θ between two vectors u and v is given by

$$\|u\|\|v\|\cos(\theta) = u \cdot v$$

+ Student Workspace

Problem 2: Estimate distance from a drawing

On graph paper, draw the vectors $u = \langle 0, -2 \rangle$ and $v = \langle 2, 1 \rangle$.
(a) Without making any calculation, draw the projection of u onto v and estimate the distance from the head of u to the line containing v.
(b) Check your answer in (a) by calculating the projection and the distance.

+ Student Workspace

Problem 3: Project a vector on a line

(a) For the vectors x and y below, use the projection of x onto y to find:
(i) the shortest distance from the head of x to the line containing y;
(ii) a nonzero vector q that is orthogonal to the line and lies in the plane containing both the line and the vector x.

```
> x := Vector([2,3,-1]);
  y := Vector([0,-4,2]);
```

(b) Replace x by z, where z is defined below. (z is orthogonal to y.) Answer (i) without doing any computation. Explain.
(c) Replace x by w, where w is defined below. (w is parallel to y.) Answer (i) without doing any computation. Explain.

```
> z := Vector([-2,1,2]);
> w := Vector([0,6,-3]);
```

+ Student Workspace

Problem 4: Project a vector on a plane

(a) For the vector $u = \langle 7, 7, 7 \rangle$ and the plane W given by $2x + y + 3z = 0$, use the projection of u onto W to find:
(i) the distance from the head of u to the plane W;
(ii) a nonzero vector q orthogonal to every vector in W.
Hint: You will have to find a pair of orthogonal vectors v_1, v_2 that span W. First find any two spanning vectors, and then use the method in Section 2 to construct an orthogonal pair from your spanning vectors.
(b) Note that your vector q is a scalar multiple of the vector $\langle 2, 1, 3 \rangle$ formed by the coefficients of x, y, z

in the equation of the plane. Why is this true?

Student Workspace

Problem 5: Construct an orthogonal basis of R^3

Use projections to find a basis $\{v_1, v_2, v_3\}$ of R^3, where $v_1 = \langle 2, 5, -4 \rangle$ and each of the vectors v_1, v_2, v_3 is orthogonal to the other two. (Such a basis is called an "orthogonal basis" of R^3.) Hint: Notice that the set of vectors $\{v_1, v_2, q\}$ in Problem 4 is an orthogonal basis of R^3; so similar techniques should work.

Student Workspace

Problem 6: Projection transformations

If v is any nonzero vector in R^n, then the matrix P that projects every vector in R^n onto the line through the origin with direction vector v is given by the formula $P = \dfrac{v\,v^T}{v^T v}$. Here's the explanation: For any vector u in R^n, its projection $P\,u$ onto the line through the origin with direction vector v is $\dfrac{u \cdot v}{v \cdot v} v$.

Rewrite this formula using the fact that $u.v = v.u = v^T u$ and $v.v = v^T v$. Also, put the scalar $\dfrac{u \cdot v}{v \cdot v}$ on the right of the vector v. Thus, $P\,u = \dfrac{v\,v^T u}{v^T v}$, and therefore the matrix P is the expression that is multiplied by u: $\dfrac{v\,v^T}{v^T v}$.

(a) Use the above formula to find the 2 by 2 matrix that projects every vector in R^2 onto the line $y = x$. (Do not use angles. However, you may check your answer against the matrix **Projectmat(Pi/4)**.)

(b) Use the above formula to find the 3 by 3 matrix that projects every vector in R^3 onto the line through the origin with direction vector $\langle 2, 1, -3 \rangle$.

(c) By a derivation similar to the one above, find the 3 by 3 matrix that projects every vector in R^3 onto the plane with normal vector $\langle 2, 1, -3 \rangle$. (Hint: If P is the desired projection matrix and u is an arbitrary vector in R^3, then $P\,u = u - p$, where p is the projection of u onto the normal vector.) You may check your answer against the matrix **Projectmat3d(<2,1,-3>)**.

Student Workspace

Linear Algebra Modules Project
Chapter 7, Module 2

Orthogonal Bases

Purpose of this module

The purpose of this module is to see how to construct and make use of an orthogonal basis of a subspace. In particular, we see how to find the projection of any vector on any subspace and thus find the shortest distance from the vector to the subspace. We also use orthogonal bases in developing the concepts of orthogonal complement and orthogonal matrix.

Prerequisites

Dot product of vectors, orthogonal vectors, length of a vector, projection of one vector on another; basis of a subspace; null space and row space of a matrix; matrix transformations.

Commands used in this module

```
> restart; with(LinearAlgebra): with(plots): with(Lamp):
  UseHardwareFloats := false: Digits := 6:
```

Tutorial

Section 1: Definition and Properties of Orthogonal Bases

We begin with two definitions:

Definition: A set of vectors $\{v_1, ..., v_k\}$ in R^n is called an ***orthogonal set*** if every pair of vectors in the set is orthogonal.

Definition: An ***orthogonal basis*** of a subspace S is a basis of S which is also an orthogonal set. That is, every pair of vectors in the basis is orthogonal.

The standard basis of R^n is an orthogonal basis of R^n. We confirm this for the standard basis of R^3:

```
> e1 := Vector([1,0,0]):
  e2 := Vector([0,1,0]):
  e3 := Vector([0,0,1]):
> Dotprod(e1,e2), Dotprod(e1,e3), Dotprod(e2,e3);
```
$$0, 0, 0$$

Exercise 1.1: Find an orthogonal basis of the subspace of R^4 in which two of the vectors are $\langle 1, 1, 1, 0 \rangle$ and $\langle 1, -1, 0, 0 \rangle$. (This can be done without any computation.)

 Student Workspace

Answer 1.1

One possible answer is $\{\langle 1, 1, 1, 0 \rangle, \langle 1, -1, 0, 0 \rangle, \langle 1, 1, -2, 0 \rangle, \langle 0, 0, 0, 1 \rangle\}$.

The following theorem provides simple formulas for computing the coordinates of a vector with respect to an orthogonal basis.

Theorem 3: If $\{v_1, \ldots, v_k\}$ is an orthogonal basis of a subspace S of R^n and u is any vector in S, then:

$$u = c_1 v_1 + \ldots + c_k v_k$$

where

$$c_1 = u.v_1/v_1.v_1, \quad c_2 = u.v_2/v_2.v_2, \quad \ldots, \quad c_k = u.v_k/v_k.v_k$$

Proof: Since u is in the span of $\{v_1, \ldots, v_k\}$, it is a linear combination of $v_1, \ldots, v_k$:

$$u = c_1 v_1 + \ldots + c_k v_k \quad [1]$$

Take the dot product of each side of equation [1] with v_1:

$$u.v_1 = c_1 v_1.v_1 + c_2 v_2.v_1 + \ldots + c_k v_k.v_1$$

Since $\{v_1, \ldots, v_k\}$ is an orthogonal set, every term on the right side is zero, except for the first term. Solving $u.v_1 = c_1 v_1.v_1$ for c_1, we get $c_1 = u.v_1/v_1.v_1$. Similarly, by taking the dot product of each side of equation [1] with $v_2, \ldots, v_k$, we derive corresponding formulas for $c_2, \ldots, c_k$.

Definition: An **orthonormal basis** of a subspace S is an orthogonal basis of S in which all of the basis vectors are unit vectors.

If we apply Theorem 3 to a subspace S that has an orthonormal basis $\{v_1, \ldots, v_k\}$, then the formulas for the coordinates of a vector u in S relative to this basis are even simpler, since the dot products in the denominators are all equal to 1. Hence we have:

$$u = c_1 v_1 + \ldots + c_k v_k$$

where $c_1 = u.v_1, c_2 = u.v_2, \ldots, c_k = u.v_k$.

Here is a useful theorem that we can see is plausible by visualizing, say, three vectors in R^3:

Theorem 4: If $\{v_1, \ldots, v_k\}$ is an orthogonal set of nonzero vectors in R^n, then $\{v_1, \ldots, v_k\}$ is a linearly independent set of vectors.

Exercise 1.2: Prove Theorem 4. Hint: You must show that the only solution to
$$0 = c_1 v_1 + \ldots + c_k v_k$$
is the trivial solution in which all the coefficients are zero. Take dot product of both sides of this equation with v_1.

[+] Student Workspace

[−] Answer 1.2

Taking the dot product of both sides of $0 = c_1 v_1 + \ldots + c_k v_k$ with v_1, we get $0 = c_1 v_1 \cdot v_1$. Therefore, after dividing by $v_1 \cdot v_1$, we have $c_1 = 0$. Similarly, taking the dot product of both sides with $v_2, \ldots, v_k$ leads to the conclusion that $c_2 = c_3 = \ldots = c_k = 0$.

[−] **Section 2: Orthogonal Projections onto Orthogonal Bases**

In this section we will use projections onto an orthogonal basis to solve the following problem:

- Given a subspace W of R^n and a point Q not in W, find the shortest distance from Q to W. Moreover, find the point in W that yields this shortest distance.

According to Theorem 2 (in Module 1), we can solve this problem by finding the projection P of Q onto W. The shortest distance from Q to W is simply the distance between Q and P. Furthermore, in Example 3A of Module 1, we saw that if W has dimension 2, we can find the projection of any vector u onto W by adding the projections of u onto an orthogonal basis of W. Let's review that example briefly:

========================

Example 2A: Let u be the vector given below, and let W be the plane spanned by the vectors v_1 and v_2 below. Find the point P in W closest to the point Q at the head of u.

```
> u := Vector([-4,5,6]):
  v1 := Vector([-3,1,-1]):
  v2 := Vector([1,4,1]):
```

Solution: The figure below shows the two vectors v_1 and v_2 that span W, the vector u (from O to Q), and the projection vector p (from O to P) of the vector u onto W.

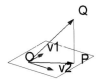

Since p is the projection of u onto W, the vector $q = u - p$ is perpendicular to every vector in W and therefore to v_1 and v_2. We chose v_1, v_2 to be <u>orthogonal</u>, which turns out to be crucial in finding the projection of u onto the plane W; we simply add the projections of u onto v_1 and u onto v_2:

$$p = \text{Project}(u, v_1) + \text{Project}(u, v_2)$$

```
> p := Project(u,v1)+Project(u,v2);
```

$$p := \begin{bmatrix} \dfrac{-16}{9} \\ \dfrac{53}{9} \\ \dfrac{2}{9} \end{bmatrix}$$

Thus $(-\dfrac{16}{9}, \dfrac{53}{9}, \dfrac{2}{9})$ is the point in W closest to the head of u.

==================

The next theorem generalizes the above result, that the projection p of u is the sum of the projections of u onto v_1 and v_2:

Theorem 5: If $\{v_1, ..., v_k\}$ is an orthogonal basis of a subspace W of R^n and u is any vector in R^n, then the projection p of u onto W is the sum of the projections of p onto $v_1, ..., v_k$:

$$p = \text{Project}(u, v_1) + ... + \text{Project}(u, v_k)$$

To prove this result, we must show that the above formula for p as a sum of projections has the two properties of a projection:

(i) p is in W and   (ii) $u - p$ is orthogonal to every vector in W.

Since each of the projections is in W, their sum is in W. So we need only verify property (ii), which is left to Problem 6.

Here's an example of the above method:

==================

Example 2B: Find the projection p of u onto the subspace W with the <u>orthogonal</u> basis $\{v_1, v_2, v_3\}$ (see below). Also find the shortest distance from the head of u to W.

```
> u := Vector([1,0,-2,1,3]):
> v1 := Vector([0,1,-1,0,0]):
  v2 := Vector([2,0,0,1,0]):
  v3 := Vector([1,1,1,-2,3]):
```

Solution: First we check that $\{v_1, v_2, v_3\}$ is indeed an orthogonal set:

```
> Dotprod(v1,v2), Dotprod(v1,v3), Dotprod(v2,v3);
```

$$0, 0, 0$$

We compute the sum of the projections of u onto v_1, v_2, v_3 to find p; then $\|u-p\|$ is the shortest distance:
```
> p := Project(u,v1)+Project(u,v2)+Project(u,v3);
```

$$p := \begin{bmatrix} \frac{63}{40} \\ \frac{11}{8} \\ \frac{-5}{8} \\ \frac{-3}{20} \\ \frac{9}{8} \end{bmatrix}$$

```
> q := u-p:
  Mag(q);
```

$$\frac{1}{10}\sqrt{895}$$

So the distance from the head of u to W is $\sqrt{895}/10$. We can also check that q is orthogonal to W:
```
> Dotprod(q,v1), Dotprod(q,v2), Dotprod(q,v3);
```

$$0, 0, 0$$

Section 3: Gram-Schmidt Process

The *Gram-Schmidt process* is a method for constructing an orthogonal basis of a subspace from any given basis. That is, suppose we are given a basis $\{u_1, ..., u_k\}$ of a nonzero subspace S of R^n. We will use this basis to construct an orthogonal basis of S. Let's look at an example first.

Example 3A: The vectors u_1, u_2 below span a two-dimensional subspace S of R^3.
```
> u1 := Vector([1,1,2]):
> u2 := Vector([-1,2,1]):
```
We will construct an orthogonal basis of S by again using a projection. We choose u_1 itself to be the first vector in our orthogonal basis. Then we find the projection p of u_2 onto u_1 and the difference vector $q = u_2 - p$. Thus $\{u_1, q\}$ is an orthogonal basis of S, since the vector $q = u_2 - p$ is orthogonal to u_1 and is a linear combination of vectors in S.
```
> p := Project(u2,u1);
```

$$p := \begin{bmatrix} \frac{1}{2} \\ \frac{1}{2} \\ 1 \end{bmatrix}$$

```
> q := u2-p;
```

$$q := \begin{bmatrix} -\frac{3}{2} \\ \frac{3}{2} \\ 0 \end{bmatrix}$$

Let's check that u_1 and q are indeed orthogonal:

```
> Dotprod(u1,q);
```

$$0$$

Here is a picture that shows both the original basis $\{u_1, u_2\}$ and the orthogonal basis $\{u_1, q\}$:

===================

Now suppose we have a 3-dimensional subspace S of R^n and a basis $\{u_1, u_2, u_3\}$ of S. To construct an orthogonal basis $\{v_1, v_2, v_3\}$ of S, we start just as we did in Example 3A: We choose v_1 to be u_1 and choose v_2 to be $u_2 - p_1$, where p_1 is the projection of u_2 onto v_1. To find v_3, we project u_3 onto the subspace spanned by the orthogonal vectors v_1 and v_2. That is, we construct

$$p_2 = \text{Project}(u_3, v_1) + \text{Project}(u_3, v_2)$$

We then choose v_3 to be $u_3 - p_2$. From the defining property (ii) of a projection, $u_3 - p_2$ is orthogonal to v_1 and v_2.

Exercise 3.1: The vectors u_1, u_2, u_3 below span a three-dimensional subspace S of R^4. Use the above method to find an orthogonal basis $\{v_1, v_2, v_3\}$ for S.

```
> u1 := Vector([1,1,2,0]):
  u2 := Vector([-1,2,1,0]):
  u3 := Vector([1,0,0,1]):
```

 Student Workspace

Answer 3.1

Let's denote the orthogonal basis that we will construct $\{v_1, v_2, v_3\}$. So

```
> v1 := u1;
```
$$v1 := \begin{bmatrix} 1 \\ 1 \\ 2 \\ 0 \end{bmatrix}$$

```
> v2 := u2-Project(u2,v1);
```
$$v2 := \begin{bmatrix} -\frac{3}{2} \\ \frac{3}{2} \\ 0 \\ 0 \end{bmatrix}$$

The projection of u_3 onto the plane Span$\{v_1, v_2\}$ is
```
> p := Project(u3,v1) + Project(u3,v2);
```
$$p := \begin{bmatrix} \frac{2}{3} \\ -\frac{1}{3} \\ \frac{1}{3} \\ 0 \end{bmatrix}$$

Therefore v_3 is
```
> v3 := u3-p;
```
$$v3 := \begin{bmatrix} \frac{1}{3} \\ \frac{1}{3} \\ -\frac{1}{3} \\ 1 \end{bmatrix}$$

Let's check our answer by computing the dot products of our three new basis vectors:
```
> Dotprod(v1,v2), Dotprod(v1,v3), Dotprod(v2,v3);
```
$$0, 0, 0$$

Furthermore, the vectors v_1, v_2, v_3 are each a linear combination of $\{u_1, u_2, u_3\}$ and hence in S. Also, since they are orthogonal and nonzero, they are linearly independent and hence a basis of S.

Here is a step-by-step description of the Gram-Schmidt process for converting any basis $\{u_1, ..., u_k\}$ of a subspace S of R^n to an orthogonal basis $\{v_1, ..., v_k\}$ of S:

1. Choose v_1 to be u_1;
2. choose v_2 to be u_2 - Project(u_2, v_1);
3. choose v_3 to be u_3 - (Project(u_3, v_1) + Project(u_3, v_2));

and so on. That is, at each step, the new vector v_j is chosen to be u_j minus the projection of u_j on the subspace spanned by the preceding basis vectors. This projection is found by adding the projections of u_j onto each of the preceding basis vectors, provided we project u_j onto <u>orthogonal</u> basis vectors.

In particular, we have proved the following theorem:

Theorem 6: Every nonzero subspace of R^n has an orthogonal basis.

The **GramSchmidt** command finds an orthogonal basis; here it is applied to the vectors in Exercise 3.1:

```
> GramSchmidt([u1,u2,u3]);
```

$$\left[\begin{bmatrix}1\\1\\2\\0\end{bmatrix}, \begin{bmatrix}-\frac{3}{2}\\\frac{3}{2}\\0\\0\end{bmatrix}, \begin{bmatrix}\frac{1}{3}\\\frac{1}{3}\\-\frac{1}{3}\\1\end{bmatrix}\right]$$

Section 4: Orthogonal Complements

Definition: Suppose S is a subspace of R^n. The ***orthogonal complement*** of S in R^n is the set of all vectors v in R^n such that $v.u = 0$ for all u in S, that is, the set of all vectors that are orthogonal to every vector in S.

=====================

Example 4A: The orthogonal complement of the line $y = 2x$ in R^2 is $y = -x/2$, the perpendicular line through the origin.

=========================

Exercise 4.1: What is the orthogonal complement of the plane $2x + y - 3z = 0$ in R^3? (No computation is needed.)

 Student Workspace

 Answer 4.1

The vectors that are orthogonal to every vector in the plane are the multiples of the normal vector $\langle 2, 1, -3 \rangle$. So the orthogonal complement of the plane is the span of this vector, which is the line perpendicular to the plane. Note that the dimensions of the plane and the orthogonal line add up to 3, the dimension of R^3.

Exercise 4.2: Find the orthogonal complement of the orthogonal complement you found in Exercise 4.1.

 Student Workspace

 Answer 4.2

The vectors that are orthogonal to the line you found in Exercise 4.1 are the vectors in the orthogonal plane, which is the plane we started with, $2x + y - 3z = 0$. More generally, the orthogonal complement of the orthogonal complement of a subspace S is always S itself.

As you may have observed in Exercise 4.1, the dimensions of the given subspace and its orthogonal complement add up to the dimension of the entire space. This is true in general:

Theorem 7: If S_1 is a subspace of R^n and S_2 is its orthogonal complement in R^n, then
$$\text{dimension}(S_1) + \text{dimension}(S_2) = n$$

The idea of the proof is to select an orthogonal basis for S_1 and an orthogonal basis for S_2, and then show that the union of the two bases is a basis for R^n.

Null Space and Row Space Revisited

The concept of orthogonal complement gives us additional geometric insight into the relationship between the null space and row space of a matrix. The following example will give the idea.

==================

Example 4B: For the matrix A below, find a basis for the null space and check that each null space basis vector is orthogonal to each row vector:

```
> A := Matrix([[0,1,-1],[-1,-4,2]]);
```

$$A := \begin{bmatrix} 0 & 1 & -1 \\ -1 & -4 & 2 \end{bmatrix}$$

Solution:

```
> N := Nullbasis(A);
  N1 := N[1]:
```

$$N := \left[\begin{bmatrix} -2 \\ 1 \\ 1 \end{bmatrix} \right]$$

```
> r1 := Row(A,1);
> r2 := Row(A,2);
```

$$r1 := [0, 1, -1]$$
$$r2 := [-1, -4, 2]$$

```
> Dotprod(r1,N1), Dotprod(r2,N1);
```

$$0, 0$$

==================

432 Chapter 7 Orthogonality

This orthogonality of the vectors is not a coincidence. A vector x is in the null space of a matrix A if and only if $A\,x = 0$. Since $A\,x$ is computed by taking the dot product of each row of A with x, we see that x is in the null space of A if and only if x is orthogonal to each row of A. This argument proves the following theorem:

> **Theorem 8:** The orthogonal complement of the row space of a matrix is its null space.

Note that we now have a new proof of Theorem 9 of Chapter 5, which says that the dimension of the null space of A is $n - r$, where n is the number of columns of A and r is the rank of A. Proof: The dimension of the row space of A is r, and so by Theorem 7 the dimension of its orthogonal complement is $n - r$. By Theorem 8 this orthogonal complement is the null space of A.

Section 5: Orthogonal Matrices

Every rotation matrix preserves the angle between any two vectors and preserves the length of any vector. For example, the vectors u and v below are orthogonal, and the rotation by the angle $\pi/3$ maps u and v to output vectors that are also orthogonal and whose lengths are the same as u and v, respectively. We confirm these facts both algebraically and geometrically:

```
> u := Vector([1,0]);
  v := Vector([0,2]);
```

$$u := \begin{bmatrix} 1 \\ 0 \end{bmatrix},\ v := \begin{bmatrix} 0 \\ 2 \end{bmatrix}$$

```
> R := Rotatemat(Pi/3);
```

$$R := \begin{bmatrix} \dfrac{1}{2} & -\dfrac{1}{2}\sqrt{3} \\ \dfrac{1}{2}\sqrt{3} & \dfrac{1}{2} \end{bmatrix}$$

```
> Dotprod(R.u,R.v);
```

$$0$$

```
> Mag(R.u), Mag(R.v);
```

$$1,\ 2$$

```
> Drawvec(u,v,[R.u,headcolor=blue,thickness=3],
    [R.v,headcolor=blue,thickness=3]);
```

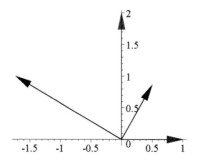

Reflection matrices also preserve angle and length. Another interesting (and, as we will see, related) property of rotation and reflection matrices is that the product of the matrix and its transpose is the identity matrix:

```
> R := Rotatemat(theta);
> RTR := Transpose(R).R;
```

$$R := \begin{bmatrix} \cos(\theta) & -\sin(\theta) \\ \sin(\theta) & \cos(\theta) \end{bmatrix}$$

$$RTR := \begin{bmatrix} \cos(\theta)^2 + \sin(\theta)^2 & 0 \\ 0 & \cos(\theta)^2 + \sin(\theta)^2 \end{bmatrix}$$

```
> map(simplify,RTR);
```

$$\begin{bmatrix} 1 & 0 \\ 0 & 1 \end{bmatrix}$$

```
> S := Reflectmat(theta);
> STS := Transpose(S).S;
```

$$S := \begin{bmatrix} \cos(2\theta) & \sin(2\theta) \\ \sin(2\theta) & -\cos(2\theta) \end{bmatrix}$$

$$STS := \begin{bmatrix} \cos(2\theta)^2 + \sin(2\theta)^2 & 0 \\ 0 & \cos(2\theta)^2 + \sin(2\theta)^2 \end{bmatrix}$$

```
> map(simplify,STS);
```

$$\begin{bmatrix} 1 & 0 \\ 0 & 1 \end{bmatrix}$$

These properties hold for a large class of interesting matrix transformations, which are defined as follows:

Definition: A square matrix U is **orthogonal** if $U^T U = I$.

Another way to state this definition is that $U^{(-1)} = U^T$ or, equivalently, $U U^T = I$. The following exercise will indicate yet another way to restate the definition of orthogonality.

Exercise 5.1: (a) (By hand) Show that the columns of the matrix R that rotates vectors through the angle $\pi/3$ (see below) are orthogonal and have length 1.
(b) (By hand) With the help of part (a), show that $R^T R = I$.

$$R = \begin{bmatrix} \dfrac{1}{2} & -\dfrac{1\sqrt{3}}{2} \\ \dfrac{1\sqrt{3}}{2} & \dfrac{1}{2} \end{bmatrix}$$

+ Student Workspace

Answer 5.1

(a) Length of column 1: $\frac{1}{4} + \frac{3}{4} = 1$. Length of column 2: $\frac{3}{4} + \frac{1}{4} = 1$. Dot product of the columns: $-\frac{\sqrt{3}}{4} + \frac{\sqrt{3}}{4} = 0$.

(b) The (1, 1) entry of $R^T R$ is the dot product of the first column of R with itself, which we saw in (a) is 1; the (1, 2) entry of $R^T R$ is the dot product of the first and second columns of R, which we saw in (a) is 0; and so on.

From Exercise 5.1 we see that the statement $U^T U = I$ is equivalent to the statement that the columns of U form an orthogonal set of unit vectors; that is, the columns form an orthonormal basis of R^n.

Yet another consequence of the definition of orthogonality is that an orthogonal matrix preserves dot products. Here is the proof, which uses the fact that $u.v = u^T v$:

$$U x . U y = (U x)^T U y = x^T U^T U y = x^T y = x.y$$

From this conclusion we can show that all orthogonal matrices preserve angles and lengths. (See Exercise 5.2 and Problem 9.)

Exercise 5.2: Use the above formula, $U x . U y = x.y$, to show that orthogonal matrices preserve lengths of vectors.

+ Student Workspace

Answer 5.2

$\|U x\|^2 = U x . U x = x.x = \|x\|^2$. So $U x$ has the same length as x.

In fact, the various properties we have been discussing are all equivalent to one another, as the following theorem says.

Theorem 9: Let U be an n by n matrix. The statement that U is orthogonal (i.e., $U^T U = I$) is equivalent to each of the following statements:

(a) $U^{(-1)} = U^T$.
(b) $U U^T = I$.
(c) The columns of U form an orthonormal basis of R^n.
(d) The rows of U form an orthonormal basis of R^n.
(e) $U x . U y = x.y$ for all vectors x and y in R^n.

Problems

Problem 1: Gram-Schmidt applied to three vectors

(a) Using the Gram-Schmidt process (but not the **GramSchmidt** command), find an orthogonal basis of the subspace of R^5 spanned by the following three linearly independent vectors:
```
> u1 := Vector([1,0,2,0,-1]):
  u2 := Vector([0,1,1,-2,0]):
  u3 := Vector([-1,2,0,1,-3]):
```
(b) Express each of the original basis vectors u_1, u_2, u_3 as a linear combination of your orthogonal basis vectors. Hint: This is most easily done by looking at the equations used in the Gram-Schmidt process. No computation should be necessary.

Student Workspace

Problem 2: Gram-Schmidt applied to same vectors, different order

(a) Using the Gram-Schmidt process (but not the **GramSchmidt** command), find an orthogonal basis of the subspace of R^5 spanned by the following three linearly independent vectors:
```
> u1 := Vector([-1,2,0,1,-3]):
  u2 := Vector([1,0,2,0,-1]):
  u3 := Vector([0,1,1,-2,0]):
```
(b) These vectors are the same vectors as in Problem 1. Why did we get a different orthogonal basis than in Problem 1?

Student Workspace

Problem 3: Gram-Schmidt applied to linearly dependent vectors

(a) Using the Gram-Schmidt process (but not the **GramSchmidt** command), find an orthogonal basis of the subspace of R^5 spanned by the following four vectors:
```
> u1 := Vector([1,0,2,0,-1]):
  u2 := Vector([0,1,1,-2,0]):
  u3 := Vector([1,-1,1,2,-1]):
  u4 := Vector([-1,2,0,1,-3]):
```
(b) Check your answer using the **GramSchmidt** command.

(c) Something quite different happened in this problem than in Problems 1 or 2. What happened and why did it happen? Explain geometrically. (Hint: Notice that u_3 is a linear combination of u_1 and u_2. Compare your orthogonal basis in this problem with your orthogonal basis in Problem 1.)

Student Workspace

Chapter 7 Orthogonality

Problem 4: Distance from a point to a subspace

Find the projection of the vector u onto the subspace S spanned by v_1, v_2, v_3, v_4 (see below). Also find the shortest distance from the head of u to the subspace S. (You may use the **GramSchmidt** command.)

```
> u := Vector([1,0,0,0,0]):
> v1 := Vector([1, 3, 1, 0, 3]):
  v2 := Vector([3, -5, 1, 2, 5]):
  v3 := Vector([-5, 1, -2, 3, -1]):
  v4 := Vector([0, -4, 4, 1, -3]):
```

+ Student Workspace

Problem 5: Orthogonal bases for null space and row space

(a) Find an orthogonal basis for the null space of the matrix A below. (You may use the **GramSchmidt** command.)

```
> A := Matrix([[2,-3,-4,-4,-3],[-9,4,-1,-1,4],[-5,1,-3,-3,1]]);
```

(b) Find an orthogonal basis for the orthogonal complement of the null space of A. (You may use the **GramSchmidt** command.)

(c) Check that the combined set of the bases in (a) and (b) is linearly independent. Why is this combined set an orthogonal basis of R^5?

+ Student Workspace

Problem 6: Defining property of a projection

If $\{v_1, v_2, v_3\}$ is an orthogonal set of vectors in R^n and u is any vector in R^n and

$$p = \text{Project}(u, v_1) + \text{Project}(u, v_2) + \text{Project}(u, v_3)$$

prove that $u - p$ is orthogonal to each of the vectors v_1, v_2, v_3. Hint: Expand and simplify $(u - p) \cdot v_1$; do the same with v_2 and v_3.

+ Student Workspace

Problem 7: Product of orthogonal matrices

If U and V are n by n orthogonal matrices, prove, using only the definition of orthogonality, that the product UV is orthogonal.

+ Student Workspace

Problem 8: Reflections in R^3

(a) Verify that every 3 by 3 reflection matrix is orthogonal. Use the command below to construct these

matrices. (The vector $\langle a, b, c \rangle$ is a normal vector to the plane of reflection.)
```
> R := Reflectmat3d(<a,b,c>);
```
(b) According to Problem 7, the product of the two reflection matrices S and T below is also orthogonal. What familiar matrix does the product $S\,T$ equal? Use geometric reasoning to determine your answer. Also, check your answer by computing the product and comparing it with your answer.
```
> S := Reflectmat3d(<0,0,1>);
  T := Reflectmat3d(<0,1,0>);
```
+ Student Workspace

Problem 9: Orthogonal matrices preserve angles

If U is any n by n orthogonal matrix and x and y are any two vectors in R^n, prove that the angle between x and y equals the angle between $U\,x$ and $U\,y$. Hint: Use the formula relating dot product and angles:

$$x \cdot y = \|x\|\|y\|\cos(\theta)$$

+ Student Workspace

Linear Algebra Modules Project
Chapter 7, Module 3

Least-Squares Solutions

Purpose of this module

The purpose of this module is to introduce the idea of a "least-squares solution" of a linear system that has no solutions. This has applications to problems in curve fitting. One such example is the problem of fitting a straight line to a set of data points that are nearly linear.

Prerequisites

Linear systems; null space and column space of a matrix; orthogonal vectors; projections.

Commands used in this module

```
> restart; with(LinearAlgebra): with(plots): with(Lamp):
  UseHardwareFloats := false: Digits := 6:
```

Tutorial

Section 1: Introduction

In real-world problems, we sometimes encounter a system of linear equations $A x = b$ that has no solutions but for which an "approximate" solution would be useful. (We will see examples of such problems in Section 3.) For instance, the following linear system has no solutions:

$$x_1 + 2 x_2 = 4$$
$$3 x_1 + 5 x_2 = 8$$
$$-x_1 + x_2 = 5$$

====================

Example 1A: We confirm that the system has no solutions by using **Matsolve**:

```
> A := Matrix([[1,2],[3,5],[-1,1]]);
  b := Vector([4,8,5]);
```

$$A := \begin{bmatrix} 1 & 2 \\ 3 & 5 \\ -1 & 1 \end{bmatrix}, \quad b := \begin{bmatrix} 4 \\ 8 \\ 5 \end{bmatrix}$$

```
> x := Matsolve(A,b);
```
Error, (in Matsolve) no solutions

====================

Although $Ax = b$ has no solution, can we find the "best approximate solution" to $Ax = b$? The criterion we will use in determining such a "solution" is to require that Ax be as close as possible to b. Another way to phrase this is to require that the distance from Ax to b, which is $\|Ax - b\|$, be as small as possible. Thus, our criterion is:

- The "best approximate solution" to $Ax = b$ is a vector x such that $\|Ax - b\|$ is as small as possible.

====================

Example 1B: Which of the two vectors x_1 and x_2 defined below is a better approximate solution to $Ax = b$, according to the above criterion?

```
> x1 := Vector([-3,3]);
> x2 := Vector([-2,3]);
```

$$x1 := \begin{bmatrix} -3 \\ 3 \end{bmatrix}, \quad x2 := \begin{bmatrix} -2 \\ 3 \end{bmatrix}$$

We compute the "error", $Ax - b$, for each of these vectors:

```
> error1 := A.x1-b;
  error2 := A.x2-b;
```

$$error1 := \begin{bmatrix} -1 \\ -2 \\ 1 \end{bmatrix}, \quad error2 := \begin{bmatrix} 0 \\ 1 \\ 0 \end{bmatrix}$$

Then we find the lengths of these vectors:

```
> Mag(error1);
  Mag(error2);
```

$$\sqrt{6}$$
$$1$$

====================

So, according to the above criterion, x_2 is a better approximate solution than x_1. However, is there a best x -- one for which $\|Ax - b\|$ is smallest? Is there more than one such vector x? And if there is a best choice for x, how do we find it? These three questions are addressed in the next section.

The proper terminology for a "best approximate solution" is a "least-squares solution." Here is why this phrase is used. When we are finding such a "solution" x, we are finding the smallest (and hence "least") value of $\|Ax - b\|$. For example, the vector $Ax - b$ for the linear system in Example 1A is

$$\begin{bmatrix} 1 & 2 \\ 3 & 5 \\ -1 & 1 \end{bmatrix} \begin{bmatrix} x_1 \\ x_2 \end{bmatrix} - \begin{bmatrix} 4 \\ 8 \\ 5 \end{bmatrix} = \begin{bmatrix} x_1 + 2x_2 - 4 \\ 3x_1 + 5x_2 - 8 \\ -x_1 + x_2 - 5 \end{bmatrix}$$

and therefore the value of $\|Ax - b\|$ is the square root of

$$(x_1 + 2x_2 - 4)^2 + (3x_1 + 5x_2 - 8)^2 + (-x_1 + x_2 - 5)^2$$

So we are finding the <u>least</u> value of this sum of <u>squares</u>, hence the phrase "least-squares."

Definition: A ***least-squares solution*** to a linear system $Ax = b$ is a vector x that yields the smallest value of $\|Ax - b\|$.

Section 2: Least-Squares Solutions and the Normal Equations

Projections are the key to finding a least-squares solution of $Ax = b$. Recall that the product Ax can be thought of as a linear combination of the columns of A:

$$\begin{bmatrix} 1 & 2 \\ 3 & 5 \\ -1 & 1 \end{bmatrix} \begin{bmatrix} x_1 \\ x_2 \end{bmatrix} = x_1 \begin{bmatrix} 1 \\ 3 \\ -1 \end{bmatrix} + x_2 \begin{bmatrix} 2 \\ 5 \\ 1 \end{bmatrix}$$

So the vector Ax is a vector in the column space of A. Therefore $Ax = b$ has a solution if and only if b is in the column space of A. We are interested in the case when b is <u>not</u> in the column space of A. In this case, we want to find x so that the distance from b to Ax (and hence the distance to the column space of A) is as small as possible. As we saw in Modules 1 and 2, the shortest distance from the head of a vector b to a subspace W is found by projecting b onto W. Here is the picture:

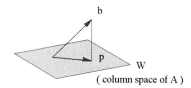

The plane W represents the column space of A; the vector not lying in the plane is the vector b; the vector in the plane is the projection p of b onto the plane; and the black line segment connecting their heads is perpendicular to the plane.

This discussion suggests the following procedure for finding a least-squares solution of $Ax = b$:

- 1. Find the projection p of the right-side vector b onto the column space of A. (Recall that we find the projection of a vector b onto a subspace W by finding an orthogonal basis of W, projecting b onto each basis vector, and adding these projections.)

- 2. Solve $Ax = p$.

Exercise 2.1: Use the above procedure to find the least-squares solution of the linear system $A\,x = b$, where A and b are defined below. Hint: The columns of A are orthogonal, which simplifies the job of finding the projection p.

```
> A := Matrix([[1,1],[2,-2],[3,1]]);
  b := Vector([3,4,2]);
```

$$A := \begin{bmatrix} 1 & 1 \\ 2 & -2 \\ 3 & 1 \end{bmatrix}, \quad b := \begin{bmatrix} 3 \\ 4 \\ 2 \end{bmatrix}$$

+ Student Workspace

− Answer 2.1

First we pull out the two columns of A and check that they are orthogonal:

```
> u := Column(A,1);
  v := Column(A,2);
```

$$u := \begin{bmatrix} 1 \\ 2 \\ 3 \end{bmatrix}, \quad v := \begin{bmatrix} 1 \\ -2 \\ 1 \end{bmatrix}$$

```
> Dotprod(u,v);
```

$$0$$

Then compute the projection p of b onto Span$\{u, v\}$, which is the column space of A:

```
> p := Project(b,u) + Project(b,v);
```

$$p := \begin{bmatrix} \dfrac{5}{7} \\ \dfrac{24}{7} \\ \dfrac{22}{7} \end{bmatrix}$$

```
> x := Matsolve(A,p);
```

$$x := \begin{bmatrix} \dfrac{17}{14} \\ \dfrac{-1}{2} \end{bmatrix}$$

So x is the least-squares solution of $A\,x = b$. Let's see what the magnitude of the error is:

```
> err := A.x-b;
  evalf(Mag(err));
```

$$err := \begin{bmatrix} -\dfrac{16}{7} \\ -\dfrac{4}{7} \\ \dfrac{8}{7} \end{bmatrix}, \quad 2.61862$$

If the columns of A are not orthogonal, we can use projections to replace them with an orthogonal basis of the column space of A, as in the next example.

====================

Example 2A: Find the least-squares solution of $A x = b$, where A and b are as in Example 1A:
```
> A := Matrix([[1,2],[3,5],[-1,1]]);
  b := Vector([4,8,5]);
```
$$A := \begin{bmatrix} 1 & 2 \\ 3 & 5 \\ -1 & 1 \end{bmatrix}, \quad b := \begin{bmatrix} 4 \\ 8 \\ 5 \end{bmatrix}$$

Solution: We orthogonalize the columns of A (see Section 3 in Module 2):
```
> u := Column(A,1);
  v := Column(A,2);
```
$$u := \begin{bmatrix} 1 \\ 3 \\ -1 \end{bmatrix}, \quad v := \begin{bmatrix} 2 \\ 5 \\ 1 \end{bmatrix}$$

```
> w := v-Project(v,u);
```
$$w := \begin{bmatrix} \frac{6}{11} \\ \frac{7}{11} \\ \frac{27}{11} \end{bmatrix}$$

So $\{u, w\}$ is an orthogonal basis of the column space of A. Then we find the projection of b onto the column space by adding the projections of b onto the orthogonal basis vectors:
```
> p := Project(b,u) + Project(b,w);
```
$$p := \begin{bmatrix} \frac{136}{37} \\ \frac{601}{74} \\ \frac{373}{74} \end{bmatrix}$$

Finally, we solve $A x = p$:
```
> x := Matsolve(A,p);
```
$$x := \begin{bmatrix} \frac{-79}{37} \\ \frac{215}{74} \end{bmatrix}$$

====================

This method is a bit tedious, and it is much more so for larger systems. We could speed it up by using the **GramSchmidt** command to compute the orthogonal basis, but that would still be tedious for large systems. Fortunately, there is a much faster method, which we describe next.

The Normal Equations

First let's recall the defining properties of a projection: The projection of a vector b onto a subspace W is the vector p that has the following two properties:

(i) p is in W; and (ii) $b - p$ is orthogonal to every vector in W.

So if W is the column space of a matrix A, property (i) tells us that $p = A\,x$ for some vector x. Also, property (ii) says that the vector $b - p$ is orthogonal to every vector in the column space of A. Since the rows of A^T are the columns of A, we can rephrase this last sentence to say that $b - p$ is orthogonal to every vector in the row space of A^T. In other words, $b - p$ is in the orthogonal complement of the row space of A^T, which is the null space of A^T (by Theorem 8 in Module 2). Thus,

$$A^T(b - p) = 0$$

Next, multiply out, and replace p by $A\,x$:

$$A^T b - A^T A\,x = 0$$

Finally, rearrange terms:

$$A^T A\,x = A^T b$$

The equations of the linear system $A^T A\,x = A^T b$ are called the *normal equations* for the original linear system $A\,x = b$. Note that the normal equations can be found simply by multiplying the original linear system on the left by A^T. So here is our much faster method for computing the least-squares solution of $A\,x = b$:

- 1. Multiply the equation $A\,x = b$ on the left by A^T. (The resulting system, $A^T A\,x = A^T b$, is the system of normal equations for the original system.)

- 2. Solve $A^T A\,x = A^T b$. (The normal equations always have a solution.)

Exercise 2.2: (a) Use the normal equations to find the least-squares solution x of the linear system in Example 1A (repeated below). (b) Verify that $A\,x - b$ is orthogonal to every vector in the column space of A.

```
> A := Matrix([[1,2],[3,5],[-1,1]]);
  b := Vector([4,8,5]);
```

$$A := \begin{bmatrix} 1 & 2 \\ 3 & 5 \\ -1 & 1 \end{bmatrix}, \quad b := \begin{bmatrix} 4 \\ 8 \\ 5 \end{bmatrix}$$

+ Student Workspace

Answer 2.2

(a) The least-squares solution to $A\,x = b$ is the solution of $M\,x = k$, where $M = A^T A$ and $k = A^T b$:

```
> M := Transpose(A).A;
> k := Transpose(A).b;
```

$$M := \begin{bmatrix} 11 & 16 \\ 16 & 30 \end{bmatrix}$$

$$k := \begin{bmatrix} 23 \\ 53 \end{bmatrix}$$

```
> x := Matsolve(M,k);
```

$$x := \begin{bmatrix} \dfrac{-79}{37} \\ \dfrac{215}{74} \end{bmatrix}$$

(b):

```
> y := A.x-b;
```

$$y := \begin{bmatrix} \dfrac{-12}{37} \\ \dfrac{9}{74} \\ \dfrac{3}{74} \end{bmatrix}$$

```
> Dotprod(y,Column(A,1));
  Dotprod(y,Column(A,2));
```

0

Uniqueness of the Least-Squares Solution

We have seen the answer to two of the three basic questions raised at the end of Section 1: Every linear system $A\,x = b$ has a least-squares solution; and we can find such a solution by solving the normal equations $A^T A\,x = A^T b$. The remaining question is whether there can be more than one least-squares solution to a linear system.

In fact, there *can* be more than one least-squares solution to a linear system. Try this example:

```
> A := Matrix([[1,2],[3,6],[-1,-2]]);
  b := Vector([1,2,-1]);
```

$$A := \begin{bmatrix} 1 & 2 \\ 3 & 6 \\ -1 & -2 \end{bmatrix}, \quad b := \begin{bmatrix} 1 \\ 2 \\ -1 \end{bmatrix}$$

We construct and solve the normal equations:

```
> M := Transpose(A).A;
  k := Transpose(A).b;
```

$$M := \begin{bmatrix} 11 & 22 \\ 22 & 44 \end{bmatrix}$$

$$k := \begin{bmatrix} 8 \\ 16 \end{bmatrix}$$

```
> x := 'x':
> Matsolve(M,k,free=x);
>
```

$$\begin{bmatrix} \frac{8}{11} - 2 x_2 \\ x_2 \end{bmatrix}$$

Notice that the columns of A are linearly dependent and hence A has a nonzero null space. Whenever a matrix A has a nonzero null space, the linear system $A x = b$ has more than one least-squares solution for the following reason. If x is a least-squares solution and n is a nonzero vector in the null space of A, then $A (x + n) = A x + A n = A x$, since $A n = 0$. So $x + n$ is another least-squares solution of $A x = b$. Conversely, if the null space of A is $\{0\}$, the least-squares solution is unique. The proof of this statement is left to Problem 4. In summary:

Theorem 10: Every linear system $A x = b$ has a least-squares solution x, and the least-squares solution is unique if and only if the null space of A is $\{0\}$.

Section 3: Fitting Curves to Data

When a scientist has collected some data, one of her first tasks is to find a pattern or a sense of order in the data. To this end, she will usually plot the data, see if the plot seems to follow a familiar shape such as a straight line, parabola, or graph of an exponential or log function, and find the function of that type that best fits the data. The most widely used method for selecting the "best fitting" function of a given type to a set of data points is the least-squares method studied in this section.

======================

Example 3A: Here is a set of data points that nearly lie on a straight line:

```
> data := [[1,3],[2,3],[3,4],[4,5],[5,7]];
```

$$data := [[1, 3], [2, 3], [3, 4], [4, 5], [5, 7]]$$

```
> p1 := pointplot(data,symbol=circle,color=red):
> display(p1,view=[-1..6,0..7]);
```

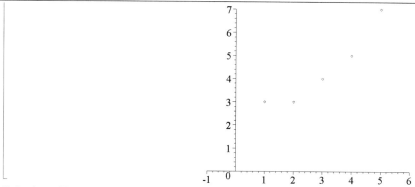

Solution: We want to find the line $y = mt + c$ that best fits the data in a least-squares sense.

```
> f := t -> m*t+c;
```
$$f := t \to m\,t + c$$

If the above five points were to lie on this line, we would have the following five equations:

```
> eqn1 := f(1)=3;
> eqn2 := f(2)=3;
> eqn3 := f(3)=4;
> eqn4 := f(4)=5;
> eqn5 := f(5)=7;
```

$$eqn1 := m + c = 3$$
$$eqn2 := 2m + c = 3$$
$$eqn3 := 3m + c = 4$$
$$eqn4 := 4m + c = 5$$
$$eqn5 := 5m + c = 7$$

This system of five linear equations in the two unknowns c and m has no solution, since the five points do not lie on a line. Instead, we will find a least-squares solution. First we write our system of five linear equations as a matrix-vector equation $Ax = b$:

```
> M := Genmatrix([eqn1,eqn2,eqn3,eqn4,eqn5],[c,m]);
```

$$M := \begin{bmatrix} 1 & 1 & 3 \\ 1 & 2 & 3 \\ 1 & 3 & 4 \\ 1 & 4 & 5 \\ 1 & 5 & 7 \end{bmatrix}$$

Maple note: The `Genmatrix` command generates an augmented matrix from a system of equations. The first argument in the command is the list of equations and the second is the list of variables used in the equations.

```
> A := M[1..5,1..2];
  b := Column(M,3);
```

$$A := \begin{bmatrix} 1 & 1 \\ 1 & 2 \\ 1 & 3 \\ 1 & 4 \\ 1 & 5 \end{bmatrix}, \quad b := \begin{bmatrix} 3 \\ 3 \\ 4 \\ 5 \\ 7 \end{bmatrix}$$

We find a least-squares solution of $A\,x = b$ by solving the normal equations:

```
> M := Transpose(A).A;
  k := Transpose(A).b;
```

$$M := \begin{bmatrix} 5 & 15 \\ 15 & 55 \end{bmatrix}$$

$$k := \begin{bmatrix} 22 \\ 76 \end{bmatrix}$$

```
> x := Matsolve(M,k);
```

$$x := \begin{bmatrix} \frac{7}{5} \\ 1 \end{bmatrix}$$

The line is therefore $y = t + \frac{7}{5}$. Let's plot the line (in blue) on top of the data points:

```
> f := t -> t+7/5;
  p2 := plot(f(t),t=0..6,color=blue):
  display([p1,p2],view=[-1..6,0..7]);
```

$$f := t \to t + \frac{7}{5}$$

The following plot shows the residuals. For each data point (t, y), the "residual" is the difference between the actual y value and the value $f(t)$ predicted by the function f.

> Residuals(f,data,view=[-1..6,0..7]);

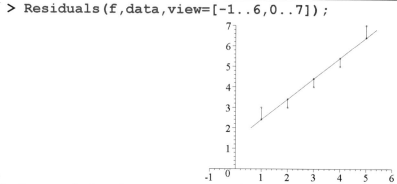

Finally, we calculate the error vector $Ax - b$ (also called the residual vector). Note that the components of this vector are the (signed) lengths of the vertical line segments in the above plot. So it is the sum of the squares of these lengths that we are minimizing when we find a "least-squares" solution.

> err := A.x-b;
> evalf(Mag(err));

$$1.09545$$

==================

Now that we see the form of the coefficient matrix A and the right-side vector b, we can construct these matrices from our data set much more quickly. We first convert the data set into a matrix N whose first column contains the t coordinates of the data and second column the y coordinates. The vector b is the second column of N (i.e., the y coordinates of the data):

> N := Matrix(data);

$$N := \begin{bmatrix} 1 & 3 \\ 2 & 3 \\ 3 & 4 \\ 4 & 5 \\ 5 & 7 \end{bmatrix}$$

> tcoords := Column(N,1);
> b := Column(N,2);

Also, observe that A is the matrix whose first column is all 1's and second column is the first column of N (i.e., the t coordinates of the data).

> A;

To construct the matrix A quickly, we apply the **Vandermat** command (short for "Vandermonde matrix") to the t coordinates of the data:

> Vandermat(tcoords,Dimension(tcoords),2);

$$\begin{bmatrix} 1 & 1 \\ 1 & 2 \\ 1 & 3 \\ 1 & 4 \\ 1 & 5 \end{bmatrix}$$

The above linear systems approach to the problem of fitting a line to a set of data points can also be used to fit other kinds of curves than straight lines. In the next example, we fit a parabola to the same set of points as in Example 3A; in fact, a parabola produces a better fit.

========================

Example 3B: Fit a quadratic function $f(t) = a_0 + a_1 t + a_2 t^2$ to the data points in Example 3A:

```
> data := [[1,3],[2,3],[3,4],[4,5],[5,7]];
```
$$data := [[1, 3], [2, 3], [3, 4], [4, 5], [5, 7]]$$

```
> f := t -> a[0]+a[1]*t+a[2]*t^2;
```
$$f := t \to a_0 + a_1 t + a_2 t^2$$

Solution: As in Example 3A, we have five equations; but now we have three unknowns, the coefficients a_0, a_1, a_2 of the quadratic function f. Here is the system of equations that we want the data to satisfy:

```
> eqn1 := f(1)=3;
  eqn2 := f(2)=3;
  eqn3 := f(3)=4;
  eqn4 := f(4)=5;
  eqn5 := f(5)=7;
```

$$eqn1 := a_0 + a_1 + a_2 = 3$$
$$eqn2 := a_0 + 2 a_1 + 4 a_2 = 3$$
$$eqn3 := a_0 + 3 a_1 + 9 a_2 = 4$$
$$eqn4 := a_0 + 4 a_1 + 16 a_2 = 5$$
$$eqn5 := a_0 + 5 a_1 + 25 a_2 = 7$$

We wrote these equations out as a reminder of the least-squares problem we are trying to solve, but we will not use them in our solution. Instead, we will use the quicker method in which we use **Vandermat** to construct the coefficient matrix A. In the above system of equations, note that the coefficient of a_0 is 1, the coefficient of a_1 is t, and the coefficient of a_2 is t^2. Thus the matrix A must now have a third column, the column of the squares of the t coordinates of the data:

```
> N := Matrix(data);
```
$$N := \begin{bmatrix} 1 & 3 \\ 2 & 3 \\ 3 & 4 \\ 4 & 5 \\ 5 & 7 \end{bmatrix}$$

```
> tcoords := Column(N,1);
> b := Column(N,2);
> A := Vandermat(tcoords,Dimension(tcoords),3);
```

$$A := \begin{bmatrix} 1 & 1 & 1 \\ 1 & 2 & 4 \\ 1 & 3 & 9 \\ 1 & 4 & 16 \\ 1 & 5 & 25 \end{bmatrix}$$

Maple note: The second and third arguments in the **Vandermat** command specify the number of rows and columns of the matrix A. The number of rows is the number of entries in the vector *tcoords*. The number of columns is the number of powers of the entries of *tcoords*. Here we need the 0th powers (all 1's), the first powers, and the second powers; hence A must have 3 columns. If we had used "4" in place of "3", A would have had a fourth column, the cubes of t (for fitting a cubic polynomial).

We construct and solve the normal equations:

```
> M := Transpose(A).A;
> k := Transpose(A).b;
  x := Matsolve(M,k);
```

$$M := \begin{bmatrix} 5 & 15 & 55 \\ 15 & 55 & 225 \\ 55 & 225 & 979 \end{bmatrix}, \quad k := \begin{bmatrix} 22 \\ 76 \\ 306 \end{bmatrix}$$

$$x := \begin{bmatrix} \dfrac{17}{5} \\ -\dfrac{5}{7} \\ \dfrac{2}{7} \end{bmatrix}$$

Then we construct our best fitting quadratic function $f(t)$ and plot it on top of the data points:

```
> f := t -> Dotprod(x,[1,t,t^2]):
  f(t);
```

$$\frac{17}{5} - \frac{5}{7}t + \frac{2}{7}t^2$$

```
> p2 := plot(f(t),t=0..6,color=blue):
> display([p1,p2],view=[-1..6,0..7]);
```

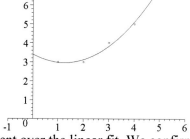

The quadratic fit is a clear improvement over the linear fit. We confirm by looking at the residuals and checking their sum of squares:

```
> err := A.x-b;
> evalf(Mag(err));
```
$$.239045$$
```
> Residuals(f,data,view=[-1..6,0..7]);
```

Problems

Problem 1: Least-squares solution of $Ax = b$

(a) Find the least-squares solution of the linear system $Ax = b$, where A and b are defined below. Rather than using the normal equations, however, compute the projection of b onto the column space of A and use the projection vector to solve the linear system.

(b) Check your answer in (a) by solving the normal equations.

```
> A := Matrix([[1,1,1],[2,1,0],[-2,-1,0],[1,0,1]]);
> b := Vector([2,5,4,1]);
```
+ Student Workspace

Problem 2: Compare linear and quadratic fits

Note: When finding the least-squares solution of $Ax = b$ for this problem, use the efficient method of Example 3B for constructing the matrices A and b.

The data points (t, y) below represent the heights in inches, t, and weights in pounds, y, of eight individuals.

(a) Find the straight line that best fits the data by solving the appropriate normal equations.
(b) Plot the line and the data points together.
(c) Find the quadratic function that best fits the data by solving the appropriate normal equations.
(d) Plot the parabola and the data points together.
(e) Compare the sum of the squares of the residuals for the results in (a) and (c). Which curve gives the better fit?

```
> hw := [[64, 141], [66, 148], [68, 157], [70, 163], [72, 172], [74, 186], [76, 206], [78, 221]];
> p1 := pointplot(hw,symbol=circle,color=red):
  display(p1);
```

+ Student Workspace

Problem 3: Cubic curve fitting

Note: When finding the least-squares solution of $A x = b$ for this problem, use the efficient method of Example 3B for constructing the matrices A and b.

Consider the following data set, which gives the number of high school soccer players (in thousands) by year (1981-1994).

```
> data :=[[81, 213], [82, 220], [83, 243], [84, 256], [85, 281], [86, 297], [87, 312], [88, 327],
  [89, 332], [90, 350], [91, 371], [92, 391], [93, 421], [94, 464]];
> p1 := pointplot(data,symbol=circle,color=red):
> display(p1);
```

(a) Find the straight line that best fits the data by solving the appropriate normal equations.
(b) Plot the line and the data points together.
(c) Find the cubic function, $f(t) = a_0 + a_1 t + a_2 t^2 + a_3 t^3$, that best fits the data by solving the appropriate normal equations.
(d) Plot the cubic and the data points together.
(e) Compare the sum of the squares of the residuals for the results in (a) and (c). Which curve gives the better fit?

+ Student Workspace

Problem 4: Uniqueness of the least-squares solution

If the null space of A is $\{0\}$, prove that $A x = b$ has only one least-squares solution. (Hint: Any least-squares solution of $A x = b$ is also a solution of $A x = p$, where p is the projection of b onto the column space of A. Why does $A x = p$ have only one solution? **Warning:** Do not assume that A has an inverse; in least-squares problems, it almost never does.)

+ Student Workspace

Problem 5: Formula for the projection

The projection of b onto the column space of A is given by the following formula (which is only valid when the null space of A is $\{0\}$):

$$p = A (A^T A)^{(-1)} A^T b$$

(a) Check the above formula for the linear system $A x = b$ in Example 1A (repeated below). That is, compute the projection p by this formula and also by projecting b onto an orthogonal basis of the column space of A.

```
> A := Matrix([[1,2],[3,5],[-1,1]]);
  b := Vector([4,8,5]);
```

(b) Use the above formula to find the 3 by 3 matrix that projects vectors onto the plane $2x - y + 3z = 0$. Check your answer against the result of the **Projectmat3d** command below.

```
> Projectmat3d(<2,-1,3>);
```

(c) Derive the above formula for p. (Hint: The least-squares solution s of $A x = b$ satisfies both of the following two equations: $A s = p$ and $A^T A s = A^T b$. Put these two equations together. **Warning:** Do not assume A has an inverse; in least-squares problems, it almost never does. However, $A^T A$ does have an inverse.)

+ Student Workspace

Problem 6: Fit a plane to multivariate data

We can also use our least-squares method to fit a surface to data points with three or more coordinates. For example, we can fit a linear function $z = f(x, y) = a_0 + a_1 x + a_2 y$ (i.e., a plane) to data points (x, y, z). Consider the following data set, from which we want to predict the volume of useable wood produced by a commercially grown tree, knowing the diameter and height of the tree. The first coordinate, x, represents the diameter of a tree (measured in inches at a height of 4 feet); the second coordinate, y, represents the height of the tree in feet; and the third coordinate, the dependent variable z, represents the volume of useable wood in cubic feet.

```
> dhv := [[14.5, 76.0, 31.4], [14.2, 75.0, 29.9], [15.7, 85.0, 40.8], [14.4, 76.0, 31.0],
  [16.8, 71.0, 35.5], [20.5, 82.0, 65.8], [14.1, 80.0, 32.7], [14.8, 69.0, 31.3], [18.9, 72.0, 48.4],
  [14.0, 66.0, 25.5]];
```

Set up and solve the appropriate normal equations. Explain where the entries of A and b come from. Note: Fitting a linear function of more than one variable to a set of data points, as you did in this problem, is usually called "multiple regression."

+ Student Workspace

Problem 7: Fault plane for earthquake data

As in Problem 6, we will fit a linear function $z = f(x, y) = a_0 + a_1 x + a_2 y$ (i.e., a plane) to data points (x, y, z). The following problem, based on actual earthquake data, will demonstrate how this method might be used in a real application.

A controversy is brewing in the country of Lampland because the proposed path of a natural gas pipeline runs near a fault plane that has caused some small earthquakes. The figure below shows a map of Lampland (a square, 250 km. on each side) and the proposed path of the pipeline. We will measure all distances in kilometers from the city of Newton, which is at the center of the country. In this coordinate system, the proposed pipeline passes through points with coordinates (0, −75, 0) and (250, −125, 0).

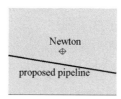

Even a small earthquake could damage the pipeline, if the fault plane ruptures the surface and if it intersects the ground ($z = 0$) on a line (called the "fault trace") that cuts through the pipeline. To determine the approximate location of the fault plane, geologists have recorded the locations of 32 seismic events (i.e., tiny movements picked up by sensitive instruments) along the subterranean ($z < 0$) fault plane. The data below are their coordinates relative to Newton. If you click on the map above and turn it, you will see these points (in red) underneath the surface.

```
> data := [[39.757, 172.063, −57.8], [15.208, 8.778, −40.8], [76.684, 74.344, −45.5],
   [36.412, 16.465, −43.2], [10.187, 47.710, −35.0], [47.028, 66.459, −46.6], [55.266, 69.793, −47.9],
   [65.724, 79.812, −46.0], [54.524, 58.679, −47.9], [73.135, 96.507, −56.2], [−20.092, 30.219, −30.8],
   [45.631, 163.161, −57.3], [75.724, 10.939, −46.0], [28.84, 199.888, −59.6], [40.205, 24.236, −44.9],
   [43.280, 64.241, −48.5], [−34.466, −25.164, −35.1], [60.461, 99.808, −50.2],
   [65.669, 102.041, −50.3], [54.530, 46.453, −46.0], [73.218, 74.278, −54.1],
   [−70.042, 93.306, −36.1], [0, 0, −35.20], [−69.320, 19.910, −42.8], [77.261, 5.389, −49.5],
   [−83.101, 109.166, −36.0], [30.240, −19.078, −39.6], [−9.504, 33.434, −34.4],
   [105.594, 52.26, −48.1], [50.785, 153.154, −54.1], [45.545, 80.909, −43.3], [5.243, −7.822, −38.7]];
```

(a) Use our least-squares method to find the plane that best fits the data points.
(b) Explain where the entries of A and b come from.
(c) Use your answer from (a) to find the fault trace (i.e., the intersection of the fault plane and the ground). On a two-dimensional plot, graph the proposed pipeline and the fault trace.
(d) Plot the fault plane together with the data points by using the **display** command below (with the equation of the fault plane suitably modified).
(e) What conclusions can you reach from your answer to part (c)? What additional information might you need to complete your risk analysis?

```
> x := 'x': y := 'y':
  fault := plot3d(-.1*x-.1*y-20,x=-250..250,y=-250..250,
```

```
      color=blue,grid=[2,2]):
  ground := plot3d(0,x=-250..250,y=-250..250,
      color=yellow,grid=[2,2]):
  pts := pointplot3d(data,symbol=circle,color=red):
  pipe := spacecurve([x,-.2*x-75,0],x=-250..250,thickness=3):
  display([fault,ground,pts,pipe],axes=none,orientation=[-90,0]);
```

Student Workspace

Problem 8: Fit a conic section

The data points below nearly lie on an ellipse, whose equation has the general form of a conic:
$$a x^2 + b y^2 + c x y + d x + e y = 1$$
(Aside: The usual form of a conic is $a x^2 + b y^2 + c x y + d x + e y + f = 0$. This form is not useful here, because a least-squares approach will produce the trivial coefficients $a = 0, ..., f = 0$. However, if the point (0, 0) does not lie on the conic, then $f \neq 0$ and so we can move f to the right side of the equation and divide by $-f$ to get the displayed equation above, which does not have a trivial solution. If a plot of the data suggests that the point (0, 0) might indeed lie on the conic, simply translate all the points by the same amount to insure this does not happen.)

(a) Use our least-squares technique to find the coefficients a, b, c, d, e. Use the function g below to create the equations, and use **Genmatrix** (as in Example 3A) to generate the coefficient matrix A.

(b) Plot your ellipse together with the data points by using the **display** command below (with the quadratic suitably modified).

```
> data := [[.1, 2.0], [.3, 1.9], [.8, 1.0], [.6, -.2], [-.2, -1.1], [-1.6, -.7], [-1.7, .4], [-.8, 1.8]];
> a := 'a': b := 'b': c := 'c': d := 'd': e := 'e':
> g := (x,y) -> a*x^2+b*y^2+c*x*y+d*x+e*y;
> p1 := pointplot(data,symbol=circle,color=blue):
  p2 := implicitplot(x^2+.5*y^2-.5*x*y+x-y=1,
      x=-3..2,y=-2..3,view=[-3..2,-2..3]):
  display({p1,p2},scaling=constrained);
```

Student Workspace

Pencil and Paper Tutorial: Gaussian Elimination

Section 1. Introduction

Gaussian elimination is the name given to a step-by-step procedure for solving systems of linear equations. In this tutorial you will apply this procedure by hand to several simple examples. Later you will use Maple to apply the same procedure to larger systems. As you work through the examples, you will be introduced to a number of new terms. A thorough understanding of these definitions is essential, since it is the foundation for all your future work in this course.

Example 1 Before we solve a system of linear equations, let's be sure what we mean by a "solution." Consider the system of linear equations:

$$\begin{cases} 2x + 6y = -6 \\ 5x + 7y = 1 \end{cases}$$

of two equations and two variables. In order to solve this system, we must find values for the unknowns x and y which satisfy both equations simultaneously. For example, the pair of values $x = 3$, $y = -2$, which we can write simply as $(3, -2)$, is a solution of this system.

Exercise 1 Use direct substitution to show that $(3, -2)$ is a solution of the above system and $(9, -4)$ is not.

Exercise 2 The graphs of the two linear equations that make up the system are shown below. Locate the points corresponding to the (x, y) pairs $(3, -2)$ and $(9, -4)$ on this graph and label the points P and Q. Note that P falls exactly at the intersection point of the two lines and Q does not. Label each line with its equation.

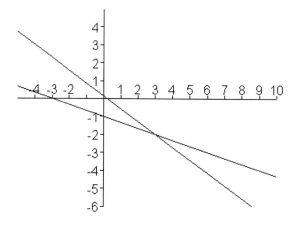

Here are our first few definitions.

Definitions:

A ***linear equation*** in the variables $x_1, x_2, ..., x_n$ is an equation of the form $a_1 x_1 + a_2 x_2 + ... + a_n x_n = b$, where the ***coefficients*** $a_1, a_2, ..., a_n$ and the right-side constant b are all real numbers. Note: From now on we will use $x_1, x_2, ..., x_n$ to designate variables instead of x, y, z etc.

A ***linear system*** is a set of one or more linear equations involving the same variables.

A ***solution*** of a linear system is a list of numbers, one for each variable, that simultaneously satisfies all of the equations of the system.

The ***solution set*** of a linear system is the set of all solutions of the system.

The first few linear systems we will consider each have exactly one solution (like the single point of intersection of the two lines). Later we will consider linear systems that have more than one solution. On the other hand, not every linear system has a solution (for example, consider two equations that represent parallel lines). If a linear system has at least one solution it is called a ***consistent system***, and if it fails to have a solution it is called an ***inconsistent system***.

Section 2.   Matrix Notation

Consider the linear system: $\begin{cases} 2x_1 + 6x_2 = -6 \\ 5x_1 + 7x_2 = 1 \end{cases}$. You will recognize this as the system from Example 1 with the variables x and y now renamed x_1 and x_2. Of course the solution set remains the same; in fact, the names of the variables are of no significance. What really matters is the particular numbers that appear as coefficients on the left sides of the equations and the constants on the right sides.

The coefficients for the system can be displayed efficiently as entries in a matrix. The ***coefficient matrix*** for the linear system is $\begin{bmatrix} 2 & 6 \\ 5 & 7 \end{bmatrix}$, a matrix with two rows corresponding to the two equations and two columns corresponding to two variables. If we include a third column with the right-side constants, we have the ***augmented matrix*** for the system: $\begin{bmatrix} 2 & 6 & -6 \\ 5 & 7 & 1 \end{bmatrix}$. Note that for an augmented matrix the number of rows again corresponds to the number of equations but the number of columns is *one more than* the number of variables.

Exercise 3 Write the coefficient and augmented matrix for each of the linear systems below.

$\begin{cases} 2x_1 + 5x_2 - 7x_3 = 0 \\ x_1 - 2x_2 + 5x_3 = 8 \\ x_2 + x_3 = 1 \end{cases}$

$\begin{cases} 2x_1 + 5x_2 - 7 = 4 \\ x_1 + 3x_2 = 9 \end{cases}$

Exercise 4 Assume that the matrix below is the augmented matrix for a linear system. Write the linear system.

$$\begin{bmatrix} 2 & 1 & 5 & -4 \\ 1 & 7 & 0 & 2 \end{bmatrix}$$

Section 3. Gaussian Elimination: What's the Big Idea?

The "big idea" behind Gaussian elimination is a familiar strategy in mathematics: replace a hard problem by an equivalent, but easier, one. In particular, Gaussian elimination provides us with a systematic way to transform any linear system into an equivalent one that is much easier to solve. By an equivalent linear system we mean one that has the same solution set as the original system. So solving the new system is equivalent to solving the original problem.

Definition: Two linear systems are ***equivalent*** if they have the same solution set.

Example 2 To make this definition more tangible, consider the linear system A shown below. Our goal is to solve this system. By applying Gaussian elimination to system A we will be able to produce system B shown on the right. We'll explain in detail exactly how this is done in the following sections.

System A: $\begin{cases} x_1 - x_2 - 2x_3 = -1 \\ 2x_1 - x_2 - x_3 = 2 \\ -3x_1 + 5x_2 + 4x_3 = 3 \end{cases}$ $\xrightarrow{\textit{Gaussian elimination}}$ System B: $\begin{cases} x_1 - x_2 - 2x_3 = -1 \\ \ldots\ldots x_2 + 3x_3 = 4 \\ \ldots\ldots\ldots x_3 = 1 \end{cases}$

Since Gaussian elimination always produces equivalent systems, systems A and B have the same solution set. How is the new system B easier to solve than the original system A?

Notice that only the first equation of system B has all three variables; x_1 has been eliminated from the second equation, and both x_1 and x_2 have been eliminated from the third equation. In fact, the third equation couldn't be simpler; it tells us that x_3 must have the value 1. With this information we can find the values for x_2 and x_3. Here's how: working backwards (i.e. working from the last equation up) starting with the value of x_3 and the second equation, we can find the value of x_2. Then, using x_2 and x_3 along with the first equation, we can find x_1. This process is called ***back substitution***.

Exercise 5 Apply back substitution to system B to find the solution. Check that the solution you find for system B is also a solution for system A.

Exercise 6 The augmented matrix for a linear system is shown below. Write the linear system that corresponds to this matrix, and then use back substitution to solve the system.

$$\begin{bmatrix} 2 & -1 & 3 & 5 \\ 0 & 1 & -4 & 6 \\ 0 & 0 & 4 & 12 \end{bmatrix}$$

Section 4. Gaussian Elimination: Three basic operations

Gaussian elimination is carried out by applying three basic operations. Each time we apply one of these operations we produce a different but equivalent system. We illustrate each operation below by applying it to the linear system A.

$$\text{System A:} \begin{cases} x_1 - x_2 - 2x_3 = -1 \\ 2x_1 - x_2 - x_3 = 2 \\ -3x_1 + 5x_2 + 4x_3 = 3 \end{cases}$$

I. The Multiply operation
If we multiply an equation by a nonzero constant, we produce an equivalent system.

Example: Multiply the first equation of system A by 5:

$$\begin{cases} x_1 - x_2 - 2x_3 = -1 \\ 2x_1 - x_2 - x_3 = 2 \\ -3x_1 + 5x_2 + 4x_3 = 3 \end{cases} \implies \begin{cases} 5x_1 - 5x_2 - 10x_3 = -5 \\ 2x_1 - x_2 - x_3 = 2 \\ -3x_1 + 5x_2 + 4x_3 = 3 \end{cases}$$

II. The Interchange operation
If we interchange two equations, we produce an equivalent system.

Example: Interchange the first and third equations of system A.

$$\begin{cases} x_1 - x_2 - 2x_3 = -1 \\ 2x_1 - x_2 - x_3 = 2 \\ -3x_1 + 5x_2 + 4x_3 = 3 \end{cases} \implies \begin{cases} -3x_1 + 5x_2 + 4x_3 = 3 \\ 2x_1 - x_2 - x_3 = 2 \\ x_1 - x_2 - 2x_3 = -1 \end{cases}$$

III. The Add operation
If we add a multiple of one equation to another, we produce an equivalent system.

Example: Multiply the first equation of system A by 10 and add this to the second equation.

$$\begin{cases} x_1 - x_2 - 2x_3 = -1 \\ 2x_1 - x_2 - x_3 = 2 \\ -3x_1 + 5x_2 + 4x_3 = 3 \end{cases} \implies \begin{cases} x_1 - x_2 - 2x_3 = -1 \\ 12x_1 - 11x_2 - 21x_3 = -8 \\ -3x_1 + 5x_2 + 4x_3 = 3 \end{cases}$$

Here's the calculation:

$$\begin{array}{rl} 10 * \text{equation one:} & 10x_1 - 10x_2 - 20x_3 = -10 \\ + \text{ equation two:} & 2x_1 - x_2 - x_3 = 2 \\ \hline \text{new equation two:} & 12x_1 - 11x_2 - 21x_3 = -8 \end{array}$$

Note: Even though equation one is used in carrying out the Add operation, only the second equation changes.

Both the Multiply and Interchange operations make only superficial changes to the linear system. Therefore it is not surprising that the solution set doesn't change. However, the third operation, the Add operation, makes a more substantial change to the system; so the fact that the solution set remains unchanged is less obvious. We will prove this fact in an appendix to this tutorial.

Section 5.   Gaussian Elimination: Solving a linear system

We are now ready to apply Gaussian elimination to solve a system. We will start by demonstrating how the three basic operations can be used to transform system A into system B from Example 2. Here's how we will do it: First we use the Add operation and equation one to eliminate the variable x_1 from equations two and three. Then we use the Add operation and equation two to eliminate the variable x_2 from equation three.

Problem: Use Gaussian elimination to transform system A to system B.

$$\text{System A:} \begin{cases} x_1 - x_2 - 2x_3 = -1 \\ 2x_1 - x_2 - x_3 = 2 \\ -3x_1 + 5x_2 + 4x_3 = 3 \end{cases} \xrightarrow{\text{Gaussian elimination}} \text{System B:} \begin{cases} x_1 - x_2 - 2x_3 = -1 \\ \ldots\ldots x_2 + 3x_3 = 4 \\ \ldots\ldots\ldots x_3 = 1 \end{cases}$$

Solution:

Step 1: Apply the Add operation to system A to eliminate the variable x_1 from the second equation.
   In particular: <u>Add –2 times the first equation to the second equation.</u>

$$\begin{cases} x_1 - x_2 - 2x_3 = -1 \\ 2x_1 - x_2 - x_3 = 2 \\ -3x_1 + 5x_2 + 4x_3 = 3 \end{cases} \implies \begin{cases} x_1 - x_2 - 2x_3 = -1 \\ \ldots\ldots x_2 + 3x_3 = 4 \\ -3x_1 + 5x_2 + 4x_3 = 3 \end{cases}$$

Step 2: Apply the Add operation to the result from Step 1 to eliminate the variable x_1 from the third equation.
In particular: <u>Add 3 times the first equation to the third equation.</u>

$$\begin{cases} x_1 - x_2 - 2x_3 = -1 \\ \ldots\ldots x_2 + 3x_3 = 4 \\ -3x_1 + 5x_2 + 4x_3 = 3 \end{cases} \implies \begin{cases} x_1 - x_2 - 2x_3 = -1 \\ \ldots\ldots x_2 + 3x_3 = 4 \\ \ldots\ldots 2x_2 - 2x_3 = 0 \end{cases}$$

Step 3: Apply the Add operation to the result from Step 2 to eliminate the variable x_2 from the third equation.
In particular: <u>Add −2 times the second equation to the third equation.</u>

$$\begin{cases} x_1 - x_2 - 2x_3 = -1 \\ \ldots\ldots x_2 + 3x_3 = 4 \\ \ldots\ldots 2x_2 - 2x_3 = 0 \end{cases} \implies \begin{cases} x_1 - x_2 - 2x_3 = -1 \\ \ldots\ldots x_2 + 3x_3 = 4 \\ \ldots\ldots\ldots -8x_3 = -8 \end{cases}$$

Step 4: Apply the Multiply operation to the result from Step 3 to simplify the third equation.
In particular: <u>Multiply the third equation by −1/8.</u>

$$\begin{cases} x_1 - x_2 - 2x_3 = -1 \\ \ldots\ldots x_2 + 3x_3 = 4 \\ \ldots\ldots\ldots -8x_3 = -8 \end{cases} \implies \begin{cases} x_1 - x_2 - 2x_3 = -1 \\ \ldots\ldots x_2 + 3x_3 = 4 \\ \ldots\ldots\ldots\ldots x_3 = 1 \end{cases}$$

Note that the resulting linear system is system B.

Exercise 7 Apply Gaussian elimination to transform the system below into an equivalent system that is easier to solve, following the strategy we used in the previous example. Then use back substitution to find the solution. Organize your work in a systematic way, indicating which operation you use at each step. When you get a solution, check it by direct substitution in the original linear system.

$$\begin{cases} x_1 + 5x_2 + 2x_3 = 1 \\ -2x_1 - 8x_2 + 2x_3 = 12 \\ x_1 + 3x_2 - 3x_3 = -9 \end{cases}$$

Section 6. Gaussian Elimination: A Matrix Approach

As you have seen, Gaussian elimination can be a tedious procedure to carry out. One way to make the process more efficient is to use augmented matrices. Since each row of the augmented matrix of a linear system corresponds to an equation, the three basic operations (Multiply, Interchange, and Add) that we applied to equations now become operations on the rows of the matrix. These three Elementary Row Operations are listed below.

Elementary Row Operations

I. Multiply Row: Multiply a row by a nonzero constant.

II. Interchange Row: Interchange two rows.

III. Add Row: Add a multiple of one row to another.

Definitions:
Two matrices A and B are ***row equivalent*** if B can be obtained from A by a sequence of elementary row operations.

When we apply Gaussian elimination to an augmented matrix by using elementary row operations, we refer to the process as ***row reduction***. That is, we use ***row reduction*** to reduce the augmented matrix to a form that is suitable for back substitution.

Example 3 Use row reduction, followed by back substitution to solve the linear system from Exercise 7 (below).

$$\begin{cases} x_1 + 5x_2 + 2x_3 = 1 \\ -2x_1 - 8x_2 + 2x_3 = 12 \\ x_1 + 3x_2 - 3x_3 = -9 \end{cases}$$

Solution: The augmented matrix for the system is shown below. To eliminate x_1 from the second equation we need to get a 0 in place of the –2.

$$\begin{bmatrix} 1 & 5 & 2 & 1 \\ \boxed{-2} & -8 & 2 & 12 \\ 1 & 3 & -3 & -9 \end{bmatrix}$$

We accomplish this by adding 2 times row 1 to row 2.

$$\begin{bmatrix} 1 & 5 & 2 & 1 \\ 0 & 2 & 6 & 14 \\ \boxed{1} & 3 & -3 & -9 \end{bmatrix}$$

To eliminate x_1 from the third equation we want to get a 0 in place of the 1.
We accomplish this by adding –1 times row 1 to row 3.

$$\begin{array}{c} \text{pivot column} \quad \text{pivot position: row 1, column 1} \\ \begin{bmatrix} \text{①} & 5 & 2 & 1 \\ 0 & 2 & 6 & 14 \\ 0 & -2 & -5 & -10 \end{bmatrix} \end{array}$$

Thus, we used the first entry of row one (circled) to produce zeros below it. This number is called a ***pivot***; its position (1, 1), that is, row 1, column 1, is called a ***pivot position***; and the column it falls in (here column 1) is called a ***pivot column***.

Our attention now turns to equations (rows) two and three. We will use equation two to eliminate x_2 from equation three. To eliminate x_1 from the third equation we want to get a 0 in place of the −2.

$$\begin{bmatrix} 1 & 5 & 2 & 1 \\ 0 & 2 & 6 & 14 \\ 0 & \boxed{-2} & -5 & -10 \end{bmatrix}$$

We accomplish this by adding 1 times row 2 to row 3. Now we are using 2 as a pivot. Its pivot position is (2, 2), i.e., row 2, column 2; and column 2 is the pivot column.

$$\begin{array}{c} \text{pivot column} \quad \text{pivot position: row 2, column 2} \\ \begin{bmatrix} 1 & 5 & 2 & 1 \\ 0 & \text{②} & 6 & 14 \\ 0 & 0 & 1 & 4 \end{bmatrix} \end{array}$$

The row reduction process is now complete. In order to apply back substitution to find the solution, we first convert back to equation format.

$$\begin{bmatrix} 1 & 5 & 2 & 1 \\ 0 & 2 & 6 & 14 \\ 0 & 0 & 1 & 4 \end{bmatrix} \text{ corresponds to the system: } \begin{cases} x_1 + 5x_2 + 2x_3 = 1 \\ \quad\quad\; 2x_2 + 6x_3 = 14 \\ \quad\quad\quad\quad\quad\; x_3 = 4 \end{cases}$$

Back substitution:

From equation three we have: $x_3 = 4$

Substituting this value into equation two: $2x_2 + 6(4) = 14$, we get $x_2 = -5$.

Substituting the values for x_2 and x_3 into equation one: $x_1 + 5(-5) + 2(4) = 1$, we get $x_1 = 18$.

So the solution of the system is: $(18, -5, 4)$.

☆**Geometry Note:** We found one solution to a system of three equations in three unknowns. Recall that a linear equation in three variables corresponds to a plane in 3-space. Therefore a geometric interpretation of our result is that the three planes corresponding to the three equations of the system intersect at the point with coordinates $(18, -5, 4)$.

Definition: The **leading entry** in a row is the leftmost nonzero entry in that row. For example, in the matrices below the leading entries for each row are circled.

$$\begin{bmatrix} ② & 6 & 0 & 5 \\ ③ & 2 & 4 & 1 \\ 0 & 0 & ② & 3 \\ ① & 0 & 0 & 1 \end{bmatrix} \quad \begin{bmatrix} ① & 7 & -5 & 0 & 5 \\ 0 & 0 & 0 & ③ & 0 \\ 0 & 0 & ⑥ & 6 & 4 \\ ② & 9 & 7 & 0 & 0 \end{bmatrix} \quad \begin{bmatrix} ② & -1 & 2 & 5 \\ 0 & ③ & 1 & 4 \\ 0 & 0 & ⑥ & 4 \end{bmatrix}$$

Example 4 A system with no solution

Consider the system of three equations below. If you were to use Maple to graph the planes corresponding to each of these equations, you would see that there is no point that all three planes have in common. Therefore there is no solution for this system. Recall that we refer to such a system as inconsistent. What happens when we apply Gaussian elimination to an inconsistent system?

$$\begin{cases} x_1 - 5x_2 + 4x_3 = -3 \\ 2x_1 - 7x_2 + 3x_3 = -2 \\ -2x_1 + x_2 + 7x_3 = -1 \end{cases}$$

The augmented matrix for the system is: $\begin{bmatrix} 1 & -5 & 4 & -3 \\ 2 & -7 & 3 & -2 \\ -2 & 1 & 7 & -1 \end{bmatrix}$

After two steps in the row reduction process we get the matrix:

$$\begin{bmatrix} 1 & -5 & 4 & -3 \\ 0 & 3 & -5 & 4 \\ 0 & -9 & 15 & -7 \end{bmatrix}$$

Now something interesting happens when we eliminate x_2 from equation three. When we add 3 times row 2 to row 3 we get:

$$\begin{bmatrix} 1 & -5 & 4 & -3 \\ 0 & 3 & -5 & 4 \\ 0 & 0 & 0 & 5 \end{bmatrix}$$

So in the process of eliminating x_2 we also lose x_3 from equation three! This time back substitution won't work. To see why, convert the matrix into equation format.

$$\begin{bmatrix} 1 & -5 & 4 & -3 \\ 0 & 3 & -5 & 4 \\ 0 & 0 & 0 & 5 \end{bmatrix} \text{ corresponds to the linear system : } \begin{cases} x_1 - 5x_2 + 4x_3 = -3 \\ \ldots\ldots 3x_2 - 5x_3 = 4 \\ \ldots\ldots\ldots\ldots\ldots\boxed{0 = 5} \end{cases}$$

Look at the third equation. Clearly no solution can exist for this system since the third equation is never true. Therefore the original system cannot have a solution.

Conclusion: If row reduction of the augmented matrix of a linear system results in a matrix with a row whose leading entry is in the rightmost column, then the system is inconsistent.

Example 5   A system with more than one solution
The linear system below has more than one solution. In fact, it has an infinite number of solutions.

$$\begin{cases} x_1 - 2x_2 + 4x_3 = 2 \\ -2x_1 + 5x_2 + 2x_3 = 4 \\ 3x_1 - 7x_2 + 2x_3 = -2 \end{cases} \quad \text{augmented matrix:} \quad \begin{bmatrix} 1 & -2 & 4 & 2 \\ -2 & 5 & 2 & 4 \\ 3 & -7 & 2 & -2 \end{bmatrix}$$

After two steps in the row reduction process we get the matrix: $\begin{bmatrix} 1 & -2 & 4 & 2 \\ 0 & 1 & 10 & 8 \\ 0 & -1 & -10 & -8 \end{bmatrix}$

And after the next step we have $\begin{bmatrix} 1 & -2 & 4 & 2 \\ 0 & 1 & 10 & 8 \\ 0 & 0 & 0 & 0 \end{bmatrix}$.

Converting this matrix back to equation form we get:

$$\begin{cases} x_1 - 2x_2 + 4x_3 = 2 \\ x_2 + 10x_3 = 8 \\ 0 = 0 \end{cases}$$

Notice that the third equation is always true regardless of the values of the variables. In particular, the third equation puts no restrictions on the value of x_3. So if we apply back substitution we can take *any* value for this variable. In this situation we refer to x_3 as a ***free variable***.

Continuing with back substitution, express the remaining variables in terms of x_3.

From equation two we have: $x_2 + 10x_3 = 8$, which we solve for x_2 to get: $x_2 = 8 - 10x_3$.

And from equation one we have: $x_1 - 2(8 - 10x_3) + 4x_3 = 2$, which yields: $x_1 = 18 - 24x_3$.

So the solution is: $\begin{cases} x_1 = 18 - 24x_3 \\ x_2 = 8 - 10x_3 \\ x_3 = x_3 \end{cases}$ where the equation $x_3 = x_3$ indicates that x_3 is free to be any real number.

Notice that for every value of x_3 we get a solution of the system. So this system has an infinite number of solutions.

☆**Geometry Note:** Suppose that we let the variable $x_3 = t$, where t represents any real number. Then the corresponding values for x_1 and x_2 are: $x_1 = 18 - 24t$ and $x_2 = 8 - 10t$. Now consider all of the solution points $(x_1, x_2, x_3) = (18 - 24t, 8 - 10t, t)$, where t ranges over all real numbers. Recall that this is the parametric description of a line in 3-space. In fact, this line is the line of intersection of the three planes corresponding to the three equations in the original system.

Section 7.   Echelon Form

We pause here to briefly look back at the last three examples. In particular, consider the new matrix that we produced each time as a result of Gaussian elimination. Note that in each case the leading entries form a step-like pattern. The fancy word for a step-like pattern is "echelon." So we say that each of these matrices is in ***echelon form***.

Appendix A Gaussian Elimination   467

Original augmented matrix   Echelon form matrix

Example 3 $\begin{bmatrix} 1 & 5 & 2 & 1 \\ -2 & -8 & 2 & 12 \\ 1 & 3 & -3 & -9 \end{bmatrix}$ → row reduction → $\begin{bmatrix} 1 & 5 & 2 & 1 \\ 0 & 2 & 6 & 14 \\ 0 & 0 & 1 & 4 \end{bmatrix}$

Example 4 $\begin{bmatrix} 1 & -5 & 4 & -3 \\ 2 & -7 & 3 & -2 \\ -2 & 1 & 7 & -1 \end{bmatrix}$ → row reduction → $\begin{bmatrix} 1 & -5 & 4 & -3 \\ 0 & 3 & -5 & 4 \\ 0 & 0 & 0 & 5 \end{bmatrix}$

Example 5 $\begin{bmatrix} 1 & -2 & 4 & 2 \\ -2 & 5 & 2 & 4 \\ 3 & -7 & 2 & -2 \end{bmatrix}$ → row reduction → $\begin{bmatrix} 1 & 2 & 4 & 5 \\ 0 & 1 & 10 & 8 \\ 0 & 0 & 0 & 0 \end{bmatrix}$

Definition: A matrix is in **echelon form** (also called **row echelon form**) if it has the following two properties:

1) All rows that consist only of zeros (if any) are grouped together at the bottom of the matrix.
2) If a row has a leading entry, then its position falls to the *right* of the leading entry of the row *above* it (if there is a row above it).

Note that all entries below a leading entry must therefore be zero.

Exercise 8 All but two of the matrices below are in echelon form. Which two are not in echelon form? Use elementary row operations to transform these two matrices into echelon form.

$\begin{bmatrix} 2 & 1 & 3 & 9 & 0 \\ 0 & 0 & 0 & 2 & 3 \\ 0 & 0 & 0 & 0 & 0 \end{bmatrix}$ $\begin{bmatrix} 1 & 1 & 1 & 1 \\ 1 & 1 & 1 & 1 \\ 0 & 0 & 1 & 1 \\ 0 & 0 & 0 & 1 \end{bmatrix}$ $\begin{bmatrix} 2 & 1 & 1 & 1 & 1 \\ 0 & 0 & 3 & 0 & 9 \\ 0 & 0 & 0 & 0 & 1 \\ 0 & 0 & 0 & 0 & 0 \\ 0 & 0 & 0 & 0 & 0 \end{bmatrix}$ $\begin{bmatrix} 2 & 0 & 0 \\ 0 & 0 & 0 \\ 0 & 0 & 3 \end{bmatrix}$ $\begin{bmatrix} 1 & 2 & 3 & 4 \end{bmatrix}$

Exercise 9 Three echelon matrices are shown below. Assume that each is the result of applying Gaussian elimination to the augmented matrix of a system of linear equations in three unknowns. For each, answer the following questions.
(a) Is the system consistent? If so, does the solution set consist of just one solution or an infinite number of solutions?
(b) If the system is consistent, find its solutions. For a system with more than one solution, write the answer as we did in Example 5.

When you are done, sketch in the step pattern for each matrix. How can you use the step pattern to answer question (a)?

$$\begin{bmatrix} 1 & -2 & 2 & 0 \\ 0 & 2 & -3 & 4 \\ 0 & 0 & 5 & 0 \end{bmatrix} \qquad \begin{bmatrix} 1 & -2 & 2 & 0 \\ 0 & 2 & -3 & 4 \\ 0 & 0 & 0 & 1 \end{bmatrix} \qquad \begin{bmatrix} 1 & -2 & 2 & 0 \\ 0 & 2 & 0 & 4 \\ 0 & 0 & 0 & 0 \end{bmatrix}$$

Once a matrix is in echelon form we can continue to apply row operations to simplify it even further into what is called *reduced* echelon form. Here is the definition:

Definition: A matrix is in reduced echelon form (also called reduced row echelon form) if it is in echelon form and additionally:

1) All leading entries are 1.
2) All entries above a leading entry are zero.

Example: The matrices below are all in reduced echelon form:

$$\begin{bmatrix} 1 & 5 & 3 & 0 & 4 \\ 0 & 0 & 0 & 1 & 3 \\ 0 & 0 & 0 & 0 & 0 \end{bmatrix} \qquad \begin{bmatrix} 1 & 0 & 3 & 0 \\ 0 & 1 & 2 & 0 \\ 0 & 0 & 0 & 1 \\ 0 & 0 & 0 & 0 \end{bmatrix} \qquad \begin{bmatrix} 1 & 0 & 9 & 4 \\ 0 & 1 & 1 & 1 \\ 0 & 0 & 0 & 0 \end{bmatrix} \qquad \begin{bmatrix} 1 & 0 & 0 & 0 & 5 \\ 0 & 1 & 0 & 0 & 4 \\ 0 & 0 & 1 & 0 & 3 \\ 0 & 0 & 0 & 1 & 2 \end{bmatrix}$$

Section 8. Systems with *m* equations and *n* unknowns

All our examples of linear systems so far have been systems of *three* equations and *three* unknowns, so-called "3 by 3 systems." In practice a linear system can have any number of equations and unknowns; i.e., *m* equations and *n* unknowns, an "*m* by *n* system." The good news is that all the machinery we have built up to solve 3 by 3 systems works for any system, regardless of the number of equations and unknowns. In the real world, it is not unusual for problems to involve thousands of equations and variables. Fortunately, we can direct computers to carry out Gaussian elimination and back substitution for us when we solve large systems. However, even in those cases we must know how to interpret the results that we get.

Example 6 The linear system below consists of 4 equations in 5 variables.

$$\begin{cases} x_1 + x_2 + 3x_4 + 3x_5 = 2 \\ -2x_1 + 2x_2 - x_3 + 2x_4 + 3x_5 = 2 \\ 4x_1 + x_3 + 4x_4 + 3x_5 = 2 \\ -7x_1 + 5x_2 - 3x_3 + 3x_4 + 6x_5 = 4 \end{cases}$$

Here is the augmented matrix:

$$\begin{bmatrix} 1 & 1 & 0 & 3 & 3 & 2 \\ -2 & 2 & -1 & 2 & 3 & 2 \\ 4 & 0 & 1 & 4 & 3 & 2 \\ -7 & 5 & -3 & 3 & 6 & 4 \end{bmatrix}$$

If we ask Maple to reduce the augmented matrix to echelon form, we get the following output:

$$\begin{bmatrix} 1 & 1 & 0 & 3 & 3 & 2 \\ 0 & 4 & -1 & 8 & 9 & 6 \\ 0 & 0 & 0 & 0 & 0 & 0 \\ 0 & 0 & 0 & 0 & 0 & 0 \end{bmatrix}$$

Let's convert the matrix back into equation form:

$$\begin{cases} x_1 + x_2 \ldots + 3x_4 + 3x_5 = 2 \\ \ldots 4x_2 - x_3 + 8x_4 + 9x_5 = 6 \\ 0 = 0 \\ 0 = 0 \end{cases}$$

With five variables and only two equations restricting the variables, we have *three* free variables. To see this more clearly, we first solve the second equation for x_2 in terms of the variables x_3, x_4 and x_5.

$$x_2 = \frac{3}{2} + \frac{1}{4}x_3 - 2x_4 - \frac{9}{4}x_5$$

Then we solve for x_1 in the first equation, substituting our expression for x_2 from the second equation.

$x_1 + (\frac{3}{2} + \frac{1}{4}x_3 - 2x_4 - \frac{9}{4}x_5) + 3x_4 + 3x_5 = 2$ which gives: $x_1 = \frac{1}{2} - \frac{1}{4}x_3 - x_4 - \frac{3}{4}x_5$

So the general solution to the system is:
$$\begin{cases} x_1 = \frac{1}{2} - \frac{1}{4}x_3 - x_4 - \frac{3}{4}x_5 \\ x_2 = \frac{3}{2} + \frac{1}{4}x_3 - 2x_4 - \frac{9}{4}x_5 \\ x_3 = x_3 \,(free) \\ x_4 = x_4 \,(free) \\ x_5 = x_5 \,(free) \end{cases}$$

Clearly we have an infinite number of solutions here since we are free to pick any values for x_3, x_4 and x_5.

Exercise 10 In Example 6 we found that the system had an infinite number of solutions. Use the general solution from that system (repeated below) to write down three different particular solutions to the original system. Take one of these solutions and check that it is in fact a solution to the original system.

$$\begin{cases} x_1 = \frac{1}{2} - \frac{1}{4}x_3 - x_4 - \frac{3}{4}x_5 \\ x_2 = \frac{3}{2} + \frac{1}{4}x_3 - 2x_4 - \frac{9}{4}x_5 \\ x_3 = x_3 \,(free) \\ x_4 = x_4 \,(free) \\ x_5 = x_5 \,(free) \end{cases}$$

Definition: Variables that are not free variables are called **leading** or **constrained variables**. In the previous example, x_3 and x_2 are constrained variables.

Exercise 11 The three matrices below are each the result of applying Gaussian elimination to the augmented matrix of a linear system. For each, answer the following questions.
(a) Is the system consistent?
(b) If the system is consistent, find its solutions. For a system with more than one solution, write the answer as we did in Example 6. Identify the free and constrained variables.

$$\begin{bmatrix} 1 & 2 & 0 & 2 & 1 \\ 0 & 0 & 1 & -2 & 2 \\ 0 & 0 & 0 & 1 & 2 \end{bmatrix} \qquad \begin{bmatrix} 1 & -3 & 2 \\ 0 & 0 & 2 \\ 0 & 0 & 0 \\ 0 & 0 & 0 \end{bmatrix} \qquad \begin{bmatrix} 1 & -2 & 3 & 2 \\ 0 & 0 & 0 & 0 \end{bmatrix}$$

Exercise 12 The purpose of this last exercise is to give you practice using all the tools that you have learned in this tutorial. Start with the linear system shown below and do the following:

a) Write the corresponding augmented matrix.
b) Use Gaussian elimination to reduce this matrix to echelon form. Identify the pivot positions. Suggestion: Use the interchange row operation first to get a convenient first row.
c) Use the echelon form to determine if the system is consistent. If the system is consistent, apply back substitution to find the solution set. Identify the constrained and free variables.
d) Use row reduction to further simplify the echelon form into reduced echelon form. To do this most efficiently, work from right to left. That is, start with the column furthest to the right that contains a leading entry. If necessary, multiply the row to create a 1. Then use row operations to create zeros above it. When you've done this, move to the next column to the left that contains a leading entry and repeat the process. Continue moving left until the matrix is in reduced echelon form.

$$\begin{cases} 2x_1 - x_2 + 7x_3 = 8 \\ x_1 + x_2 - x_3 = 1 \\ -x_1 + x_2 - 5x_3 = -5 \end{cases}$$

Appendix. Why Gaussian Elimination Produces Equivalent Systems

In this appendix, we complete the discussion begun in Section 4; that is, we explain why the three basic operations on a linear system (Multiply, Interchange, and Add) leave the solution set unchanged.

We will focus on the Add operation, since the effect of the other two operations is easy to understand. We begin by recalling the example in Section 4. Specifically, we multiply the first equation of the linear system A by 10 and add it to the second equation, thus producing a new linear system B:

$$A: \begin{cases} x_1 - x_2 - 2x_3 = -1 \\ 2x_1 - x_2 - x_3 = 2 \\ -3x_1 + 5x_2 + 4x_3 = 3 \end{cases} \implies B: \begin{cases} x_1 - x_2 - 2x_3 = -1 \\ 12x_1 - 11x_2 - 21x_3 = -8 \\ -3x_1 + 5x_2 + 4x_3 = 3 \end{cases}$$

If (x_1, x_2, x_3) is any solution of the system A, we see that (x_1, x_2, x_3) is also a solution of system B, as follows. (x_1, x_2, x_3) is a solution of the first and third equations of system B, since these equations are also in system A. Also, the following calculation shows that (x_1, x_2, x_3) is a solution of the second equation in system B:

$$\begin{array}{ll} 10 * \text{equation one}: & 10x_1 - 10x_2 - 20x_3 = -10 \\ + \quad \text{equation two}: & 2x_1 - x_2 - x_3 = 2 \\ \hline \text{new equation two}: & 12x_1 - 11x_2 - 21x_3 = -8 \end{array}$$

This reasoning is valid for every possible application of the Add operation on any linear system. That is, suppose $(x_1, ..., x_n)$ is a solution of any linear system L. Also, suppose we replace the jth equation of L by the sum of the jth equation and c times the ith equation $(i \neq j)$, thus producing a new linear system M. Then $(x_1, ..., x_n)$ is also a solution of the system M, since only the jth equation has been changed and since $(x_1, ..., x_n)$ satisfies

$$\begin{array}{l} c * \text{equation } i \text{ of } L \\ + \quad \text{equation } j \text{ of } L \\ \hline \text{equation } j \text{ of } M \end{array}$$

However, to see that systems L and M have exactly the same solution set, we must also show that every solution of system M is also a solution of system L. In fact, we can obtain system L from system M by the Add operation in which we use $-c$ instead of c:

$$\begin{array}{l} -c * \text{equation } i \text{ of } M \text{ (also of } L) \\ + \quad \text{equation } j \text{ of } M \\ \hline \text{equation } j \text{ of } L \end{array}$$

Therefore, by our reasoning in the preceding paragraph, every solution of system M is also a solution of system L.

We will say that the two Add operations, one using c and the other $-c$, are the *reverse* of one another. For example, the reverse of the above Add operation that changed system A to system B adds -10 times equation one to equation two:

$$B: \begin{cases} x_1 - x_2 - 2x_3 = -1 \\ 12x_1 - 11x_2 - 21x_3 = -8 \\ -3x_1 + 5x_2 + 4x_3 = 3 \end{cases} \implies A: \begin{cases} x_1 - x_2 - 2x_3 = -1 \\ 2x_1 - x_2 - x_3 = 2 \\ -3x_1 + 5x_2 + 4x_3 = 3 \end{cases}$$

Similarly, the Multiply and Interchange operations leave the solution set of a linear system unchanged. Specifically, the reverse of the operation in which the ith equation is multiplied by the nonzero constant c is the operation in which the same i^{th} equation is multiplied by $1/c$. Also, the reverse of the operation in which equations i and j ($i \neq j$) are interchanged is exactly the same Interchange operation.

Therefore, if we apply any sequence of Multiply, Interchange, and Add operations to a linear system, the solution set of the linear system will be unchanged.

Finally, let's reformulate our conclusions in terms of row operations on the augmented matrix of a linear system. Note that applying the Multiply Row, Interchange Row, and Add Row operations to the augmented matrix has the same effect as applying the corresponding Multiply, Interchange, and Add operations to the linear system itself. For example, here is the Multiply Row operation corresponding to the Multiply operation at the beginning of this section. We multiply the first row of the augmented matrix by 10 and add it to the second row:

$$\begin{matrix} x_1 - x_2 - 2x_3 = -1 \\ 2x_1 - x_2 - x_3 = 2 \\ -3x_1 + 5x_2 + 4x_3 = 3 \end{matrix} \quad \begin{bmatrix} 1 & -1 & -2 & -1 \\ 2 & -1 & -1 & 2 \\ -3 & 5 & 4 & 3 \end{bmatrix} \implies \begin{bmatrix} 1 & -1 & -2 & -1 \\ 12 & -11 & -21 & -8 \\ -3 & 5 & 4 & 3 \end{bmatrix} \quad \begin{matrix} x_1 - x_2 - 2x_3 = -1 \\ 12x_1 - 11x_2 - 21x_3 = -8 \\ -3x_1 + 5x_2 + 4x_3 = 3 \end{matrix}$$

Conclusion: Suppose we form the augmented matrix of a linear system L and apply a sequence of row operations to its augmented matrix. Then the linear system M corresponding to the new augmented matrix has the same solution set as the system L. We have therefore proved Theorem 1 (see the beginning of Module 2).

In particular, suppose we use the method of Gaussian elimination to produce a sequence of row operations that transforms the augmented matrix of system L to echelon form. Then the system M corresponding to this echelon form can be solved easily by back substitution. Furthermore, by our above conclusion, in solving system M we have solved the original system L.

Pencil and Paper Tutorial: Determinants and Cofactors

Section 1. Introduction

The determinant of a square matrix A, denoted $\text{Det}(A)$, is a single number. That single number happens to provide quite useful information about the matrix, and it also has convenient algebraic properties. However, we leave these aspects of the determinant to Module 7 of Chapter 3. Our goal here is merely to define the determinant, which is a necessary preparation for the module.

We define the determinant by showing how to calculate it. We will start with small matrices and work our way up. If A is a 1 by 1 matrix, $A = [a]$, then $\text{Det}(A) = a$. The determinant of a 2 by 2 matrix is also easy to define and calculate:

$$\text{If } A = \begin{bmatrix} a & b \\ c & d \end{bmatrix} \text{ then } \text{Det}(A) = ad - bc.$$

For example, if $A = \begin{bmatrix} 2 & 3 \\ 4 & 7 \end{bmatrix}$ then $\text{Det}(A) = 14 - 12 = 2$.

Before we consider the formula for calculating the determinant of a 3 by 3 matrix, we introduce some notation that will facilitate the discussion. We will use the notation $a_{i,j}$ to denote the entry in row i and column j of the matrix A. So for the matrix $A = \begin{bmatrix} 2 & 3 \\ 4 & 7 \end{bmatrix}$ we have $a_{1,1} = 2$, $a_{1,2} = 3$, $a_{2,1} = 4$ and $a_{2,2} = 7$.

Section 2.   Minors

If we cross out row 1 and column 1 of the matrix A below, we are left with a 2 by 2 matrix in the lower-right corner. This submatrix of A is called the (1, 1) *minor* of A.

$$A = \begin{bmatrix} 1 & 3 & 5 \\ 4 & 2 & -3 \\ 1 & -1 & -2 \end{bmatrix} \Rightarrow \begin{bmatrix} * & * & * \\ * & 2 & -3 \\ * & -1 & -2 \end{bmatrix} \Rightarrow \begin{bmatrix} 2 & -3 \\ -1 & -2 \end{bmatrix}$$

We will use the notation $A_{1,1}$ to denote this minor; so $A_{1,1} = \begin{bmatrix} 2 & -3 \\ -1 & -2 \end{bmatrix}$. Similarly, we have minors associated with each of the other 8 entries of A; for example:

$$\text{Crossing out row 1 and column 2 of } A: \begin{bmatrix} * & * & * \\ 4 & * & -3 \\ 1 & * & -2 \end{bmatrix} \Rightarrow A_{1,2} = \begin{bmatrix} 4 & -3 \\ 1 & -2 \end{bmatrix}$$

Crossing out row 2 and column 2 of A: $\begin{bmatrix} 1 & * & 5 \\ * & * & * \\ 1 & * & -2 \end{bmatrix} \Rightarrow A_{2,2} = \begin{bmatrix} 1 & 5 \\ 1 & -2 \end{bmatrix}$

In general, the *(i, j) minor* of a matrix A, denoted $A_{i,j}$, is the submatrix that remains when we delete row i and column j of A.

Exercise 1 Write down the (2, 3) minor of the matrix M.

$$M = \begin{bmatrix} 2 & 3 & 6 & 5 \\ 6 & 2 & 9 & 0 \\ 7 & 0 & -4 & 6 \\ -2 & 3 & 7 & 5 \end{bmatrix}$$

Section 3. Determinant of a 3 by 3 Matrix

The formula for the determinant of a 3 by 3 matrix is as follows:

$$\text{If } A = \begin{bmatrix} a & b & c \\ d & e & f \\ g & h & i \end{bmatrix} \text{ then } \text{Det}(A) = aei - afh - bdi + bfg + cdh - ceg \quad [1]$$

As you can see, the formula is substantially more complicated than the 2 by 2 case. For larger matrices, the formulas become much longer yet. A more efficient scheme for writing these formulas makes use of minors. All that's needed is a bit of algebra.

Note what happens when we group the terms on the right side of formula [1] two at a time and then factor:

$$aei - afh - bdi + bfg + cdh - ceg = a(ei - fh) - b(di - fg) + c(dh - eg)$$

Observe that each factor in parentheses on the right is the determinant of a minor:

$$ei - fh = \text{Det}(A_{1,1}), \quad di - fg = \text{Det}(A_{1,2}), \quad dh - eg = \text{Det}(A_{1,3})$$

If we make these substitutions and replace a, b, c by $a_{1,1}, a_{1,2}, a_{1,3}$, respectively, we get:

$$\text{Det}(A) = a_{1,1} \text{Det}(A_{1,1}) - a_{1,2} \text{Det}(A_{1,2}) + a_{1,3} \text{Det}(A_{1,3}) \quad [2]$$

This is called the *cofactor expansion across row 1 of A*. Note the alternation in signs for the three terms $(+, -, +)$.

Example 1 Using formula [2], find the determinant of $A = \begin{bmatrix} 1 & 3 & 5 \\ 4 & 2 & -3 \\ 1 & -1 & -2 \end{bmatrix}$.

Solution: The entries of the first row are $a_{1,1} = 1$, $a_{1,2} = 3$, $a_{1,3} = 5$. The minors are:

$$A_{1,1} = \begin{bmatrix} 2 & -3 \\ -1 & -2 \end{bmatrix}, \quad A_{1,2} = \begin{bmatrix} 4 & -3 \\ 1 & -2 \end{bmatrix}, \quad A_{1,3} = \begin{bmatrix} 4 & 2 \\ 1 & -1 \end{bmatrix}$$

So we have $\text{Det}(A) = 1\,\text{Det}(A_{1,1}) - 3\,\text{Det}(A_{1,2}) + 5\,\text{Det}(A_{1,3}) = 1(-7) - 3(-5) + 5(-6) = -22$.

Section 4.   Cofactors

We have referred to formula [2] as a "cofactor" expansion. We now define this term.

   Definition: If A is a square matrix and $A_{i,j}$ is the *(i, j)* minor of A, then $C_{i,j}$, the ***(i, j) cofactor*** of A, is defined as:

$$C_{i,j} = \text{Det}(A_{i,j}) \qquad \text{if } i+j \text{ is an even number}$$

$$C_{i,j} = (-1)\,\text{Det}(A_{i,j}) \qquad \text{if } i+j \text{ is an odd number}$$

We can also write this in a single formula:

$$C_{i,j} = (-1)^{(i+j)}\,\text{Det}(A_{i,j})$$

We can now rewrite formula [2] using cofactors as

$$\text{Det}(A) = a_{1,1}C_{1,1} + a_{1,2}C_{1,2} + a_{1,3}C_{1,3}$$

Or, using summation notation, we have the more compact formula

$$\text{Det}(A) = \sum_{j=1}^{3} a_{1,j}C_{1,j} \qquad\qquad [3]$$

The cofactor expansion for calculating the determinant of a matrix can be based on <u>any</u> row or column of the matrix. For example, the cofactor expansion <u>across row 2</u> of the matrix A is:

$$\text{Det}(A) = \sum_{j=1}^{3} a_{2,j}C_{2,j} = a_{2,1}C_{2,1} + a_{2,2}C_{2,2} + a_{2,3}C_{2,3}$$

Exercise 2 (a) Using formula [1] for the determinant of a 3 by 3 matrix, regroup and factor out $d, e,$ and f, the entries of row 2. Show that this factorization leads to the above formula for the cofactor expansion across row 2. (b) Use this formula to recompute the determinant of the matrix in Example 1:

$$\begin{bmatrix} 1 & 3 & 5 \\ 4 & 2 & -3 \\ 1 & -1 & -2 \end{bmatrix}$$

Section 5. Determinant of an *n* by *n* Matrix

The general definition of the determinant of an n by n matrix follows the pattern we observed in the 3 by 3 case. That is, we define the determinant of an n by n matrix in terms of the determinants of minors of A, each of size $n-1$. For example, we define the determinant of a 4 by 4 matrix using four minors of size 3 by 3. Since we define the determinant of any n by n matrix in terms of the determinants of $n-1$ by $n-1$ matrices, this is called a "recursive" definition.

The determinant of an n by n matrix A is defined recursively by its cofactor expansion across row 1:

$$\text{Det}(A) = \sum_{j=1}^{n} a_{1,j} C_{1,j}$$

where, as defined in Section 4, $C_{1,j} = (-1)^{(1+j)} \text{Det}(A_{1,j})$.

Note the similarity to formula [3]. Furthermore, just as we saw for 3 by 3 matrices, we can regroup terms to obtain other cofactor expansions:

Appendix B Determinant and Cofactors

Theorem 10: The determinant of an n by n matrix A can be calculated using the cofactor expansion across any row or down any column:

The cofactor expansion across row i is: $\quad \text{Det}(A) = \sum_{j=1}^{n} a_{i,j} C_{i,j}$

The cofactor expansion down column j is: $\quad \text{Det}(A) = \sum_{i=1}^{n} a_{i,j} C_{i,j}$

where $C_{i,j} = (-1)^{(i+j)} \text{Det}(A_{i,j})$ and $A_{i,j}$ denotes the (i, j) minor of A.

As a practical matter, the cascading series of calculations required to apply these formulas to even a moderately large matrix makes the cofactor expansion unworkable for hand calculations and highly inefficient for computers as well. However, the determinants of some special types of matrices are easy to calculate with cofactor expansions:

Exercise 3 Calculate the determinant of the matrix A below. Exploit the fact that A has lots of zeros by expanding along a row or column that has many zeros.

$$A = \begin{bmatrix} 3 & 4 & 0 & -1 \\ 5 & 6 & 0 & 0 \\ -3 & 0 & 0 & 3 \\ 0 & 0 & 5 & -2 \end{bmatrix}$$

Exercise 4 Here is one more example on which to practice cofactor expansions. Evaluate the determinant of the matrix B below.

$$B = \begin{bmatrix} 3 & 4 & 5 & 6 \\ 0 & -2 & 1 & -1 \\ 0 & 0 & 1 & 7 \\ 0 & 0 & 0 & 2 \end{bmatrix}$$

Answers to Exercises

Answer 1:

$$M_{2,3} = \begin{bmatrix} 2 & 3 & 5 \\ 7 & 0 & 6 \\ -2 & 3 & 5 \end{bmatrix}$$

Answer 2:

(a) If $A = \begin{bmatrix} a & b & c \\ d & e & f \\ g & h & i \end{bmatrix}$ then $\text{Det}(A) = aei - afh - bdi + bfg + cdh - ceg$

Now regroup:

$$\text{Det}(A) = -bdi + cdh + aei - ceg - afh + bfg$$

and factor:

$$\text{Det}(A) = d(-1)(bi - ch) + e(ai - cg) + f(-1)(ah - bg)$$

Rewriting in terms of cofactors, we have $\text{Det}(A) = a_{2,1}C_{2,1} + a_{2,2}C_{2,2} + a_{2,3}C_{2,3}$.

(b) If $A = \begin{bmatrix} 1 & 3 & 5 \\ 4 & 2 & -3 \\ 1 & -1 & -2 \end{bmatrix}$ then $\text{Det}(A) = -4\,\text{Det}(A_{2,1}) + 2\,\text{Det}(A_{2,2}) + 3\,\text{Det}(A_{2,3})$.

So $\text{Det}(A) = -4(-6+5) + 2(-2-5) + 3(-1-3) = 4 - 14 - 12 = -22$.

Answer 3:

First expand along column 3, since it has three zero entries:

$$\text{Det}(A) = -5\,\text{Det}(A_{4,3}) = -5\,\text{Det}\left(\begin{bmatrix} 3 & 4 & -1 \\ 5 & 6 & 0 \\ -3 & 0 & 3 \end{bmatrix}\right)$$

Then expand along row 3:

$$\text{Det}(A) = -5\left(-3\text{Det}\left(\begin{bmatrix} 4 & -1 \\ 6 & 0 \end{bmatrix}\right) + 3\text{Det}\left(\begin{bmatrix} 3 & 4 \\ 5 & 6 \end{bmatrix}\right)\right) = -5(-3(0+6) + 3(18-20)) = (-5)(-24) = 120.$$

Answer 4:

Expand along column 1 again and again:

$$\text{Det}\left(\begin{bmatrix} 3 & 4 & 5 & 6 \\ 0 & -2 & 1 & -1 \\ 0 & 0 & 1 & 7 \\ 0 & 0 & 0 & 2 \end{bmatrix}\right) = 3\,\text{Det}\left(\begin{bmatrix} -2 & 1 & -1 \\ 0 & 1 & 7 \\ 0 & 0 & 2 \end{bmatrix}\right) = 3(-2)\,\text{Det}\left(\begin{bmatrix} 1 & 7 \\ 0 & 2 \end{bmatrix}\right) = 3(-2)(1)(2) = -12.$$

The Lamp Library

Getting help with a command in the Lamp Library

The Lamp Library includes a complete help system that is integrated within Maple's online Help Browser. You can access help on the Lamp commands in a number of ways:

- You can access a help page from a worksheet by typing a question mark followed by the name of the command. For example, to access help on the command Matsolve, type: ?Matsolve

- You can access a help page from the Help Browser by clicking on the name of the command. As you read from left to right, the Lamp commands are found under: Mathematics...Linear Algebra...Lamp Library...

- Finally, you can access the help page for any command by clicking anywhere within the word and either pressing the F1 key (if you are using Windows) or clicking on the command name that now appears in the help menu.

- Keep in mind that, once you are within Maple's help system, you can use the Help toolbar to navigate back and forth between related help pages. For example, if you click on a hyperlink in one page, you can return to the previous page by using the back arrow. Here are the most useful toolbar buttons:

    Display the last help page viewed.

    Go forward in the help page history.

    Go to previous help topic alphabetically.

The Lamp environment

The first step in working through a module from "Linear Algebra: Modules for Interactive Learning Using MAPLE" is to load the "Lamp environment." This environment activates all the commands used in the modules and sets defaults for some important computing variables. This is accomplished by executing the following two lines that you see at the start of every module:

```
> restart: with(LinearAlgebra): with(plots): with(Lamp):
UseHardwareFloats := false: Digits := 6:
```

Here is what these two lines do:

- The **restart** command causes the Maple kernel to clear its internal memory so that it acts as if you had just started Maple.
- **with(LinearAlgebra)** loads the LinearAlgebra package.
- **with(plots)** loads the plots package.
- **with(Lamp)** loads the Lamp Library.
- **UseHardwareFloats := false** and **Digits := 6** instructs Maple to use its software floating-point computation environment to perform all floating-point operations; then it sets the default number of digits to 6.

Every time you open a module, you must first execute the two lines above for the module to function properly. Likewise, if you open a new Maple worksheet and want access to the commands and settings of the Lamp environment, you will want to enter the two initialization lines above.

Additionally, the Lamp Library instructs Maple to (1) use lower-case i as the symbol for $\sqrt{-1}$ (in place of Maple's default choice I), (2) put no restriction on the size of matrices that can be displayed, and (3) display output in mathematical notation. Specifically, Lamp's initialization procedure has the following three commands:

> `interface(imaginaryunit=i):`
> `interface(rtablesize=infinity):`
> `interface(prettyprint=3):`

Shortcut for loading the Lamp environment

As an alternative to entering the two initialization lines described above, we have provided a shortcut for setting the Lamp environment in one step. Simply enter the following command:
> `read(lampstart):`

This shortcut, besides executing the same commands as above (except for **restart**), enables users to employ lower-case names for all commands. So you can then enter "drawvec" instead of "Drawvec," "vector" instead of "Vector," etc. Students find this a convenience since it relieves them of having to remember which commands require capitalization. Of course, you are still free to use the upper-case version.

Note 1: This shortcut does not include a **restart**.. If you want to clear Maple's internal memory in addition to setting the Lamp environment, enter this sequence of commands:
> `restart: read(lampstart):`
Note 2: Maple already has a command called "expand." This is not the same as the Lamp command "Expand." So, for this one command, lower-case cannot be substituted for upper-case.

Glossary of Linear Algebra Definitions

Note: the 3-digit code following each definition gives the location, [Chapter.Module.Section], in which the definition is first used. For example, [2.4.3] means Chapter 2, Module 4, Section 3. "A" stands for appendix.

algebraic multiplicity of an eigenvalue:

    The number of times that a factor $\lambda - r$ occurs in the characteristic polynomial is called the ***algebraic multiplicity*** of the eigenvalue r. [6.2.1]

basis for a subspace:

    Let S be any subspace of R^n. A set of vectors $\{v_1, ..., v_p\}$ in S is called a ***basis*** of S if $\{v_1, ..., v_p\}$ spans S and $\{v_1, ..., v_p\}$ is linearly independent. [5.2.1]

characteristic polynomial of a matrix:

    The polynomial $p(\lambda) = \text{Det}(A - \lambda I)$ is called the ***characteristic polynomial*** for the matrix A. The equation $p(\lambda) = 0$ is called the ***characteristic equation*** of A. [6.2.1]

cofactor of a matrix:

    If A is a square matrix and $A_{i,j}$ is the (i, j) minor of A, then $C_{i,j}$, the ***(i, j) cofactor*** of A, is defined as:

$$C_{i,j} = (-1)^{(i+j)} \text{Det}(A_{i,j}) \qquad [3.7.1]$$

column space of a matrix:

    The ***column space*** of A is the span of the column vectors of A, which is a subspace of R^m. [5.3.1]

consistent linear system:

    A system of linear equations is ***consistent*** if it has at least one solution. [1.2.A]

coordinates relative to a basis:

    Suppose $B = \{v_1, ..., v_p\}$ is a basis of a subspace S and w is a vector in S. Then the weights $c_1, ..., c_p$ in the linear combination $w = c_1 v_1 + ... + c_p v_p$ are called the ***coordinates of w relative to the basis B***, or more succinctly, the ***B-coordinates of*** w. [5.2.3]

dimension of a subspace:

    Let S be any subspace of R^n other than the zero subspace $\{0\}$. If S has a basis consisting of p vectors, then p is called the ***dimension*** of S. By convention, we say that the dimension of the zero subspace, $\{0\}$, is 0. [5.2.1]

distance between vectors:

If u and v are vectors in R^n then the ***distance between u and v*** is the magnitude of the vector $u - v$, i.e., $\|u - v\|$. [7.1.1]

dot product:

If $u = \langle u_1, ..., u_n \rangle$ and $v = \langle v_1, ..., v_n \rangle$ are vectors in R^n, then the ***dot product*** of u and v, denoted $u.v$, is the sum of the products of the corresponding components of u and v:

$$u.v = u_1 v_1 + ... + u_n v_n \quad [7.1.1]$$

echelon form of a matrix: (see *row echelon form*):

eigenspace of a matrix:

If A is a square matrix with eigenvalue λ, then the null space of the matrix $A - \lambda I$ is called the ***eigenspace*** associated with the eigenvalue λ. [6.1.1]

eigenvalue of a matrix:
eigenvector of a matrix:

If A is a square matrix, we say that a nonzero vector x is an ***eigenvector*** of A and a scalar λ is the associated ***eigenvalue*** if the following equation holds:

$A x = \lambda x$

In other words, when x is multiplied by A, the result is just a scalar multiple of x. [6.1.1]

elementary matrix:

An ***elementary matrix*** is a matrix that is created by performing a single elementary row operation on an identity matrix. [3.6.1]

equivalent linear systems:

Two systems of linear equations in n unknowns are ***equivalent*** if they have the same set of solutions.[1.2.A]

geometric multiplicity of an eigenvalue:

The dimension of the eigenspace associated with an eigenvalue λ is called the ***geometric multiplicity*** of λ. (That is, the dimension of the null space of $A - \lambda I$ is the geometric multiplicity of λ.) [6.1.1]

homogeneous linear system:

A system of linear equations is ***homogeneous*** if all the constants on the right side are zero.

inconsistent linear system:

A system of linear equations is *inconsistent* if it has no solutions.[1.2.A]

inverse of a matrix:
invertible matrix:

The *inverse* of an *n* by *n* matrix A is an *n* by *n* matrix B such that: $AB = I$ and $BA = I$ where I is the *n* by *n* identity matrix. The notation for the inverse of A is $A^{(-1)}$. The matrix A is said to be *invertible* if it has an inverse (synonym: *nonsingular*). [3.5.1]

kernel of a matrix transformation:

The *kernel* of T is the set of vectors x in R^n such that $T(x) = 0$ (where 0 denotes the zero vector in R^m) ; in other words, the kernel is the set of "zeros" of T. [4.2.2]

leading entry:

The first nonzero entry in each nonzero row of a matrix in row echelon form is called the *leading entry* of that row.[1.2.A]

least-squares solution of a linear system:

A *least-squares solution* to a linear system $Ax = b$ is a vector x that yields the smallest value of $\|Ax - b\|$. [7.3.1]

length of a vector:

The *length* (or *magnitude*) of a vector $u = \langle u_1, ..., u_n \rangle$ in R^n is defined to be the square root of $u.u$:

$$\|u\| = \sqrt{u.u} \qquad [7.1.1]$$

linear combination of vectors:

Given a set of vectors $v_1, v_2, ..., v_k$ in R^n, a *linear combination* of these vectors is a vector of the form $c_1 v_1 + c_2 v_2 + ... + c_k v_k$ (where the scalar weights $c_1, c_2, ..., c_k$ are real numbers). [2.2.3]

linear dependence relation for a set of vectors:
linearly dependent set of vectors:
linearly independent set of vectors:

A set of vectors $\{v_1, v_2, ..., v_k\}$ in R^n is *linearly dependent* if the vectors satisfy a *linear dependence relation*

$$c_1 v_1 + c_2 v_2 + ... + c_k v_k = 0$$

where at least one of the weights $c_1, c_2, ..., c_k$ is not zero. (The 0 in the above equation is the zero vector in R^n.) The set of vectors is ***linearly independent*** if it is not linearly dependent, that is, if there is no linear dependence relation satisfied by the vectors. [2.4.2]

matrix product:

Suppose A is a matrix with m rows and n columns and B is a matrix with n rows and r columns. The ***matrix product $A\,B$*** is defined column-by-column as follows. The first column of $A\,B$ is the matrix A times the first column of B; the second column of $A\,B$ is A times the second column of B; and so on. Thus each column of $A\,B$ is the matrix-vector product formed by multiplying the matrix A by the corresponding column of B.

We can express this definition in symbols: Let $B_1, ..., B_r$ denote the columns of the matrix B. Then

$$A\,B = A\,[B_1, B_2, ..., B_r] = [A\,B_1, A\,B_2, ..., A\,B_r] \qquad [3.2.1]$$

matrix-vector product:

Suppose A is a matrix with m rows and n columns, and x is a vector with n components. The ***matrix-vector product*** $A\,x$ is defined to be the linear combination of the columns of A in which the weights are the components of x.

We can express this definition in symbols: Let $A_1, A_2, ..., A_n$ denote the columns of the matrix A, and $x_1, x_2, ..., x_n$ denote the components of the vector x. Then

$$A\,x = [A_1, A_2, ..., A_n] \begin{bmatrix} x_1 \\ \cdot \\ \cdot \\ x_n \end{bmatrix} = x_1\,A_1 + x_2\,A_2 + ... + x_n\,A_n \qquad [3.1.1]$$

minor of a matrix:

The ***(i, j) minor*** of a matrix A, denoted $A_{i,j}$, is the submatrix that remains when we delete row i and column j of A. [3.7.1]

n-space:

The set of all vectors with n components is denoted by R^n and is called ***n-space***. Such vectors are added componentwise, and scalars are multiplied by such vectors componentwise. [2.2.3]

nonsingular matrix:

A square matrix A is ***nonsingular*** if and only if A is invertible. [3.5.1]

null space of a matrix:

The ***null space*** of A is the solution set of $A\,x = 0$, which is a subspace of R^n. [5.3.1] *orthogonal* vectors:

Vectors u and v in R^n are ***orthogonal*** if $u.v = 0$; i.e., their dot product is zero. [7.1.1]

orthogonal basis of a subspace:

An ***orthogonal basis*** of a subspace S is a basis of S which is also an orthogonal set. That is, every pair of vectors in the basis is orthogonal. [7.2.1]

orthogonal complement of a subspace:

Suppose S is a subspace of R^n. The ***orthogonal complement*** of S in R^n is the set of all vectors v in R^n such that $v.u = 0$ for all u in S, that is, the set of all vectors that are orthogonal to every vector in S. [7.2.4]

orthogonal matrix:

A square matrix U is ***orthogonal*** if $U^T U = I$. [7.2.5]

orthogonal set of vectors:

A set of vectors $\{v_1, ..., v_k\}$ in R^n is called an ***orthogonal set*** if every pair of vectors in the set is orthogonal. [7.2.1]

orthonormal basis of a subspace:

An ***orthonormal basis*** of a subspace S is an orthogonal basis of S in which all of the basis vectors are unit vectors. [7.2.1]

permutation matrix:

A ***permutation matrix*** is a matrix obtained by rearranging (i.e., permuting) the columns of an identity matrix. [3.2.2]

pivot columns:
pivot position:

The first nonzero entry in each nonzero row of a matrix in row echelon form is called the leading entry of that row; the locations of the leading entries are called the ***pivot*** positions, and the columns containing the leading entries are called the ***pivot columns***.[1.2.A]

powers of a matrix:

If A is a square matrix, we define A^2 to be the product $A\,A$. Higher powers are defined similarly: A^n is the product of n copies of A. [3.2.2]

projection of a vector onto a subspace:

If u is any vector in R^n and W is any subspace of R^n, the ***projection of u onto W*** is the vector p that has the following two properties:

(i) p is in W and (ii) $u - p$ is orthogonal to every vector in W. [7.1.3]

range of a matrix transformation:

The ***range*** of T is the set of vectors $T(x)$, where x ranges over all vectors in R^n; in other words, the range is the set of all output (i.e., image) vectors of T in R^m. [4.2.2]

rank of a matrix:

The ***rank*** of a matrix A is the number of nonzero rows in any row echelon form of A; i.e., the dimension of the row space of A. [1.2.6] also [5.3.1]

row echelon form of a matrix:
reduced row echelon form of a matrix:

A matrix is in ***row echelon form*** if:
(1) all rows that consist entirely of zeros are grouped together at the bottom of the matrix; and
(2) the first (counting left to right) nonzero entry in each nonzero row appears in a column to the right of the first nonzero entry in the preceding row (if there is a preceding row).

A matrix is in ***reduced row echelon form*** if:
(1) the matrix is in ***row echelon form***; and
(2) the first nonzero entry in each nonzero row is the number 1; and
(3) the first nonzero entry in each nonzero row is the only nonzero entry in its column. [1.2.A]

row equivalent matrices:

Two m by n matrices A and B are ***row equivalent*** if B can be obtained from A by a sequence of elementary row operations.[1.2.A]

row operations:

The elementary ***row operations*** performed on a matrix are:
(1) replace row j by row $j + (c$ times row $k)$
(2) replace row k by c times row k $(c \neq 0)$
(3) interchange row k and row j [1.2.A]

row space of a matrix:

The ***row space*** of A is the span of the row vectors of A, which is a subspace of R^n. [5.3.1]

similar matrices:

Two *n* by *n* matrices *A* and *B* are ***similar*** if there is an invertible matrix *P* such that:

$$P^{-1} A P = B, \text{ or equivalently, } A = P B P^{-1} \quad [6.4.2]$$

singular matrix:

A matrix that does not have an inverse is called a ***singular*** matrix (synonym: ***noninvertible***). [3.5.1]

span of a set of vectors:

The ***span*** of the set of vectors $\{v_1, ..., v_k\}$ is the set of all linear combinations of $v_1, ..., v_k$ and is denoted by Span$\{v_1, ..., v_n\}$. [2.2.3]

stochastic matrix:

A vector whose entries are non-negative numbers adding to 1 is called a probability vector. A square matrix *P* whose columns are probability vectors is called a ***stochastic matrix*** (or a ***Markov matrix***). [3.4.3]

subspace:

Let S be a set of vectors in R^n. S is a ***subspace*** of R^n if:
- (1) *k u* is in S whenever *u* is a vector in S and *k* is a scalar; and
- (2) *u* + *v* is in S whenever *u* and *v* are vectors in S; and
- (3) The zero vector of R^n is contained in S   [5.1.1]

symmetric matrix:

A matrix *A* is ***symmetric*** if it equals its transpose; i.e., $A = A^T$.

transpose of a matrix:

The ***transpose*** of a matrix *A* is the matrix obtained from *A* by making its rows into its columns (and hence its columns into its rows). [3.3.3]

List of Linear Algebra Theorems

 Theorems for Chapter 1

Theorem 1: Suppose we are given two systems of linear equations, both with m equations and n unknowns. If their augmented matrices are row equivalent, then the two systems have exactly the same solution set.

Theorem 2: Every matrix is row equivalent to a matrix in row echelon form.

Theorem 3: Suppose we are given a consistent linear system with m equations and n unknowns. If a row echelon form of the augmented matrix of this system has exactly r nonzero rows (i.e., rows that are not made up entirely of zeros), then the solution set of the system has exactly $n - r$ free variables.

 Theorems for Chapter 3

Theorem 1: Suppose A is a m by n matrix with columns $A_1, A_2, \ldots, A_n$ and b is a vector in R^m. Then
- the vector equation $x_1 A_1 + x_2 A_2 + \ldots + x_n A_n = b$,
- the matrix-vector equation $A x = b$, and
- the linear system with coefficient matrix A and right-side vector b

all have the same set of solutions.

Theorem 2: If p is a particular solution of the linear system $A x = b$ and $\text{Span}\{v_1, \ldots, v_k\}$ is the solution set of the corresponding homogeneous system $A x = 0$, then $p + \text{Span}\{v_1, \ldots, v_k\}$ is the solution set of $A x = b$.

Theorem 3: Matrix algebra obeys the following rules (where uppercase letters denote matrices and lowercase letters denote scalars; also, the letter O denotes the zero matrix, and the letter I denotes the identity matrix):

$$A + B = B + A \quad \text{(commutative law for addition)}$$
$$(A + B) + C = A + (B + C) \quad \text{(associative law for addition)}$$
$$(A B) C = A (B C) \quad \text{(associative law for multiplication)}$$
$$A (B + C) = A B + A C \quad \text{(first distributive law)}$$
$$(A + B) C = A C + B C \quad \text{(second distributive law)}$$

$$A + O = A \quad \text{(identity law for addition)}$$
$$A I = I A = A \quad \text{(identity law for multiplication)}$$
$$A - A = O \quad \text{(inverse law for addition)}$$

And here are rules involving scalar multiplication:

$$(c\,d)\,A = c\,(d\,A)$$
$$c(A\,B) = (c\,A)\,B = A\,(c\,B)$$
$$(c + d)\,A = c\,A + d\,A$$
$$c\,(A + B) = c\,A + c\,B$$

Theorem 4: The transpose operation for matrices obeys the following rules of algebra (where uppercase letters denote matrices and the lowercase letter is a scalar):

$$(A + B)^T = A^T + B^T$$
$$(c\,A)^T = c\,A^T$$
$$(A^T)^T = A$$
$$(A\,B)^T = B^T A^T$$

Theorem 5: Suppose A and B are n by n invertible matrices. Then:
(a) A has only one inverse matrix;
(b) $A^{(-1)}$ is invertible and $(A^{(-1)})^{(-1)} = A$;
(c) $A B$ is invertible and $(A B)^{(-1)} = B^{(-1)} A^{(-1)}$.

Theorem 6 (Invertibility Theorem): Suppose A is a n by n matrix. Then the statement that A is invertible is equivalent to each of the following statements:
(a) the only solution to the homogeneous linear system $A x = 0$ is the trivial solution $x = 0$;
(b) the columns of A are linearly independent;
(c) the linear system $A x = b$ is consistent for every vector b in R^n;
(d) the columns of A span R^n.

Theorem 7: A square matrix A is invertible if and only if it is row equivalent to the identity matrix.

Theorem 8: Suppose A and B are n by n matrices such that $A B = I$. Then A and B are both invertible, and $A^{(-1)} = B$.

Theorem 9: Suppose A and B are matrices where B is obtained from A by applying a single elementary row operation to A. Then $B = EA$, where E is the elementary matrix obtained by applying the same elementary row operation to the identity matrix.

Theorem 10: The determinant of an n by n matrix A can be calculated using the cofactor expansion across any row or down any column:

The cofactor expansion across row i is: $\quad \text{Det}(A) = \sum_{j=1}^{n} a_{i,j} C_{i,j}$

The cofactor expansion down column j is: $\quad \text{Det}(A) = \sum_{i=1}^{n} a_{i,j} C_{i,j}$

where $C_{i,j} = (-1)^{(i+j)} \text{Det}(A_{i,j})$ and $A_{i,j}$ denotes the (i,j) minor of A.

Theorem 11: The determinant of an n by n upper triangular matrix is the product of its diagonal entries.

Theorem 12: If A and B are n by n matrices, then $\text{Det}(AB) = \text{Det}(A)\text{Det}(B)$.

Theorem 13: If A is an n by n matrix, then $\text{Det}(A^T) = \text{Det}(A)$.

Theorem 14: A square matrix A is invertible if and only if $\text{Det}(A) \neq 0$. Equivalently, A is singular if and only if $\text{Det}(A) = 0$.

Theorem 15: If a matrix B is obtained from an n by n matrix A by:
 (a) interchanging two rows of A, then $\text{Det}(B) = -\text{Det}(A)$;
 (b) multiplying one row of A by a scalar k, then $\text{Det}(B) = k\,\text{Det}(A)$;
 (c) adding a multiple of one row of A to another row of A, then $\text{Det}(B) = \text{Det}(A)$.

Theorem 16: **(Cramer's Rule)** If A is an n by n invertible matrix and b is a vector in R^n, then the components of the unique solution x of the matrix-vector equation $Ax = b$ are given by:

$$x_i = \frac{\text{Det}(Ai\_b)}{\text{Det}(A)} \quad \text{for } i = 1 \,..\, n$$

Theorems for Chapter 5

Theorem 1: If $v_1, v_2, ..., v_p$ are vectors in R^n, then $\text{Span}\{v_1, v_2, ..., v_p\}$ is a subspace of R^n.

Theorem 2: Let A be any m by n matrix. The set of solutions of the homogeneous equation $Ax = 0$ is a subspace of R^n.

Theorem 3: Suppose a subspace S is spanned by $\{v_1, v_2, ..., v_p\}$ and that $\{w_1, w_2, ..., w_q\}$ is another set of vectors in S. If $p < q$, then $\{w_1, w_2, ..., w_q\}$ is linearly dependent.

Theorem 4: If $\{v_1, v_2, ..., v_p\}$ and $\{w_1, w_2, ..., w_q\}$ are bases of a subspace S, then $p = q$.

Theorem 5: If a nonzero subspace S is spanned by $\{v_1, v_2, ..., v_p\}$, then some subset of $\{v_1, v_2, ..., v_p\}$ is a basis of S.

Theorem 6: Suppose S is a subspace of dimension p. Then:
 (a) Any set of more than p vectors in S is linearly dependent.
 (b) Any set of fewer than p vectors in S will not span S.
 (c) Any set of exactly p vectors in S is linearly independent if and only if it spans S.

Theorem 7: If S is a nonzero subspace of R^n, then S has a basis.

Theorem 8: The column space and row space of a matrix A have the same dimension. This dimension is equal to r, the rank of A.

Theorem 9: The dimension of the null space of a matrix A is $n - r$, where n is the number of columns of A and r is the rank of A.

Theorem 10: Suppose A is an m by n matrix and T is the corresponding transformation from R^n to R^m defined by $T(x) = Ax$. Then (a) the kernel of T is the null space of A, and (b) the range of T is the column space of A.

Theorems for Chapter 6

Theorem 1: A number λ is an eigenvalue of the square matrix A if and only if the equation $(A - \lambda I)x = 0$ has a nontrivial solution x. The nontrivial solutions are the eigenvectors associated with λ.

Theorem 2: An n by n matrix has at most n distinct eigenvalues.

Theorem 3: A number λ is an eigenvalue of a square matrix A if and only if the matrix $A - \lambda I$ is singular.

Theorem 4: λ is an eigenvalue of the square matrix A if and only if $\mathrm{Det}(A - \lambda I) = 0$.

Theorem 5: The geometric multiplicity of an eigenvalue is always less than or equal to its algebraic multiplicity.

Theorem 6: The sum of the geometric multiplicities of the eigenvalues of an n by n matrix is at most n.

Theorem 7: The geometric multiplicity of the eigenvalue λ of an n by n matrix A is $n - r$, where r is the rank of $A - \lambda I$.

Theorem 8: Suppose A is a square matrix and $\{v_1, ..., v_m\}$ is a set of eigenvectors of A constructed as follows: Select a basis for each eigenspace of A and take all these basis vectors collectively. Then $\{v_1, ..., v_m\}$ is linearly independent.

Theorem 9: Let A be an n by n matrix with n linearly independent eigenvectors $v_1, ..., v_n$ and associated eigenvalues $\lambda_1, ..., \lambda_n$. Then, for any vector x_0 in R^n, we have the eigenvector decomposition

$$A^k x_0 = a_1 \lambda_1^k v_1 + ... + a_n \lambda_n^k v_n \qquad [4]$$

where the scalars $a_1, ..., a_n$ are determined by the equation $x_0 = a_1 v_1 + ... + a_n v_n$.

Theorem 10: If A is an n by n matrix with n linearly independent eigenvectors, then A can be diagonalized. Specifically, if P is a matrix whose columns are n linearly independent eigenvectors of A, then $P^{(-1)} A P = K$, where K is the diagonal matrix whose diagonal entries are the corresponding eigenvalues.

Theorem 11: If A and B are similar matrices, they have the same characteristic polynomial and hence the same eigenvalues with the same algebraic multiplicities.

Theorem 12: If A and B are similar matrices, their eigenvalues have the same geometric multiplicities.

Theorem 13: Suppose A is an a square matrix with real entries. If A has a complex eigenvector w and associated eigenvalue λ, then $\overline{(w)}$ is also an eigenvector of A and its associated eigenvalue is $\overline{(\lambda)}$. That is, if $A\,w = \lambda\,w$, then $A\,\overline{(w)} = \overline{(\lambda)}\,\overline{(w)}$.

Theorem 14: If $K = \begin{bmatrix} a & -b \\ b & a \end{bmatrix}$ is any scale-rotation with $b \neq 0$, then its eigenvalues are $a+b\,i$ and $a - b\,i$ with corresponding eigenvectors $\begin{bmatrix} i \\ 1 \end{bmatrix}$ and $\begin{bmatrix} -i \\ 1 \end{bmatrix}$, respectively. Furthermore, we can factor the scale-rotation K as a product of a dilation and a rotation

$$K = \begin{bmatrix} r & 0 \\ 0 & r \end{bmatrix} \begin{bmatrix} \cos(\theta) & -\sin(\theta) \\ \sin(\theta) & \cos(\theta) \end{bmatrix}$$

where the numbers r and θ come from the polar coordinates of $a + b\,i = r\cos(\theta) + r\sin(\theta)\,i$.

Theorems for Chapter 7

Theorem 1: (The Pythagorean Theorem for R^n) If u and v are orthogonal vectors in R^n, then the square of $\|u+v\|$ equals the sum of the squares of $\|u\|$ and $\|v\|$; that is:

$$\|u+v\|^2 = \|u\|^2 + \|v\|^2$$

Theorem 2: If u is any vector in R^n and W is any nonzero subspace of R^n, then the projection p of u onto W is the vector in W that is closest to u. That is, $\|u - p\| < \|u - w\|$ for every vector w in W ($w \neq p$).

Theorem 3: If $\{v_1, \ldots, v_k\}$ is an orthogonal basis of a subspace S of R^n and u is any vector in S, then:

$$u = c_1 v_1 + \ldots + c_k v_k$$

where

$$c_1 = u.v_1/v_1.v_1, \quad c_2 = u.v_2/v_2.v_2, \quad \ldots, \quad c_k = u.v_k/v_k.v_k$$

Theorem 4: If $\{v_1, \ldots, v_k\}$ is an orthogonal set of nonzero vectors in R^n, then $\{v_1, \ldots, v_k\}$ is a linearly independent set of vectors.

Theorem 5: If $\{v_1, ..., v_k\}$ is an orthogonal basis of a subspace W of R^n and u is any vector in R^n, then the projection p of u onto W is the sum of the projections of p onto $v_1, ..., v_k$:

$$p = \text{Project}(u, v_1) + ... + \text{Project}(u, v_k)$$

Theorem 6: Every nonzero subspace of R^n has an orthogonal basis.

Theorem 7: If S_1 is a subspace of R^n and S_2 is its orthogonal complement in R^n, then
$$\text{dimension}(S_1) + \text{dimension}(S_2) = n$$

Theorem 8: The orthogonal complement of the row space of a matrix is its null space.

Theorem 9: Let U be an n by n matrix. The statement that U is orthogonal (i.e., $U^T U = I$) is equivalent to each of the following statements:
(a) $U^{(-1)} = U^T$.
(b) $U U^T = I$.
(c) The columns of U form an orthonormal basis of R^n.
(d) The rows of U form an orthonormal basis of R^n.
(e) $Ux \cdot Uy = x \cdot y$ for all vectors x and y in R^n.

Theorem 10: Every linear system $Ax = b$ has a least-squares solution x, and the least-squares solution is unique if and only if the null space of A is $\{0\}$.

Index

Algebraic multiplicity, 327
Augmented matrix, 457
Basis, 268-275
    coordinates relative to, 273-275
    linear dependence relation and, 271-272
    method for finding, 270-272, 284-292
    theorems on, 275
Car on curved path, 251-255
Cayley-Hamilton theorem, 337
Characteristic equation, 327
Characteristic polynomial, 326
    similar matrices and, 368
Clock command, 316
Coefficient matrix, 457
Cofactor, 176-178, 476-478
Coffee blends, 57-58
Column space, 279
    basis of, 287-291
    dimension of, 280, 292
Complex number, 380
    absolute value, 381
    argument, 382
    conjugate, 381
    geometry of, 381
    imaginary part, 380
    real part, 380
Complex vector, 383
    conjugate, 384
    imaginary part, 383
    real part, 383
Computer graphics, 238-255
Concrete application, 60-61, 91-92
Conic section, 33, 455
Convex combinations, 61-62
Coordinates relative to a basis, 342, 356
Coordinates, 51, 273-275
Cramer's rule, 189-191
Cubic spline, 28-31, 34-35
Curve fitting, 445, 452
Decomposition of solution set, 70-73, 105-107
    linear independence, 87-88, 92

basis and, 270, 284-285
Determinant, 152, 176-191
    2 by 2, 474
    3 by 3, 475-476
    algebraic properties, 178-180
    cofactor of, 176-178, 476-478
    Cramer's rule, 189-191
    elementary matrix and, 193-194
    geometric properties, 185-189
    matrix inverse and, 180-181, 185
    matrix transformation and, 221-222
    n by n, 477-478
    row operation and, 181-185, 194
    upper triangular matrix and, 177-178
Diagonal matrix, 116
Diagonalizing a matrix, 360
    geometric meaning, 370
Diagonalizing powers of a matrix, 363
Dilation matrix, 199-203
Dimension, 268-275
    column space, 280, 292
    null space, 280, 286
    row space, 280, 292
    theorems on, 275
Direction vector, 43, 63, 67, 71
Distance between vectors, 412
Dot product, 98, 410
    matrix multiplication and, 98, 113
    rules of algebra, 412
Dynamical system, 340, 348
    attracted by origin, 350, 399
    complex eigenvalues, 392
    foxes and rabbits, 354
    Leslie model, 357
    predator-prey, 354
    repelled by origin, 399
    saddle point, 351
    spiral trajectory, 353
    three countries, 359
    trajectory of, 348
Earthquake data, 453

Echelon form, 16-18, 20, 170-174, 466-467
Eigenspace, 317
    geometric multiplicity, 317
Eigenvalue, 308-312
    algebraic multiplicity, 327
    characteristic polynomial, 326
    complex, 380, 385
    of diagonal matrix, 312
    Evalues command, 319
    geometry and, 313, 319
    of letter matrices, 338-339
    of matrix transformations, 321
    of nilpotent matrix, 323
    number of, 313
    patterns and, 333
    of power of a matrix, 323
    similar matrices and, 368
    of a symbolic matrix, 336
    of the transpose, 337
    of upper triangular matrix, 322
Eigenvector decomposition, 343
Eigenvector, 308
    basis for R^n, 340
    characteristic polynomial, 326
    Clock command, 316
    complex, 380, 385
    diagonalization by, 361
    eigenvector decomposition, 343, 403
    Evectors command, 330
    geometric multiplicity, 317, 332
    geometry of, 313-317, 319
    Headtail command, 315
    Mapcolor command, 324
    matrix transformations and, 321
    visualizing in R^3, 324
Elementary row operation, 14, 20
Foxes and rabbits, 354
Gambler's ruin, 147-148
Gaussian elimination, 456, 458, 460-461, 472-473
Geometric multiplicity, 317, 332
    rank and, 335
Graham-Schmidt process, 427, 429
Gridgame command, 51
Headtail command, 315

Healthy-sick workers, 147
Homogeneous coordinates, 238-251
    translation and, 241, 247
Hypercube, 420
Identity matrix, 114
Inverse of a matrix, 121-122, 150-161
Invertibility Theorem, 154
Kernel of a transformation, 233-235, 281-284
    subspace and, 262
Lampland migration, 134-142, 344
Leading entry, 20, 464
Least-squares solution, 438-440, 444, 448
Length of a vector, 411
Leslie population model, 357, 402
Line
    cartesian equation, 2
    graph of, 3
    parametric representation, 43-45, 63-66
Linear combination, 49, 56
Linear dependence relation, 78, 81
    basis and, 271-272
Linear equation, 457
Linear system, 457
    augmented matrix, 457
    coefficient matrix, 457
    consistent, 20, 457
    equivalent, 20, 458
    homogeneous, 10
    inconsistent, 20, 457
    matrix inverse and, 156-161
    matrix-vector equations and, 99
    particular solution of, 71, 105
    solution of, 12, 70-73, 457
    solution set, 70-73, 105-107
    three unknowns, 5-9
    two unknowns, 1-5
    vector equations and, 99
Linearly independent set, 80, 81
    basis and, 268
    decomposition and, 87, 92, 270, 284-285
    matrix inverse and, 158-61
    matrix-vector equation and, 103
Magnitude of a vector, 411
Markov chain, 142, 344

Markov matrix, 142, 364, 376
Matrix
- augmented, 12
- backward identity, 122-123
- diagonal, 116
- diagonalization of, 360
- dilation, 199-203
- elementary, 166-174, 192-193
- identity, 114
- Jordan, 120, 122
- letter matrix, 123, 295
- Markov, 142
- minor, 176-178, 474-475
- multiplication symbol, 96, 112
- nilpotent, 323
- noncommutativity of, 126
- nonsingular, 153
- permutation, 115
- power, 120
- product, 111
- projection, 217-219, 231-233, 241, 378
- rank of, 19, 161, 280, 292
- reflection, 214-217, 229-231, 241
- regular stochastic, 145
- rotation, 209-212, 228, 239, 248
- row equivalent, 20, 170-174, 462
- rules of algebra for, 104-105, 124-130
- shear, 205-207, 336, 378
- similar matrices, 360, 365
- singular, 153
- square root of, 376
- square, 116
- stochastic, 142
- symmetric, 133
- transpose, 127-129
- upper triangular, 322
- -vector equation, 95, 99-104
- vector- product, 117

Matrix inverse, 121-122, 150-161, 170-174
- by hand, 162-163
- determinant and, 180-181, 185
- elementary matrix and, 167-168
- kernel and, 234
- linear independence and, 158-161
- linear system and, 158-161
- range and, 234
- span and, 158-161

Matrix transformation, 198
- animation of, 212-214, 244-245
- composite of, 207-209
- determinant and, 221-22
- kernel of, 233-235, 281-284
- range of, 233-235, 281-284

Migration matrix, 137-142, 344
Minimal spanning set, 270
Minor, 176-178, 474-475
Nilpotent matrix, 323
Nonsingular matrix, 153
Normal equations, 440, 443, 447
Null space, 279, 431
- basis of, 284-287
- dimension of, 280, 286

Oil pipeline construction, 301-303
Orthogonal
- basis, 422-424
- complement, 430, 432
- matrix, 432-434
- projection, 425
- vectors, 425

Orthogonal set of vectors, 423
Orthonormal basis, 424
Parametric representation
- of line, 43-45, 63-66
- of plane, 67-70

Permutation matrix, 115
Pivot, 15-16, 20, 463
- column, 20, 463

Plane
- cartesian equation, 5
- graph of, 6
- parametric representation of, 67-70

Planes at three hubs, 147
Polynomial
- fitting to data, 24-28, 445-451

Population model, 359
Position vector, 44, 63, 71
Predator-prey model, 354
Probability vector, 142

Projection
- formula, 415
- onto a line, 414, 421
- onto a plane, 417, 421
- onto a subspace, 417, 420
- vector, 415

Projection matrix, 217-219, 231-233, 241
Pythagorean Theorem for R^n, 413
Random walks, 143-145, 148
Range of a transformation, 233-235, 281-284
Rank of a matrix, 19, 161, 280, 292
Reduced row echelon form, 18-20, 170-174, 468
Reflection matrix, 214-217, 229-231, 241
Residual, 447
Rotation matrix, 209-212, 228, 239, 248
Row echelon form, 16-18, 20, 170-174, 466-467
Row operation, 14, 20, 166-174, 462
- determinant and, 181-185, 194

Row space, 279, 431
- basis of, 291-292
- dimension of, 280, 292

Scale-rotations, 388
Scaling matrix, 203-205, 225-227
Shear matrix, 205-207
Similar matrices, 365
- characteristic polynomial and, 368
- eigenvalues and, 368
- geometry of, 367, 370

Singular matrix, 153
Ski jump, 33
Solution set, 12, 70-73, 457
Span, 49, 56
- basis and, 268
- matrix inverse and, 158-161
- matrix-vector equation and, 102
- solution set and, 105-107
- subspace and, 262

Spanning set, 270
Spanning tree, 303
Sphere intersecting planes, 7
Spline, 28-31, 34-35
Spotted owl population dynamics, 407
Steady-state vector, 139-142
Stochastic matrix, 142
- regular, 145

Subspace, 260-264
- basis of, 268-275
- closure properties, 261
- dimension of, 268-275
- geometry of, 263-264
- kernel and, 262
- span and, 262

Symmetric matrix, 133
Temperature estimation, 32-33, 35
Traffic flow, 296-301
Transpose of a matrix, 127-129
Unit span, 52
Vandermonde matrix, 448
Variable
- constrained, 17, 470
- free, 17, 466, 470
- leading, 17, 470

Vector
- addition, 38-39
- complex entries, 383
- convex combination of, 61-62
- distance between, 412
- geometry of, 37-43
- in 3-space, 54-56
- in n-space, 56-58
- in the plane, 37-47
- length, 411
- linear combination of, 49, 56
- linear dependence relation for, 78, 81
- linear independent set of, 80, 81
- magnitude, 411
- probability, 142
- scalar multiplication, 42-43
- span of, 49, 56
- steady-state, 139-142
- translation by, 39-42

Vector form
- of line, 43-45
- of plane, 67-70

Weights (i.e., coefficients), 49

LAMP Installation Instructions for Maple 6.01 (or later) on a Macintosh system.

1. Exit Maple before installing the LAMP materials.

2. In the Macintosh folder on the LAMP CD-ROM, double-click on the self-extracting archive LampInstaller. A window will come up that shows where the LAMP folders and files will be installed. Although you can change this location, we recommend that you accept the default, lamp6, as this will save you a step later in the installation procedure.

3. Your final step will be to set up the file Maple 6 Init Preferences, a new copy of which is now in the lamp6 folder. (This file will tell Maple where to find the Lamp library.) First determine where Maple is installed on your system. It is usually in the folder Maple 6 at the top level of your hard drive. Within the Maple 6 folder, open the Global Preferences folder and find the existing copy of Maple 6 Init Preferences. Drag and drop the new copy from the lamp6 folder onto this existing copy, unless you have already made some custom changes to Maple 6 Init Preferences. (In the latter case, see "Custom Installation" below.)

If you did not choose lamp6 as the location of the LAMP folders and files, or if lamp6 is not on a hard drive with the name Macintosh HD, you must change Maple 6 Init Preferences to specify where you put the Lamp library. Do so by double-clicking on this file, and edit the line that says

lamppath := "Macintosh HD:lamp6:lamplib":

For example, if you put the LAMP folders and files in the folder My folder on the hard drive My HD, the above line should be

lamppath := "My HD:My folder:lamplib":

This completes the installation.

LAMP folders and files:
Within the lamp6 folder are two folders: lampmod, which contains the twenty-nine LAMP modules and twenty-nine worksheets of problems, and lamplib, which contains the four Lamp library files, maple.hdb, maple.ind, maple.lib, and maple.rep, and also a text file, lamp.txt. The lamp6 folder also contains the file Maple 6 Init Preferences.

Custom Installation:
If have made custom changes to Maple 6 Init Preferences, copy the contents of the Maple 6 Init Preferences file in lamp6 to the end of the Maple 6 Init Preferences file in the Global Preferences folder in the Maple 6 folder. (You can use SimpleText to do so.)

Testing the Installation:
In a Maple worksheet, enter the Maple command

with(Lamp);

If the result is a list of Lamp library procedures, the installation has succeeded. If, instead, an error message appears, type

libname;

The first name in the output should be Macintosh HD:lamp6:lamplib (or whatever folder you put the library in). If it is not, then Maple is not reading Maple 6 Init Preferences. Check that you do have a file of that name in the Global Preferences folder in the Maple 6 folder on your hard drive, and check that it contains the above definition of lamppath.

If the first name in the output of libname is the correct folder name but Maple does not find the Lamp library, check that the folder name assigned to lamppath in Maple 6 Init Preferences is where you put the Lamp library files.

LAMP Installation Instructions for Maple 6 in Windows 95, 98, 2000, and NT

1. Exit Maple before installing the LAMP materials.

2. In the Windows folder on the LAMP CD-ROM, double-click on the self-extracting archive install.exe. A window will come up that shows where the LAMP folders and files will be installed. Although you can change this location, we recommend that you accept the default, c:\lamp6, as this will save you a step later in the installation process.

3. Your final step will be to copy the file maple.ini from the lamp6 folder to a Maple folder. (maple.ini is an initialization file that will tell Maple where to find the Lamp library.) To find out where maple.ini goes, determine where Maple 6 is installed on your system. It is usually in the folder

  c:\Program Files\Maple 6

Within that folder, open the folder Bin.wnt, which is the folder from which Maple is started. (The last three letters, wnt, may vary with the version of Windows.)

If you do not already have a file called maple.ini in Bin.wnt, copy maple.ini from lamp6 and paste it into the Bin.wnt folder.

On the other hand, if you do already have a file called maple.ini in Bin.wnt, copy the contents of the maple.ini file in lamp6 to the end of your existing maple.ini file. (You can use NotePad to do so.)

This completes the installation.

LAMP folders and files:
Within the lamp6 folder are two folders: lampmod, which contains the twenty-nine LAMP modules and twenty-nine worksheets of problems, and lamplib, which contains the four Lamp library files, maple.hdb, maple.ind, maple.lib, and maple.rep, and also a text file, lamp.txt. The lamp6 folder also contains the file maple.ini.

Custom Installation:
If you did not choose c:\lamp6 as the location for the LAMP folders and files, you must change maple.ini to specify where you put the Lamp library. Do so by editing the line of maple.ini that says

lamppath := "c:\\lamp6\\lamplib\\":

For example, if you put the LAMP folders and files in c:\myfolder, the above line should be

lamppath := "c:\\myfolder\\lamplib\\":

NOTE: The slashes are DOUBLE BACK SLASHES.

Testing the Installation:
In a Maple worksheet, enter the Maple command

with(Lamp);

If the result is a list of Lamp library procedures, the installation has succeeded. If, instead, an error message appears, type

libname;

The first name in the output should be c:\lamp6\lamplib\ (or whatever folder you put the library in). If it is not, then Maple is not reading maple.ini. Check that you have put maple.ini in the correct folder.

If the first name in the output of libname is the correct folder name but Maple does not find the Lamp library, check that the folder name assigned to lamppath in maple.ini is where you put the Lamp library files.

LAMP Installation Instructions for Maple 6 in Unix

1. Exit Maple before installing the LAMP materials.

2. Put the LAMP CD in your CD-ROM drive and mount it at an appropriate point in your file system. You may need to log in as root in order to do this. Here's what this step looks like on a typical Linux system:

```
$ /bin/su
Password:
# /bin/mount -r /dev/cdrom /mnt/cdrom
```

We'll assume that you have mounted the CD-ROM drive at /mnt/cdrom. If you use a different mount point, make appropriate adjustments in the following instructions.

3. Determine where Maple 6 is installed and move into that directory.

```
# cd /usr/local/maple
```

We'll assume that Maple 6 is in /usr/local/maple. Again, if it is located somewhere else on your system, make appropriate changes in the following instructions.

4. The LAMP materials are in the tape archive file Unix/lamp6.tar on the CD. Use the tar command to extract the LAMP files from this archive:

```
# /bin/tar xvf /mnt/cdrom/Unix/lamp6.tar
```

The LAMP files are placed in /usr/local/maple/lib/lamplib and /usr/local/maple/lib/src/lampmod. The tar command creates these automatically if they do not already exist.

Here is a list of the files in the tape archive:

lib/lamplib/maple.hdb
lib/lamplib/maple.ind
lib/lamplib/maple.lib
lib/lamplib/maple.rep

lib/lamplib/lamp.txt
lib/src/lampmod/README
lib/src/lampmod/init
lib/src/lampmod/*.mws (58 Maple worksheets with .mws extension)

Notice that the filenames are relative to the current working directory when the tar command is given; this is why it is important to cd first into the directory in which Maple is installed.

5. One of the files you extracted from the archive, lib/src/lampmod/init, is an initialization file that contains Maple instructions for loading the Lamp library whenever Maple is started.

5a. This file also presupposes that Maple 6 was installed in /usr/local/maple. If this is not the case, edit the line

lamppath := "/usr/local/maple/lib/lamplib/":

to replace /usr/local/maple with the full pathname of the directory in which Maple 6 was installed.

5b. The following command adds the instructions for your system-wide Maple initialization file:

/bin/cat /usr/local/maple/lib/src/lampmod/init >> /usr/local/maple/lib/init

(If you are using the C shell and you do not already have a system-wide Maple initialization file, you'll have to create an empty file before the preceding command will work. `/bin/touch /usr/local/maple/lib/init' will do the job.)

Alternatively, you can have each LAMP user append this file to his or her personal Maple initialization file, by giving the command

$ /bin/cat /usr/local/maple/lib/src/lampmod/init >> ~/.mapleinit

(Again, a C shell user who does not already have a personal Maple initialization file must precede this command with `/bin/touch ~/.mapleinit'.)

6. Unmount the CD-ROM drive and remove the CD.

/bin/umount /mnt/cdrom

This completes the installation.

Testing the Installation:
As an ordinary user, start Maple. In a Maple worksheet, enter the command

> with(Lamp);

If the result is a list of LAMP library procedures, the installation has succeeded. If, instead, an error message appears, type

> libname;

The first filename in the output should be the directory into which the library was unpacked — typically, /usr/local/maple/lib/lamplib. If it is not, then Maple is not reading the LAMP initialization; make sure that you have appended the contents of /usr/local/maple/lib/src/lampmod/init either to your system-wide Maple initialization file or to your personal Maple initialization file, as described in step 5b above.

If the first filename is correct, but Maple is nevertheless not finding the Lamp library, check the specified directory to make sure that the Lamp library files are actually in it.

The system-wide Maple initialization file /usr/local/maple/lib/init should have permissions 644 (in particular, it should be readable by all users), the LAMP directories /usr/local/maple/lib/lamplib and /usr/local/maple/lib/src/lampmod should have permissions 755, and the files in those directories should have permissions 644.